**W9-CEV-510**

DISCARD

DATE DUE

BRODART, CO.   Cat. No. 23-221-003

Language of the Earth

Second Edition

EDITED BY FRANK H.T. RHODES,
RICHARD O. STONE AND
BRUCE D. MALAMUD

Blackwell
Publishing

Editorial material and organization © 2008 by Blackwell Publishing Ltd

BLACKWELL PUBLISHING
350 Main Street, Malden, MA 02148-5020, USA
9600 Garsington Road, Oxford OX4 2DQ, UK
550 Swanston Street, Carlton, Victoria 3053, Australia

The right of Frank H.T. Rhodes, Richard O. Stone and Bruce D. Malamud to be
identified as the authors of the editorial material in this work has been asserted in
accordance with the UK Copyright, Designs, and Patents Act 1988.

All rights reserved. No part of this publication may be reproduced, stored in a retrieval system,
or transmitted, in any form or by any means, electronic, mechanical, photocopying, recording or
otherwise, except as permitted by the UK Copyright, Designs, and Patents Act 1988,
without the prior permission of the publisher.

Designations used by companies to distinguish their products are often claimed as trademarks.
All brand names and product names used in this book are trade names, service marks, trademarks,
or registered trademarks of their respective owners. The publisher is not associated
with any product or vendor mentioned in this book.

This publication is designed to provide accurate and authoritative information in regard to the subject matter
covered. It is sold on the understanding that the publisher is not engaged in rendering professional services.
If professional advice or other expert assistance is required, the services of a competent professional should be sought.

First published 2008 by Blackwell Publishing Ltd
First edition published by Pergamon Press 1981

1   2008

Library of Congress Cataloging-in-Publication Data

Language of the earth. – 2nd ed. / edited by Frank H.T. Rhodes, Richard O. Stone and Bruce D. Malamud.
p. cm.
Includes bibliographical references and index.
ISBN 978-1-4051-6067-4 (hardcover : alk. paper)
1. Geology. 2. Geologists. I. Rhodes, Frank Harold Trevor. II. Stone, Richard O., 1920–1978.
III. Malamud, B.D. (Bruce D.)

QE26.3.L35 2008
550–dc22
2007045041

ISBN: 978-1-4051-6067-4 (hardback)
A catalogue record for this title is available from the British Library.

Set in 10.5 / 12pt DanteMT
by SPi Publisher Services, Pondicherry, India
Printed and bound in Singapore
by Markono Print Media Pte Ltd

The publisher's policy is to use permanent paper from mills that operate a sustainable forestry policy,
and which has been manufactured from pulp processed using acid-free and elementary chlorine-free practices.
Furthermore, the publisher ensures that the text paper and cover board used have met acceptable
environmental accreditation standards.

For further information on
Blackwell Publishing, visit our website at
www.blackwellpublishing.com

ACC Library Services
Austin, Texas

Contents

11. Art 244

PART 4: THE CROWDED PLANET 257

12. Human History 259

13. Resources 277

14. Benevolent Planet 289

Preface

Every life is a continuing encounter with Earth. Every breath is a transaction with our planet. Every meal is an assimilation of terrestrial products. Every item we use, touch, or manufacture is a piece of Earth. Every element in our bodies, atom for atom, comes from Earth's crust. Every calorie of energy we use comes directly from Earth, and indirectly from Earth's parent star, the Sun.

We are formed of the dust of the Earth, just as the ancient scripture affirms. But that dust is star dust. We are not only Earth's children; we are the grandchildren of a star. "Dust we may be," Maurice Boyd has remarked, "but the dust of a star, and troubled by dreams." And in those dreams rest our greatness and our hope.

This is an anthology, a book of writings about our parent planet, Earth. It is not a book of science, though some writers are scientists. It is not didactic, though we hope it will be instructive. It is not comprehensive, though it covers an expansive range of topics. Our purpose in writing is to inspire interest, rather than to tell the whole story; to whet the appetite, rather than "provide all the data"; to ignite the imagination, rather than instruct in detail. Certainly we hope students of Earth science will find the book useful, whether as required supplementary reading in formal courses, or as a diversion from a surfeit of scientific literature. But our intended audience is wider: because we are all Earth's children, we hope that these reports on the parent planet will be of interest to the general reader and will be read as letters from home. This is a book for browsing, for tasting, for reflecting.

These accounts of the home planet, their style and their viewpoints, are as varied as their authors. We hope that by using the form of an anthology, based as it is on the writings of authors of many backgrounds, periods and interests, we can capture a sense of the fascination and mystery of this ancient and beautiful, blue planet. We hope that the reader will catch a glimpse of its contradictory moods: its benevolence and its terror, its resilience and its fragility, its tranquility and its episodic violence, its regularity and its unpredictability.

In this respect, the emphasis of the present book differs from that of the first edition. That work was directed chiefly to students of Earth science; this volume, though retaining that goal, is consciously more general in its scope and more expansive in its range of topics. The first edition aimed to provide a context for that particular category of knowledge which we identify as Earth science, to show not only its range and scope, but also its flavor, style and implications; to show all knowledge as provisional, rather than infallible, as refinable rather than complete and finished.

Our purpose in the present volume is still to achieve those goals, but also something more. We hope to show not only the intimacy of science with other expression and study, but also to give a sense of the individuality, insight and intuition that underlie all encounters with our parent planet and all descriptions of it. For scientific knowledge is similar to all other knowledge; in spite of its public verifiability, it grows by private insight and personal intuition. "If you want to know the essence of scientific method," Einstein is said to have remarked, "don't listen to what a scientist may tell you. Watch what he does." So this book is a window on the world of people "doing" – not just geologists, but sculptors and soldiers, artists and aviators, politicians and poets, prophets and prospectors, novelists and naturalists.

We have greatly expanded the extracts included in the present volume, retaining – we hope – the best of the first edition, but omitting twenty of the original authors, shortening some extracts, and adding another fifty-five. Though we have made a particular effort to include contemporary literature and topics of current interest, we have also tried to achieve a reasonable balance with the older literature. We have added Jane Hirshfield, John McPhee and Carl Sagan, for example, but we have kept William Wordsworth, Voltaire and James Hutton.

The thread that gives continuity is their common concern with the Earth. Nor is the span of time and place any less comprehensive than that of occupation. We have deliberately selected writings from the time of the fifth or sixth century BC when the book of Job was written, to the heyday of the Victorian era, when William Buckland fascinated and dominated the Oxford scene; from the medieval ages of wonder in the created universe, to the novels of John Fowles written in the 1970s, to books published in the new millennium. In the most literal sense our extracts involve the comprehension of the whole Earth. Our authors' accounts range from dripping caves, deep below the Pyrenees, to the deserts of North Africa, from the Alps to the Andes, from mountain tops to the depths of the ocean, from the Antarctic ice sheet to the surface of the moon. Their writings concern both social and scientific topics, as well as personal interests and political goals. All reveal some aspect of our planet in the life of humankind.

We have arranged our selections in various categories, in order to give some structure and coherence to the book, and also to allow the reader to use it more selectively. These categories are generally self-evident, but a few articles could have fitted into some other category just as easily as the one in which they stand. Most of the articles that we include are fragments of much longer articles, essays, or books, and even the extracts themselves, we have generally abridged. We regret the need to make these abbreviations, and we have tried very carefully to preserve the sense and style of the original authors. Our reluctance to make the abridgement is exceeded only by our conviction that diversity and variety are of more importance than comprehensive quotation of a smaller number of authors. In the book as it now stands, our task of selection has been painfully difficult; we have collected many outstanding articles for which we have been unable to find space. With each extract, we have retained the essential footnotes and references, though we have shortened or deleted footnotes and references where these did not seem to us essential to the main argument. We hope that our readers will wish to explore the new worlds of information represented in some of the extracts that are included, just as we also hope that they may develop a sufficiently strong taste for some of the present authors to encourage them to explore their other writings.

The first edition of this book was planned with Richard Stone, Professor of Geology at the University of Southern California. He and the first editor, Frank Rhodes, worked together from 1972 to 1978, when Richard Stone died from a disease against which he had fought with great courage. Frank Rhodes finished work on the book, which was published in 1981.

Though the book had been out of print for almost a decade, the third editor, Bruce Malamud, while a PhD student at Cornell discussed the book and found that it occupied a respectable place on a list of the "Great Books of Geology" selected by the readers of the *Journal of Geological Education* ["Geologists Select Great Books of Geology", 1993, vol. 41, p. 26]. It was this that led to his suggestion of a new edition and we are grateful for the decision of Ian Francis of Blackwell Publishing to publish it. Frank Rhodes rewrote and expanded all the introductions and we have worked together on the preparation of the book, supported and helped by several people whose efforts we are anxious to acknowledge: Joy Wagner, John Briggs, Joel Haenlein, and Helen Sullivan have helped us greatly with the typing and editorial work.

We are conscious that in aiming at a wider readership, we face a delicate task in balancing the technical and the nontechnical content. In this we have deliberately leaned towards general intelligibility, partly because we are persuaded that such broader contextual understanding and interest are beneficial to the specialist and partly because we believe the substance of this book is of vital concern and significance for the more general reader. If war is too important to be left to the generals, then the understanding and care of the Earth are far too important to be left to the Earth scientists.

For each of us – specialist or not – is a custodian of this beautiful planet, which is our home. That is why our response to the issues raised in this book is of more than casual importance. It is, of course, of vital concern as a basis for the vocational skills and technical knowledge required of those pursuing courses and careers in Earth science. But, in a far wider sense, it is an important component in adding to the richness and exuberance of life for all of us. We have attempted to use a concern with the Earth to introduce the reader to a growing range of experiences and involvements ranging from sculpture to literature, and from architecture to history. Through such interests, it is possible to develop a new perspective which enriches the quality of human experience.

Our response to the Earth is a factor of profound significance. We share our over-crowded, polluted, plundered planet with more than six billion other members of the human race, as well as a couple of million other species of animals and plants. The next 50 years will call for critical decisions in the problems of conservation, energy policy, urban development, transportation, mineral and water resources, land use, atmospheric protection and utilization of the oceans. The terms of our survival will depend in large part on our careful comprehension of the language of the Earth and our stewardship of this rare planet. Earth's fitness as a home for humans and other species is the end product of its long and complex history. Soil, air, water, food, fuels, minerals, and microbes are the products of Earth processes acting over incalculable periods, and of the life, activity, and death of countless organisms that have gone before us. Each of us, in our own turn, absorbs, utilizes and recreates this sustaining environment.

"Speak to the Earth, and it shall teach thee," Job was told 25 centuries ago. To those with the patience to master its language, the Earth still responds. To those with the insight to comprehend its moods and reflect on its mysteries, the Earth still teaches.

Frank H.T. Rhodes
Bruce D. Malamud
Cornell University, Ithaca, NY
King's College, London

Preface from the first edition

One of the problems with our conventional styles of teaching and conventional patterns of learning at the introductory undergraduate level is that the "subject" – whatever it may be – all too easily emerges as given, frozen, complete, canned. Add to this quizzes, multiple-choice exams, and a single textbook, and the pursuit of knowledge sometimes becomes a kind of catechism – the recital of prepared answers to a limited set of questions ("You lost one point because you missed the hardness of hornblende," or some other equally inconsequential fragment of information). It is against such a limited view of knowledge that students so frequently react, and so they should.

But it need not be so. In every department there are still successful teachers who can win the interest and enthusiasm of once uninterested and unenthusiastic students. They do so partly by the example of their own commitment to learning, and partly by revealing knowledge as a continuing personal quest. For no knowledge, and least of all scientific knowledge, exists as a finished corpus of categorized facts. Knowledge exists because there are people; it is the accumulated personal experience of our race. It becomes meaningful, useful and intelligible as we grasp, not only its content, but also its basis, its implications, its relationship and its limitations. Its coherence and significance lie in its relatedness to the whole of the rest of our human experience.

The aim of this book is to provide such a context for the particular category of knowledge which we identify as earth science. Our intention is not didactic. We make no attempt to cover the ground in the sense of describing the current content of each area of earth science. Our aim is to illustrate the scope and range of the science and to convey its flavor and style, rather than catalog its contents; to display all our knowledge as provisional rather than infallible, as refinable rather than complete and finished; to show the inspiration and sweeping implications of earth science, rather than representing it as an isolated area of study.

We hope that by using the form of an anthology, based as it is on the writings of authors representing many countries, periods and viewpoints, we can show something of the individuality which lies at the heart of science. The categorization of science as less humane than other areas of human knowledge and as dehumanizing in its effects is a verdict reached too readily by many of our contemporaries. For scientific knowledge, in spite of its public verifiability, grows by private insight and personal intuition. "If you want to know the essence of scientific method," Einstein is said to have remarked, "don't listen to what a scientist may tell you. Watch what he does." So this book is a window on the world of people "doing" – not just geologists – but sculptors and soldiers, artists and aviators, politicians and poets, prophets and prospectors, novelists and naturalists.

The thread that gives continuity is their common concern with the Earth. Nor is the span of time and place any less comprehensive than that of occupation. We have deliberately selected writings from the time of the 5th or 6th century BC, when the book of Job was written, to the heyday of the Victorian era, when William Buckland fascinated and dominated the Oxford scene; from the medieval ages of wonder in the created universe, to the novels of John Fowles written in the 1970s. In the most literal sense, geology involves the comprehension of the whole Earth. Our authors' accounts range from dripping caves, deep below the Pyrenees, to the deserts of North Africa, from the Alps to the Andes, from mountain tops to the depths of the ocean, from the Antarctic ice sheet to the surface of the moon. Their writings concern both social and scientific topics, as well as personal and political goals. All reveal some aspect of geology in the life of mankind.

We have arranged our selections in various categories, in order to give some structure and coherence to the book, and also to allow the reader to use it more selectively. These categories are generally self-evident, but a few articles could have fitted into some other category just as easily as the one in which they stand. Most of the articles that we include are mere fragments of much longer articles, essays, or books, and even the extracts themselves, we have generally abridged. We regret the need to make these abbreviations, and we have tried very carefully to preserve the sense and style of the original authors. Our reluctance to make the abridgement is exceeded only by our conviction that diversity and variety are of more importance than comprehensive quotation of a smaller number of authors. In the book as it now stands, our task of selection has been painfully difficult; we have collected many outstanding articles for which we have been unable to find space. With each extract, we have retained the essential footnotes and references, though we have shortened or deleted footnotes and references where these were not essential to the main argument. We hope that our readers will wish to explore the new worlds of information represented in some of the references that are quoted, just as we also hope that they may develop a sufficiently strong taste for some of the present authors to encourage them to explore their other writings.

We hope that the book will provide useful supplementary reading for those who are enrolled in introductory earth science courses. We see it – not merely as a collection of required readings – but as an anthology for browsing. We hope that the book might even stand by itself, and have some interest for those who, though having no formal concern with courses or teaching in earth science, possess a curiosity about the planet which is our home, and about our varying responses to it. This response is of more than casual importance. It is, of course, of vital concern as a basis for the vocational skills and technical knowledge required of those pursuing courses in earth science. But, in the far wider sense, it is an important component in adding to the richness and exuberance of life. We have attempted to use a concern with the Earth to introduce the reader to a growing range of experiences and involvements ranging from sculpture to literature, and from architecture to history. Through such interests, it is possible to develop a new perspective which enriches the quality of human experience.

Our response to the Earth is a factor of profound significance. We share our over-crowded, polluted, plundered planet with 3.5 billion other members of our race. The next

30 years will call for critical decisions in the problems of conservation, energy supplies, mineral and water resources, land use, atmospheric protection and utilization of the oceans. The terms of our survival will depend in large part on our careful comprehension of the language of the earth and our stewardship of this rare planet. Earth's fitness as a home for humans is the end product of its long and elaborate history. Soil, air, and water, food and fuels, minerals and microbes are the products of earth processes acting over incalculable periods, and of the life, activity, and death of countless organisms. Each of us, in his own turn, absorbs, utilizes and recreates this sustaining environment.

"Speak to the Earth, and it shall teach thee," Job was told 25 centuries ago. To those with the patience to master its language, the Earth still responds. To those with the insight to comprehend its moods and reflect on its mysteries, the Earth still teaches.

Dick Stone and I began work on this book in 1972. We worked steadily at it, fitting it in as best we could between other pressing commitments. Most of the writing and selection of materials was finished by 1976.

Because of increasing publishing costs, we were then faced with the need to reduce the length of the book. This took much longer than either of us had anticipated. Dick Stone became ill with a disease against which he fought with courage and hope, but from which he died on July 23, 1978. I have completed the book, but its essential form and much of its content are as we both planned it.

Frank H.T. Rhodes
Cornell University, Ithaca, New York

Acknowledgments from the first edition

I am most grateful to several individuals whose willing cooperation and enthusiastic interest have contributed significantly to the completion of this book. My greatest debt is to my late colleague, Dick Stone, who derived great pleasure from the selection and preparation of material for the book. The breadth of the topics we cover in the book reflects the scope and quality of his own interests.

To those authors and publishers who have given permission to reproduce their work, to extract from it and otherwise abridge it, I have a particular debt of gratitude. In some cases, this permission was given without the requirement of payment of copyright fees. All the original sources of information are fully acknowledged.

Mrs. Jean Schleede and Mrs. Margaret Gihingham typed earlier versions of the manuscript, while Mrs. Marcia Parks, Mrs. Clara Pierson and Mrs. Joy Wagner assisted in its final preparation. Their careful typing, checking and verification of sources have been especially important, and I am most grateful for their help.

Frank H.T. Rhodes

Part 1

The Earth Experienced

To live is to experience the Earth, for we are Earth's children. Every atom of our bodies comes from the Earth. Every breath we take inhales and modifies Earth's fragile atmospheric envelope.

The food and water by which our life is sustained are Earth's products. Every object we touch, the clothes we wear, the cities we build, the houses where we live, the vehicles that we drive, the energy on which we depend – every substance we use and enjoy, everything from computer chips to cathedrals, comes from Earth's materials. The light by which we see, the energy source for all the teeming world of plants on which all other life depends, radiates from our grandparent: Earth's parent star, the Sun.

And we are children who never grow up, who remain lifelong dependents, who never leave home, who – like it or not – never escape our parent's apron strings. How remarkable, then, that most of us give our parent so little attention, with scarcely a passing thought for Earth's continuing support and well-being. How strange that we know and care so little for our family history.

The following section (Part 1) gives a parental portrait: the Earth experienced in its changing moods, its spasmodic violence, and its secret places. Reporters, historians, explorers, Earth scientists, and those better known for other achievements, provide us with a series of letters from home, reminding us of the old homestead, the place where we grew up and where – though sometimes forgetful of it – we still live.

1. Eyewitness Accounts of Earth Events

Science, it is generally asserted, is concerned with facts. But ultimately there is nothing in Nature labeled "fact." Facts represent human abstractions, and our recognition and understanding of facts are based upon individual perception and experience. To this extent geology includes both contrived encounters, which we call experiments, which are based on studies such as those in the laboratory where conditions and materials are carefully controlled, and also other encounters and experiences that can be gained only by firsthand contact with the Earth as it is, which generally means the rocks of the Earth's crust. In this sense, the work of the field geologist provides the basic link between our knowledge and the materials and processes on which it is based.

The role of experiment on Earth science is limited by the problems of scaling down the Earth to appropriate models, the problem of physical states existing in the Earth but unattainable in the laboratory, and the time element, which is rarely attainable under laboratory conditions. For these and other reasons, laboratory experimentation plays a lesser role in geology than in most other sciences, although there is a rather different kind of "experimentation" which is also possible in the field, especially with such natural geological phenomena as the various agents of erosion and deposition, hot springs, geysers, and volcanoes, to take some obvious examples.

Geology has a twofold concern. It is concerned first with the present configuration and composition of the Earth, and second with the interpretation of the past history of the Earth. Field observations are of critical importance, because the recognition of earlier events in the history of the Earth depends heavily on an analysis of present composition and configuration of Earth materials, and the identification of the action of processes and laws observable in our existing environment. Here, again, time and scale are important elements. Catastrophes do indeed take place, but most processes that influence the crust of the Earth act almost imperceptibly. Unfortunately, nearly all the processes that we can observe (as opposed to those we can infer) are rapid ones, which play only a local or limited role in the overall development of the Earth's crust. In spite of this limitation, such events and processes are still of great significance because they allow us to observe and compare multiple sequences of events in which various components and conditions may be analyzed, and in which a measure of prediction and experimentation is possible. Even the speed and extent of these processes have great significance.

Because most centers of population are deliberately established in areas remote from scenes of frequent terrestrial violence and instability, field observation frequently takes the geologist into distant and isolated areas. The areas included in the present accounts range from the Sahara Desert to the Antarctic, and the accounts themselves present fascinating differences in character and style.

1-1　Los Angeles Against the Mountains – John McPhee

John McPhee was born in Princeton in 1931, educated at Deerfield, Princeton and Cambridge and has been a staff writer for The New Yorker *for forty years, as well as teaching a popular writing class at Princeton. The range of McPhee's books and other writing is extraordinary: basketball, tennis, prep school, orange farming, environmental activism, nuclear terrorism, Alaska and, most notably, a series of four books –* Basin and Range *(1986),* In Suspect Terrain *(1983),* Rising from the Plains *(1986), and* Assembling California *(1993) – which,*

through detailed discussion of road cuts on Interstate 80, geologists in the field and the emergence of plate tectonics theory, give an account of the shaping and character of the North American continent. These four books, compiled in a single volume Annals of the Former World *(1998), led to the award of the Pulitzer Prize in 1999. The present extract, taken from the book* The Control of Nature *(1989), describes the devastating effects of a torrential debris flow in Shields Creek on the slopes of the San Gabriel Mountains.*

In Los Angeles versus the San Gabriel Mountains, it is not always clear which side is losing. For example, the Genofiles, Bob and Jackie, can claim to have lost and won. They live on an acre of ground so high that they look across their pool and past the trunks of big pines at an aerial view over Glendale and across Los Angeles to the Pacific bays. The setting, in cool dry air, is serene and Mediterranean. It has not been everlastingly serene.

On a February night some years ago, the Genofiles were awakened by a crash of thunder – lightning striking the mountain front. Ordinarily, in their quiet neighborhood, only the creek beside them was likely to make much sound, dropping steeply out of Shields Canyon on its way to the Los Angeles River. The creek, like every component of all the river systems across the city from mountains to ocean, had not been left to nature. Its banks were concrete. Its bed was concrete. When boulders were running there, they sounded like a rolling freight. On a night like this, the boulders should have been running. The creek should have been a torrent. Its unnatural sound was unnaturally absent. There was, and had been, a lot of rain.

The Genofiles had two teen-age children, whose rooms were on the uphill side of the one-story house. The window in Scott's room looked straight up Pine Cone Road, a cul-de-sac, which, with hundreds like it, defined the northern limit of the city, the confrontation of the urban and the wild. Los Angeles is overmatched on one side by the Pacific Ocean and on the other by very high mountains. With respect to these principal boundaries, Los Angeles is done sprawling. The San Gabriels, in their state of tectonic youth, are rising as rapidly as any range on earth. Their loose inimical slopes flout the tolerance of the angle of repose. Rising straight up out of the megalopolis, they stand ten thousand feet above the nearby sea, and they are not kidding with this city. Shedding, spalling, self-destructing, they are disintegrating at a rate that is also among the fastest in the world. The phalanxed communities of Los Angeles have pushed themselves hard against these mountains, an aggression that requires a deep defense budget to contend with the results. Kimberlee Genofile called to her mother, who joined her in Scott's room as they looked up the street. From its high turnaround, Pine Cone Road plunges downhill like a ski run, bending left and then right and then left and then right in steep christiania turns for half a mile above a three-hundred-foot straightaway that aims directly at the Genofiles' house. Not far below the turnaround, Shields Creek passes under the street, and there a kink in its concrete profile had been plugged by a six-foot boulder. Hence the silence of the creek. The water was now spreading over the street. It descended in heavy sheets. As the young Genofiles and their mother glimpsed it in the all but total darkness, the scene was suddenly illuminated by a blue electrical flash. In the blue light they saw a massive blackness, moving. It was not a landslide, not a mudslide, not a rock avalanche; nor by any means was it the front of a conventional flood. In Jackie's words, "It was just one big black thing coming at us, rolling, rolling with a lot of water in front of it, pushing the water, this big black thing. It was just one big black hill coming toward us."

In geology, it would be known as a debris flow. Debris flows amass in stream valleys and more or less resemble fresh concrete. They consist of water mixed with a good deal of solid material, most of which is above sand size. Some of it is Chevrolet size. Boulders bigger than cars ride long distances in debris flows. The dark material coming toward the Genofiles was not only full of boulders; it was so full of automobiles it was like bread dough mixed with raisins. On its way down Pine Cone Road, it plucked up cars from driveways and the street. When it crashed into the Genofiles' house, the shattering of safety glass made terrific explosive sounds. A door burst open. Mud and boulders poured into the hall. We're going to go, Jackie thought. Oh, my God, what a hell of a way for the four of us to die together.

The parents' bedroom was on the far side of the house. Bob Genofile was in there kicking through white satin draperies at the panelled glass, smashing it to provide an outlet for water, when the three others ran in to join him. The walls of the house neither moved nor shook. As a general contractor, Bob had built dams, department stores, hospitals, six schools, seven churches, and this house. It was made of concrete block with steel reinforcement, sixteen inches on center. His wife had said it was stronger than any dam in California. His crew had called it "the fort." In those days, twenty years before, the Genofiles' acre was close by the edge of the mountain brush, but a developer had come along since then and knocked down thousands of trees and put Pine Cone Road up the slope. Now Bob Genofile was thinking, I hope the roof holds. I hope the roof is strong enough to hold. Debris was flowing over it. He told Scott to shut the bedroom door. No sooner was the door closed than it was battered down and fell into the room. Mud, rock, water poured in. It pushed everybody against the far wall. "Jump on the bed," Bob said. The bed began to rise. Kneeling on it – on a gold velvet spread – they could soon press their palms against the ceiling. The bed also moved toward the glass wall. The two teen-agers got off, to try to control the motion, and were pinned between the bed's brass railing and the wall. Boulders went up against the railing, pressed it into their legs, and held them fast. Bob dived into the muck to try to move the boulders, but he failed. The debris flow, entering through windows as well as doors, continued to rise. Escape was still possible for the parents but not for the children. The parents looked at each other and did not stir. Each reached for and held one of the children. Their mother felt suddenly resigned, sure that her son and daughter would die and she and her husband would quickly follow. The house became buried to the eaves. Boulders sat on the roof. Thirteen automobiles were packed around the building, including five in the pool. A din of rocks kept banging against them. The stuck horn of a buried car was blaring. The family in the darkness in their fixed tableau watched one another by the light of a directional signal, endlessly blinking. The house had filled up in six minutes, and the mud stopped rising near the children's chins.

1-2   The Night the Mountain Fell – Gordon Gaskill

This report by Gordon Gaskill of the 1963 dam failure at the Vajont Dam in the Italian Alps, describes one of the most terrible disasters of modern times. Some three years before the disaster, a great dam, 858 feet high, had been built across the narrow valley to impound a deep reservoir. Repeated, but very small Earth movements were recorded in the ensuing period, but on the night of October 9, after torrential rains, a rockslide of some 600 million tons occurred high in the valley above the reservoir and threw a huge volume of rock material into the reservoir, creating a great surge *of water that rose 300 feet above the top of the dam. As it swept through the valley below, some 2,000 people lost their lives. Even in the absence of the reservoir, the rock fall was of such proportions that it would probably have caused considerable damage, but the rock fall itself was facilitated by the construction of the dam. The bedrock consisted of interbedded limestones and clays, which had been weakened by deformation, and further weakened by water percolating through it from the reservoir. This extract is from an article which appeared in* Readers Digest *(1965).*

They say the animals knew. In that last peaceful twilight – Wednesday, October 9, 1963 – hares grew suddenly bold and, oblivious of passing men and automobiles, raced silently, intently, down the paved road – away from the lake. As darkness gathered, cows milled uneasily in their stalls, dogs whimpered, chickens stirred in their pens, unwilling to sleep. A couple watching television was irritated by the unnatural, noisy fluttering of their caged canary. Then the fluttering abruptly stopped: in its strange panic the bird had caught its head in the cage bars and strangled to death. Husband stared at wife: "Something's going to happen! The dam . . . ?"

Life or death this night in the small town of Longarone, in northeast Italy, would turn on a simple, single fact: how high up the hillsides of the river valley below the Vajont Dam you happened to be. All but those in the highest parts would soon die.

An engaged couple, due to marry in six days, had a slight difference of opinion. Giovanna wanted to go to the movies at Belluno, the provincial capital about 12 miles away, but her fiancé, Antonio, felt too tired and begged off. They separated for the night, he to his higher home, she to a lower one. Next morning he would be digging in the muddy waste where Giovanna, her family and home had disappeared, repeating endlessly, "If only I had taken her to the movies. . . . If only I had taken her to the movies."

A teen-age boy astride his motorbike fidgeted with embarrassment as, from a window, his mother tried to talk him out of riding off to another village to see a girl. But from inside the house his father, remembering his own salad days, called out indulgently, "Oh, let him go!" The mother sighed, gave in and the boy rode off to safety – never to see home or parents again.

Visiting Longarone were three Americans of Italian descent, all staying in a low-lying little hotel. One of them, John De Bona, of Riverside, Calif. retired to his room – and never would be seen again. Two others, Mr. and Mrs. Robert De Lazzero, of Scarsdale, N.Y. had panted uphill nearly 150 steps to have dinner with his great-aunt Elisabetta, and two cousins. Shortly before ten, dinner over, they were about to walk back down to the hotel when Aunt Elisabetta said, "Don't go yet. See, I've saved a special bottle of wine for you." Somewhat reluctantly, they stayed on a little longer. They were lucky.

For the clocks of Longarone would never strike 11 on this night. Just before that hour Longarone and the hamlets clustering near it would be erased from the earth, and more than 2000 people would die in perhaps the world's most tragic dam disaster.

Four years later the great new Vajont Dam had been both the pride and the fear of people living around Longarone, a sub-Alpine town not far south of the Austrian border. Flung across the nearby Vajont gorge – so deep and narrow that sunlight touched its bottom only fleetingly at noon – the dam was the highest arch dam in all the world, a showpiece for visiting foreign engineers, a magnet for tourists. Its graceful curving wall of tapered concrete rose 858 feet above its base – nearly five times as high as Niagara Falls, 132 feet higher than America's highest dam, Hoover. Its impounded lake, not yet full, would provide enormous amounts of electricity to bring industry, jobs and prosperity to mountain folk for miles around.

Still, many feared it.

Ever since 1959 a swelling chorus of protest had demanded that the dam be stopped or that absolute assurance be given that it was safe. But the site had been approved and all work had been carefully supervised by some of Italy's most respected geologists and engineers. Chief among them was the very father of the Vajont project, Dr. Carlo Semenza, an internationally known engineer who had built dams in several countries. He and others concluded that there might be a little minor land slippage at first – as with nearly all artificial lakes – but nothing to worry about.

People living nearby weren't so sure. Above all, they distrusted the stability of Monte Toc, which anchored the dam's left shoulder and hung nearly 4000 feet above the new lake. They gave it an ominous nickname, *la montagna che cammina* – "the mountain that walks." The village of Erto, just above the lake, felt itself especially menaced and made most of the early protests.

With construction having begun in 1956, all was ready for a partial test-filling of the lake by March 1960. The results were worrisome. Even modest amounts of water produced, on November 4, 1960, an alarming crack in the earth high up on Monte Toc – a gaping split more than a foot wide and about 8000 feet long. At the same time, a half-million tons of earth and rock slipped into the still-small lake, churning up waves six feet high.

Disappointed, Società Adriatica di Elettricità (SADE), the power company building the dam, lowered the lake and, its timetable wrecked, turned to two years of expensive testing. Also, it performed extensive remedial work: strengthening the dam, digging a large bypass tunnel, sealing suspected rock fractures with pressurized concrete.

But, less than six months after the warning slide, Dr. Semenza himself began to lose hope. In an April 1961 letter, not divulged until after the disaster, he wrote to an engineering friend: "The problems are probably too big for us, and there are no practical remedies to take." He died six months afterward. Yet neither he nor others sharing his worries ever dreamed of any serious danger to *human life*. They feared only that landslides might so clog the basin as to make it useless for water storage.

One test (No. 19) on SADE's elaborate scale model (1/200 actual size) of the entire dam and basin was to have especially disastrous consequences. The formal report on this test said that *if the lake's level was 75 feet below maximum*, this would be "absolutely safe, even in the face of the most catastrophic landslide that can be foreseen." Such a fantastic slide, said the test, would churn up dangerous waves about 80 feet high on the lake.

The safety measures to counteract this seemed obvious: if a slide seemed imminent, (a) get the lake down at least 75 feet below maximum, and (b) evacuate everybody from the shoreline belt so that the expected waves could spend their rage harmlessly. As for people in Longarone, 1½ miles *below* the dam, there was no reason to worry. With the lake down to this "safety level," only about five feet of water, a harmless trickle, could possibly go *over* the dam.

In April 1963 the situation seemed ripe for a new test-raising of the water. By now a new factor had entered the picture. SADE and many other private power companies had been forcibly nationalized under a new state power board, ENEL (Ente Nazionale per l'Energia Elettricà). The exit valves were closed, and the water started up again.

As the waters crept higher, the old disheartening signals appeared. Another long, frightening crack split the earth high on Monte Toc. From July to September, small earth tremors shook the area. Strange rumbling came from deep in the earth; the lake water "boiled up" ominously.

SADE earlier had implanted dozens of sentinel bench marks on the mountain's flank. These, watched regularly by optical instruments so sensitive that they noted even a hairbreadth of movement, would signal any tendency of the earth to slip downward. And signal they did. Quiet for months, they suddenly began reporting an ever-rising tendency of Monte Toc's earthen flank to slip . . . 6 . . . 8 . . . 12 . . . 22 millimeters per 24 hours – edging up toward the highest "danger reading" of 40, registered three years earlier at the time of the first slide.

Perturbed, ENEL–SADE halted the test-raising with the waters still 41 feet below maximum, hoping that the earth would settle down to a new stability. Unfortunately, it did not. And, to make matters worse, thrice-normal rains, the heaviest in 20 years, had made the earth unusually sodden.

Nino A. Biadene, ENEL–SADE deputy director-general for technical matters, from the head office in Venice, now declared that there would be no question of letting the water go any higher, even though it was authorized. On the contrary, he would order the water lowered if the alarm signals continued.

They did continue – and got worse. Thus, on September 26 – with disaster 13 days away – Biadene gave the emergency order: "Take the water down!" Instantly, the great exit valves were opened, and water began rushing out. But not *too* fast, for this would too quickly remove the water-cushion supporting Monte Toc's soaked earth and make it even more unstable. . . .

Next day was golden and merry in Longarone. Most of the harvesting had been done. The year's business had been good: new factories had been coming in; ever more tourists had come to visit the dam; everybody had jobs and fatter paychecks. Best of all, Longarone's famed ice-cream makers, who fanned out over Europe every March to make and sell their delicacy, were now streaming home to spend the winter with their families. It was a gay time of homecoming, of matchmaking and marrying, of meeting old friends in the many bar-cafés which served as clubs, of swapping stories of the season's business over a glass of *vino*.

True, the great dam above cast a certain shadow on the merriment. Word had leaked out that the slippage rate was high today. A truck driver named Antonio Savi said that he had driven over the paved road on Monte Toc and that it was buckling so much he refused to go there again. But this seemed familiar stuff somehow.

Hundreds of men stayed at the bar-cafés later than usual that evening. They were ardent soccer fans, and at 9:55 began a television relay from Madrid of an important game between the Spanish Real Madrid team and the Scottish Glasgow Rangers. Most would never live to know who won the game (the Spaniards, 6-0), and later it was bitterly cursed for luring so many to death. In fact it probably made little difference in the final death toll; it doomed some, but saved others.

Meanwhile, up on Monte Toc the sentinel bench marks had gone wild. Far in the past was the old danger-reading of 40. Now the signals were reading 190 . . . 200!

Just before 9 p.m. – with disaster barely 100 minutes away – Biadene in Venice decided that it would be wise to institute *below* the dam and around Longarone some of the same precautions already in force up around the lake itself. By phone he ordered a subordinate engineer in Belluno to have police bar all traffic on the below-dam roads in and around Longarone. And from the dam went out a series of phone calls to people and establishments down by the Piave – to the sawmill, the spinning mill, the quarry, a tavern – passing on the message: "Maybe a little water over the dam tonight . . . nothing to get alarmed about."

At 10:39 the mountain fell. Not all of it, but a greater single mass than has fallen in Europe since prehistoric times – with a shock so great that it simulated genuine earthquake effects on seismographs in five countries. About 600 million tons came down – roughly equivalent to a football field piled with earth and rock to a height of *40 miles*. It didn't fall slowly, by inches, as predicted, disintegrating as it went. Instead, the mountain split away cleanly, as if cut by a knife, and fell straight into the lake.

Don Carlo Onorini, parish priest of Casso village perched high on a mountain just across the lake, happened to be watching. In the bright glare of the floodlights he saw the mountainside suddenly slip loose "with a sound as if the end of the earth had come." A muddy flood leaped up toward him and he saw, just below him, the mountainside clawed away, a church and some lower houses vanish, before the flood fell back into the valley. An enormous blue-white flash filled the sky as the great 20,000-volt high-tension lines shorted, fused and broke, plunging the valley into darkness.

All around the lakeshore the tormented water raced – not 80 feet high, but in places, clawing up to *800 feet* above lake level. It thundered at the dam – and the dam held. But the water went over the dam not five feet high but up to 300 feet high, and smashed to the bottom of the gorge 800 feet below. There it was constricted as in a deadly funnel, and its speed fearfully increased. It shot out of the short gorge as from a gun barrel and spurted across the wide Piave riverbed, scooping up millions of deadly stones. Ahead of it raced a strange icy wind and a storm of fragmented water, like rain, but flying *upward*. By now it was more than a wave, more than a flood. It was a tornado of water and mud and rocks, tumbling hundreds of feet high in the pale moonlight, leaping straight at Longarone.

In the next minutes – about six – it thundered far up the hillside where Longarone stood, then recoiled into the Piave Valley with a fearful sucking noise, as if a mile-wide sink were emptying. In those six minutes Longarone vanished from the earth.

Almost none of the survivors – even those watching from so high up that they were never in danger – could give any coherent account of what they saw. One man remembers "a great milky cloud settling over our town." Another, "a huge grayish-silverish mass that seemed so big it hardly seemed to move at all. Then I saw things swirling in the mass – bodies, lumber, automobiles." Most remember the strangely cold wind and the horrible noise "like a thousand express trains rushing on us . . . a noise so great the ears refused to hear it."

At one bar, somebody yelled, "The dam's broken! Run for your lives!" Those spry enough made it; those too old or too dazed died. In another bar, those who jumped out uphill windows made it; those who went out the front door did not.

A girl of 22, Maria Teresa Galli, was just closing her balcony shutters when she felt a great cold wind, and somehow the house seemed to dissolve around her. Some great force, part wind, part water, picked her up and whirled her along as she thought dazedly: "I'm flying ... walking ... swimming!" Two hundred yards away, an old couple Arduino Burrigana and his wife Gianna, watched from the top floor as the flood invaded the ground floor – and dropped some dark bundle which emitted a groan. It was Maria Teresa Galli, fainting with shock, bruised, but little harmed.

A paralyzed man, helpless in his chair, called out in panic to his wife, "What is it? What is it?" She stepped out on the balcony to look. He felt the house tremble to some great shock, and called out, "Where are you?" She never answered. A passing edge of the wave had flicked her away.

The visiting American couple from Scarsdale had nearly finished the special bottle of wine when the roar came. One of the cousins pulled open the door, stared out, slammed it shut again and cried, "We're all dead!" Water poured over them, and the man remembers dazed thoughts passing through his mind like, "What's the use ... with a thousand feet of water over us?" Yet in a moment, miraculously, the water retreated. He was safe, and so were his wife and cousins – except for fractures – but it had been too much for Aunt Elisabetta. She lay dead.

It took some time before the outer world learned how enormous the tragedy had been. The flood had isolated Longarone in a sea of mud. The first reporter waded in about 2:30 a.m., and well before dawn in came 1000 of Italy's famed mountain troops, the *Alpini*, the vanguard of nearly 10,000 rescue workers – soldiers, police, firemen, Red Cross, Boy Scouts, volunteers of all kinds. ...

The great flood of death was matched by an equal flood of compassion and help from all Italy. While no single Italian admitted to any blame for the disaster, somehow all Italy felt responsible. ...

All around the world the news produced horror and help. From Australia to Canada, many Italian communities – often led by local Italian-language newspapers – gave money. James Bez, 58, an unemployed worker in Stamford, Conn., who had been born in Longarone and lost nearly all his family there, personally collected $350 in cash plus 40 boxes of clothing, which a Greek ship carried to Italy free.

One day recently, along with an engineer who had no connection with the dam, I climbed a narrow, muddy road, still being rebuilt, up to the village of Casso – where, on that terrible night, the parish priest had looked across to see the mountain fall. Here, 800 feet above the lake, the whole scene lay clear before us.

Below us, to the right stood the great dam, still intact save for a few minor bruises at the top. Directly below us lay the huge mass of the slide, looking as if it had been there forever, its trees and bushes still growing, already being called Monte Nuovo – "the new mountain." It nudges up against the dam and is, in effect, a new natural dam about 1½ miles thick with earth and rock, towering nearly 300 feet higher than the man-made dam, now forever useless.

The lake has shrunk to about half its former size. But Italy will soon need all the electric power it can get, and there is an unspoken hope that later, when fear and feeling have cooled, some absolutely safe way may be found to go on using what is left of the lake for precious water storage.

As we stood at Casso looking down, my engineer companion pointed with his pipestem to the great dam, still a proud work of man. "In building things these days," he said, "man can calculate stresses and strains just about 100 percent – as that dam proves. But so far, not even the best experts, using the best equipment, can be absolutely sure what goes on deep down in the earth. These days engineering is pretty much an exact science. Geology isn't – not yet."

1-3   The Turtle Mountain Slide –
R.G. McConnell and R.W. Brock

Landslides are rapid down-slope movements of rock debris. They represent a particular example of the universal down-slope movement of weathered material. This "mass wasting" takes place under the influence of gravity, and involves down-slope movement ranging from sudden rockfalls to slow hill creep, which proceeds almost imperceptibly over many years. Landslides are facilitated by certain structural rock conditions, such as faults, or bedding planes inclined parallel to the slope of the ground, interbedding of shales with more brittle strata, excessively heavy rainfall which may lubricate joints and bedding planes, and by artificial disturbance of the geologic environment.

A landslide took place in 1903 near the small mining village of Frank, Alberta, and killed 70 people. It is estimated that about 40 million cubic yards of rock avalanched down the face of the 3,000 feet high Turtle Mountain at a speed reaching 60 miles per hour, and struck the valley bottom with such force that it spread across the two-mile width of the floor and was carried 400 feet up the other side. Turtle Mountain consisted of jointed limestone, in which, although the stratification was perpendicular to the face of the mountain, the joints were parallel to it. The opening of these joints by weathering facilitated the movement. The account that follows, published soon after the event took place, is taken from a 1904 Canadian Department of the Interior report by Richard McConnell (1857–1942) and Reginald Brock (1874–1935).

At dawn, on April 29, 1903, a huge rock mass, nearly half a mile square and probably 400 to 500 feet thick in places, suddenly broke loose from the east face of Turtle mountain and precipitated itself with terrific violence into the valley beneath, overwhelming everything in its course. The great mass, urged forward by the momentum acquired in its descent, and broken into innumerable fragments, ploughed through the bed of Old Man river, and carrying both water and underlying sediments along with it, crossed the valley and hurled itself against and up the opposite terraced slopes to a height of 400 feet. Blocks of limestones and shale, mingled with mud, now cover the valley to a depth of from 3 to probably 150 feet, over an area of 1.03 square miles.

The number of people killed by the slide is not known exactly, but it is given at about 70. The property destroyed includes the tipple and plant at the mouth of the Canadian American Coal and Coke Company's mine, the company's barn and seven cottages at the east end of the town of Frank, half a dozen outlying houses, with some shacks and camps, besides a considerable number of horses and cattle and a couple of ranches. The track of the Crow's Nest Railway was hopelessly buried for a distance of nearly 7,000 feet and the lower mile of the Frank and Grassy Mountain Railway met a similar fate. The people occupying the houses in the track of the slide were all swept away with it and destroyed, with the exception of a few near the edge of the slide, who escaped in some almost miraculous way, which they themselves cannot explain.

The slide occurred about 4:10 a.m., at a time when most of the inhabitants of the valley were asleep and before full daylight. The statements of the few eye-witnesses throw little light on the character of the slide, but the following notes obtained from them are not without interest:—

Karl Cornelianson was awakened by the noise. He rushed to the door of his house, which looked out over the first terrace flat and the base of the second. His first thought was that there had been an explosion at the mine and his first look was in that direction. Seeing nothing there, he glanced round to the terrace flat in time to see the rock debris hurl itself against the slope of the second terrace, and its momentum spent, fall back to the lower level. His impression was that an explosion had taken place directly in front of him. The edge of the slide was about a quarter of a mile from his door.

Mr. McLean, who kept a boarding house in Frank, was already up. Hearing the noise, he rushed to the door in time to see the slide rush by only a few feet in front of him. The passage of the slide was so instantaneous that he thought an eruption had taken place directly in front of him.

A freight train was shunting on the mine siding at the time of the disaster. They had taken some cars of coal from the tipple at the mouth of the tunnel and were slowly backing up for another load, when the engineer heard the rocks breaking away from the mountain above. He immediately changed to full speed ahead and ran out of danger. The conductor saw the men at the tipple become alarmed and start to run, but they were overtaken by the slide and perished. Immediately after everything was shrouded in a cloud of dust.

Mr. Warrington, who was sleeping in one of the cottages destroyed, was awakened by a noise which he thought was caused by hail. He jumped out of bed and then realized that it was something more serious, but before he had time to become alarmed the house began to rock, and the next thing he was conscious of was finding himself in the lee of some rocks, forty feet from where the house had stood. His bed was some twenty feet farther on. His thigh was broken and he was otherwise injured by small fragments of debris being forced into his body. He pulled himself out with his arms, and was trying to work his way to some children, whose cries he heard, when the first rescue party arrived.

1-4 Candide – Voltaire

Earthquakes are movements of the Earth's crust, generally caused by movement of rock masses relative to one another. They range in intensity from minor tremors, so slight that they can be detected only by sensitive seismographs, to earthquakes so violent that major areas are destroyed. The great loss of life caused by some earthquakes is often the result of secondary causes, such as fires from broken gas mains (which may run out of control because of broken water mains), great tidal waves (tsunamis) from submarine earthquakes, and flooding caused by landslides. Some 900,000 are said to have perished in the Shantzee Earthquake of 1556, for example.

The two accounts which we include of the Lisbon Earthquake are quite different. While neither represents an eyewitness description – and both are reflective, as well as descriptive – the first account, by Voltaire, is taken from Candide *(1759), a novel which parodied contemporary virtue and philosophy. Voltaire – the pseudonym of François Marie Arouet (1694–1778) – was born in Paris, educated by Jesuit teachers, studied law and epitomized the Enlightenment movement. One of the most influential figures of the eighteenth century and a prolific writer, his work includes poetry, drama, scientific, philosophical, biographical and historical work. In and out of favor at the French court, he lived* at various times in England, Germany, and Switzerland. In Candide, Dr. Pangloss, Candide's tutor, and a "brutal sailor" suffer storm and shipwreck and are finally cast ashore in Portugal, only to find themselves then involved in the great Lisbon Earthquake. Voltaire wrestles with the problem of death, including death resulting from "acts of God" (the drowning of the Anabaptist) and the injustice of death at the hands of other men. Pangloss, still arguing not only for a "train of sulphur running underground from Lima to Lisbon," but also for the ultimate harmony of things ("It is impossible that things should not be what they are; for all is well."), is hanged by the Portuguese because of his comments on the earthquake. "Must I see you hang without knowing why" pleads Candide of his dead master. "Was it necessary that you should be drowned in port?" he groans over his dead Anabaptist friend. In these questions, Voltaire echoes a baffled and angry agony that swept Europe after the Lisbon Earthquake of 1755.

Half the enfeebled passengers, suffering from that inconceivable anguish which the rolling of the ship causes in the nerves and in all the humors of bodies shaken in contrary directions, did not retain strength enough even to trouble about the danger. The other half screamed and prayed; the sails were torn, the masts broken, the vessel leaking. Those worked who could, no one cooperated, no one commanded. The Anabaptist tried to help the crew a little: he was on the main deck; a furious sailor struck him violently and stretched him on the deck; but the blow he delivered gave him so violent a shock that he fell head-first out of the ship. He remained hanging and clinging to part of the broken mast. The good Jacques ran to his aid, helped him to climb back, and from the effort he made was flung into the sea in full view of the sailor, who allowed him to drown without condescending even to look at him. Candide came up, saw his benefactor reappear for a moment and then be engulfed for ever. He tried to throw himself after him into the sea; he was prevented by the philosopher Pangloss, who proved to him that the Lisbon roads had been expressly created for the Anabaptist to be drowned in them. While he was proving this *a priori*, the vessel sank, and everyone perished except Pangloss, Candide and the brutal sailor who had drowned the virtuous Anabaptist; the blackguard swam successfully to the shore and Pangloss and Candide were carried there on a plank. When they had recovered a little, they walked toward Lisbon; they had a little money by the help of which they hoped to be saved from hunger after having escaped the storm. Weeping the death of their benefactor, they had scarcely set foot in the town when they felt the earth tremble under their feet; the sea rose in foaming masses in the port and smashed the ships which rode at anchor. Whirlwinds of flame and ashes covered the streets and squares; the houses collapsed, the roofs were thrown upon the foundations, and the foundations were scattered; thirty thousand inhabitants of every age and both sexes were crushed under the ruins. Whistling and swearing, the sailor said: "There'll be something to pick up here." "What can be the sufficient reason for this phenomenon?" said Pangloss. "It is the last day!" cried Candide. The sailor immediately ran among the debris, dared death to find money, found it, seized it, got drunk, and having slept off his wine, purchased the favors of the first woman of good-will he met on the ruins of the houses and among the dead and dying. Pangloss, however, pulled him by the sleeve. "My friend," said he, "this is not well, you are disregarding universal reason, you choose the wrong time." "Blood and 'ounds!" he retorted, "I am a sailor and I was born in Batavia; four times have I stamped on the crucifix during four voyages to Japan; you have found the right man for your universal reason!" Candide had been hurt by some falling stones; he lay in the street covered with debris. He said to Pangloss: "Alas! Get me a little wine and oil; I am dying." "This earthquake is not a new thing," replied Pangloss. "The town of Lima felt the same shocks in America last year; similar causes produce similar effects; there must certainly be a train of sulphur underground from Lima to Lisbon." "Nothing is more probable," replied Candide; "but, for God's sake, a little oil and wine." "What do you mean, probable?" replied the philosopher; "I maintain that it is proved." Candide lost consciousness, and Pangloss brought him a little water from a neighboring fountain. Next day they found a little food as they wandered among the ruins and gained a little strength. Afterward they worked like others to help the inhabitants who had escaped death. Some citizens they had assisted gave them as good a dinner as could be expected in such a disaster; true, it was a dreary meal; the hosts watered their bread with their tears, but Pangloss consoled them by assuring them that things could not be otherwise. "For," said he, "all this is for the best; for, if there is a volcano at Lisbon, it cannot be anywhere else; for it is impossible that things should not be where they are; for all is well." A little, dark man, a familiar of the Inquisition, who sat beside him, politely took up the conversation, and said: "Apparently, you do not believe in original sin; for, if everything is for the best, there was neither fall nor punishment." "I most humbly beg your excellency's pardon," replied Pangloss still more politely, "for the fall of man and the curse necessarily entered into the best of all possible worlds." "Then you do not believe in free-will?" said the familiar. "Your excellency will pardon me," said Pangloss; "free-will can exist with absolute necessity; for it was necessary that we should be free; for in short, limited will . . ." Pangloss was in the middle of his phrase when the familiar nodded to his armed attendant who was pouring out port or Oporto wine for him.

After the earthquake which destroyed three-quarters of Lisbon, the wise men of that country could discover no more efficacious way of preventing a total ruin than by giving the people a splendid *auto-da-fé*. It was decided by the University of Coimbre that the sight of several persons being slowly burned in great ceremony is an infallible secret for preventing earthquakes. Consequently they had arrested a Biscayan convicted of having married his fellow-godmother, and two Portuguese who, when eating a chicken, had thrown away the bacon; after dinner they came and bound Dr. Pangloss and his disciple Candide, one because he had spoken and the other because he had listened with an air of approbation; they were both carried separately to extremely cool apartments, where there was never any discomfort from the sun; a week afterward each was dressed in a sanbenito and their heads were ornamented with paper mitres; Candide's mitre and sanbenito were painted with flames upside down and with devils who had neither tails nor claws; but Pangloss's devils had claws and tails, and his flames were upright. Dressed in this manner they marched in procession and listened to a most pathetic sermon, followed by lovely plain-song music. Candide was flogged in time to the music, while the singing went on; the Biscayan and the two men who had not wanted to eat bacon were burned; and Pangloss was hanged, although this is not the custom. The very same day, the earth shook again with a terrible clamour. Candide, terrified, dumbfounded, bewildered, covered with blood, quivering from head to foot, said to himself: "If this is the best of all possible worlds, what are the others? Let it pass that I was flogged, for I was flogged by the Bulgarians, but, O my dear Pangloss! The greatest of philosophers! Must I see you hanged without knowing why! O my dear Anabaptist! The best of men! Was it necessary that you should be drowned in port! O Mademoiselle Cunegonde! The pearl of women! Was it necessary that your belly should be slit!" He was returning, scarcely able to support himself, preached at, flogged, absolved and blessed, when an old woman accosted him and said: "Courage, my son, follow me."

1-5 The Lisbon Earthquake – James R. Newman

The second account of the Lisbon earthquake, by James Roy Newman (1907–1966) – educated at City College of New York and Columbia University, American lawyer, diplomat, mathematician, journalist, and author – catches the setting of the earthquake and the mood of the continent. The catastrophe, which lasted only ten minutes, brought great physical damage and destruction. But in those ten minutes, "an era came to an end." Europe was never quite the same again, for that brief interval changed the course of human thought. The stability of seventeenth century philosophy, the ordered domestic tranquility of the prosperous, and the confidence of the common man, all tumbled before what many regarded as the wrath and the judgment of God. It was this prevailing mood of optimism that Voltaire attacked, although the extension of his own view, into a sustained cynicism, as others from Rousseau to Newman have pointed out, is, itself, too extreme. If the "all is well" brand of optimism leads to a life of shallow unconcern and indifference, it is also not clear that an "all is meaningless" brand of pessimism leads to a life which is more productive. The quest and debate raised by the Lisbon earthquake still continue. Our extract here is taken from his book Science and Sensibility *which he first published in 1961.*

At about 9:30 a.m. on All Saints' Day, Saturday, November 1, 1755, a massive earthquake convulsed the great city of Lisbon in Portugal. There were three distinct shocks separated by intervals of about a minute. "The first alarm" – I quote from Sir Thomas Kendrick's description – "was a rumbling noise that many people said sounded like that of exceptionally heavy traffic in an adjacent street, and this was sufficient to cause great alarm and make the buildings tremble; then there was a brief pause and a devastating shock followed, lasting over two minutes, that brought down roofs, walls, and façades of churches, palaces and houses and shops in a dreadful deafening roar of destruction. Close on this came a third trembling to complete the disaster, and then a dark cloud of suffocating dust settled foglike on the ruins of the city. It had been a clear, bright morning, but in a few moments the day turned into the frightening darkness of night." The earthquake lasted about ten minutes; but in that ten minutes an era came to an end. After the shaking of Lisbon, Europe was not the same, as after Hiroshima the world was changed.

For the bicentenary of the disaster, the director and principal librarian of the British Museum prepared a scholarly and entertaining study based upon the immense literature occasioned by the event. Since the fall of Rome in the fifth century no other happening had so shocked Western civilization. Not only in Portugal but throughout Europe there was an outpouring of diagnosis and commentary. Scientists offered explanations, preachers sermonized, philosophers speculated, poets lamented, theologians moralized and exhorted. One opinion evoked another and controversies multiplied. Sir Thomas reports the earthquake in vivid detail, but this is only the setting for his book. Primarily he is concerned with the effect of the calamity on men's minds: how they interpreted its meaning, what light it shed on the deep question of man's place in the world, of crime and punishment and the wrath of God, of duty to one's fellows, of competing faiths and creeds, of optimism and man's prospects. Physical damage is soon repaired and the loss of human life easily forgotten; but the earthquake shaped the future by altering the course of thought.

For a few minutes of that dreadful day in November, Lisbon seemed – to observers in ships on the Tagus and on the higher ground around the city – to "sway like corn in the wind." Then avalanches of falling masonry hid the ruins under a cloud of dust. But the violent tremors, which collapsed some of the finest buildings and hundreds of the smaller houses and shops, were only the first ordeal. Within a quarter of an hour of the triple shock, fires broke out in different parts of the town, and as the conflagration mounted, a third disaster occurred. The waters of the Tagus "rocked and rose menacingly, and then poured in three great towering waves over its banks, breaking with their mightiest impact on the shore between the Alcántara docks and [the palace square] the Terreiro do Paço."

Some 10,000 to 15,000 persons, it is said, lost their lives in the earthquake. Original estimates were wildly exaggerated, and the official statistics compiled with great difficulty afterward were not reliable. But the careful historian, Moreira de Mendonça, thought that not more than 5,000 of Lisbon's 275,000 persons were killed on the first of November, the casualties being doubled or trebled during the course of the month. The loss of property was severe, the fire doing the most damage. Much could have been salvaged from the shattered buildings, but the flames destroyed pictures, furniture, tapestries, jewelry and plate, and enormous stocks of merchandise. The foreign traders alone – principally British and Hamburg merchants – lost about £10,000,000 worth of goods. Among the irreplaceable treasures incinerated in a single building, the palace of the Marquês de Louriçal, were two hundred pictures including works by Titian, Correggio, and Rubens, a library of 18,000 printed books, 1000 manuscripts, and a huge collection of maps and charts relating to the Portuguese voyages of discovery and colonization in the East and in the New World. Seventy thousand books perished in the burning of the King's palace.

Measured by its death roll and damage, Lisbon was not one of the greatest disasters of its kind. India, China, and Japan have suffered much more terrible earthquakes. The Kwanto quake of 1923, which almost leveled Tokyo and Yokohama, killed 100,000 persons. Nonetheless, the earthquake of November 1, 1755, was in every sense a colossal seismic disturbance. It was felt over a large area. It shook heavily the whole southwestern corner of Portugal; there was a tremendous upheaval in North Africa in the area of Fez and Meknes, with much loss of life; Spain and France experienced shocks, as did Switzerland and northern Italy; tidal waves were recorded in England and Ireland at 2 p.m. and in the West Indies at 6 p.m. Over almost all Europe, including Scandinavia, the water in rivers, canals, and lakes was suddenly agitated. But while the ravages and astonishing side-phenomena of this cruel day caused widespread alarm, there were other factors that had a special turning effect on men's minds.

Consider, first, how "men of action" responded to the disaster. It must be said that they came off well. Lisbon was, of course, in a state of terror and panic. Corpses were everywhere. Men, women, and children crept out of the rubble bleeding and mangled, and searched frantically for others in their families. Suicidal attempts were made to escape from the upper stories of partly wrecked houses. The air was filled with screams and groans, and with the piteous cries of wounded horses and dogs. As flames roared through the town, it was soon enveloped in an impenetrable sulphurous cloak. And the earthquake continued. Brief aftershocks, some quite violent, although they did little damage, kept hysteria alive. (In the week after November 1, there were nearly thirty earthquakes, and by August, 1756, some five hundred aftershocks had been recorded.) There was reason to believe, as many thought, that God would not desist until the city had been razed.

But the men of action wasted no time on such feckless speculations. Foremost among these energetic leaders was King José's secretary of state, Sebastião José

Carvalho e Mello, the future Marquês de Pombal – the name by which he is remembered. Pombal, who was dictator of Portugal throughout the reign of José I, was troublesome, touchy, and aggressive, but a man of exceptional ability. Where others, on learning of the catastrophe, wrung their hands and looked to heaven for succor, he saw the need for firm and immediate terrestrial steps to avert social chaos. When the unhappy King asked despairingly what was to be done, Pombal is said to have replied, "Bury the dead and feed the living."

This he did and much more. . . .

It cannot be said that the response of the men of science, such as they were, compared favorably with that of the men of action. Lisbon needed some plain truths about the earthquake to counteract the monstrous exaggerations. But the very suggestion that the earthquake was a natural phenomenon, like an eclipse or a storm, shocked the devout and enraged their religious instructors. Even honest men deemed it prudent to hedge. A physician, José Alvares da Silva, was one of the first to venture the opinion that while the earthquake might be a judgment of God, it could also be naturally explained. He put forth several possible theories, among them the notion that compressed air was responsible for the quake, and also that electricity – an increasingly fashionable phenomenon – was an important factor. It is our duty, he said, to find out how nature works before looking to supernatural causes; undoubtedly Lisbon is a wicked city, but to compare it with Babylon is absurd, and if God intends to set an example there are much more deserving candidates in other countries. A noteworthy aspect of da Silva's essay is his stress on the spiritual enrichment of the world, which the researchers of such men as Descartes and Newton had effected. He proposed that their approach be emulated, and the physical sciences encouraged. God cannot be expected to help man if he is too witless or too wayward to help himself. . . .

The theme of God's anger and of retribution was endlessly repeated in sermons, tracts and poetry throughout Europe. In Lisbon, especially when the exhortations were insanely ferocious, the effect was to make people apathetic and seriously to hamper the work of recovery. Should a man rebuild and start afresh, or should he, as the saintly and crazy Jesuit Gabriel Malagrida passionately urged, "set all this miserable worldly business aside and seek in which might well be his last hours to save his soul"? These extremes could not be reconciled; there was no conceivable compromise here between the men of action and the more fervid men of God. But it was a conflict not confined to Lisbon: its repercussions sounded everywhere.

"One generation passeth away, and another generation cometh: but the earth abideth forever," said the preacher. But now man could not be sure that the earth would abide. We must recall the setting for this upheaval in thought. Looking backward in time from the first half of the eighteenth century, it is astonishing to see the change in outlook that had taken place in two hundred years. Science, philosophy, politics, technology, trade and commerce had transformed society. This life was no longer to be regarded as a mere preparation for the hereafter. Man was to enjoy what he had. He saw himself and the world about him as no medieval thinker could have imagined, as few of even the most ebullient philosophers of the Renaissance would have dared hope. Sir Thomas writes: "It is said that the first half of the eighteenth century with its enlightenment, its optimism, its cult of happiness and its content with the *status quo* was a fortunate age, so much so that it might be preferred to all other times in the past as the one in which a sensible man might elect to live." It was in all things an "age of stability." For the wealthy and educated, Basil Willey has said, it was "the nearest approach to earthly felicity ever known to man"; and even the common man could share in this cozy feeling. Thus the Lisbon earthquake could scarcely have come as a more terrible shock. Suddenly the whole edifice of confidence began to crumble; in ten awful minutes the world's sense of security was swallowed up. There is nothing like an earthquake to make men feel helpless, to remind them of their mortality, of the vanity of all they covet and acquire, of the ridiculous insubstantiality of what they have made and built. If the ground itself will not stand firm, what remains? Even in our time earthquakes strike terror; how much more shattering the experience must have been in an age when earthquakes were not understood as natural events but regarded as occurrences "instinct with deity."

Voltaire was the foremost figure who seized upon the Lisbon earthquake as an opportunity to attack the climate of optimism. The chief target of his attack was what he called the *tout est bien* philosophy expressed twenty years earlier in Alexander Pope's "Essay on Man." We must admit, said Voltaire, that there is injustice and evil in the world, that there is inexcusable suffering, that there are inexplicable calamities. It is stupid and self-deluding to pretend that every misfortune is a benefit in disguise. It is folly to believe that Providence will assure safe-conduct to the virtuous. Man is "weak and helpless, ignorant of his destiny, and exposed to terrible dangers, as all must now see." Optimism must be replaced by realism; at best, by an "apprehensive hope that Providence will lead us through our dangerous world to a happier state."

The poem was widely read and discussed. Voltaire himself was a little apprehensive as to the offense it might give in religious circles and for that reason tinkered with some of the lines. It was scarcely calculated to nourish the belief in a kind and loving God. Nor, on the other hand, did it support a blind faith in a just God whose ways might be hard but whose over-all scheme could not be questioned. Rousseau was one of those who strongly opposed Voltaire's pessimism; Kant, while less cheery, took the position that "the only possible theodicy is a practical act of faith in divine justice."

In 1759 Voltaire published his immortal satire *Candide*, which was far more influential than his poem. It blew the *tout est bien* philosophy to bits. . . .

Optimism never recovered from Lisbon and *Candide*. "There was no more to be said; the case was finished and the case was lost." Not, of course, that it vanished all at once. "A doctrine," as the noted French literary historian Paul Hazard wrote, "lives on for a long time, even when wounded, even when its soul has fled." But within a few years the wounds proved fatal, and a French poet could say that the age of optimism had degenerated into the Dark Ages.

1-6 The Temblor – Mary Austin

The second earthquake described in the present book is more familiar than that of Lisbon. It concerns the great San Francisco earthquake of 1906. Mary Austin (1868–1934) describes the event with clarity and perception, as well as with an obvious feeling of tenderness and understanding toward those individuals who suffered. Her own impressions convey the intensely personal details that remain after the terror of such a catastrophe has subsided. Mary Austin was a writer whose work was a lively blending of her interests in nature, feminism, Native American customs, social justice, and western life. Born in Illinois and educated at Blackburn College, she followed her family to California, where she began a long series of books and articles. An unhappy marriage lay behind her early writing and she produced a succession of novels, westerns, poems and mythological, political and theological works, many having as a theme the link between the individual and the land. Our extract is taken from her essay The Temblor from The California Earthquake of 1906, edited by David Starr Jordan (1907).

There are some fortunes harder to bear once they are done with than while they are doing, and there are three things that I shall never be able to abide in quietness again – the smell of burning, the creaking of house-beams in the night, and the roar of a great city going past me in the street.

Ours was a quiet neighborhood in the best times; undisturbed except by the hawker's cry or the seldom whistling hum of the wire, and in the two days following April eighteenth, it became a little lane out of Destruction. The first thing I was aware of was being wakened sharply to see my bureau lunging solemnly at me across the width of the room. It got up first on one castor and then on another, like the table at a séance, and wagged its top portentously. It was an antique pattern, tall and marble-topped, and quite heavy enough to seem for the moment sufficient cause for all the uproar. Then I remember standing in the doorway to see the great barred leaves of the entrance on the second floor part quietly as under an unseen hand, and beyond them, in the morning grayness, the rose tree and the palms replacing one another, as in a moving picture, and suddenly an eruption of nightgowned figures crying out that it was only an earthquake, but I had already made this discovery for myself as I recall trying to explain. Nobody having suffered much in our immediate vicinity, we were left free to perceive that the very instant after the quake was tempered by the half-humorous, wholly American appreciation of a thoroughly good job. Half an hour after the temblor people sitting on their doorsteps, in bathrobes and kimonos, were admitting to each other with a half twist of laughter between tremblings that it was a really creditable shake.

The appreciation of calamity widened slowly as water rays on a mantling pond. Mercifully the temblor came at an hour when families had not divided for the day, but live wires sagging across housetops were to outdo the damage of falling walls. Almost before the dust of ruined walls had ceased rising, smoke began to go up against the sun, which, by nine of the clock, showed bloodshot through it as the eye of Disaster.

It is perfectly safe to believe anything any one tells you of personal adventure; the inventive faculty does not exist which could outdo the actuality; little things prick themselves on the attention as the index of the greater horror.

I remember distinctly that in the first considered interval after the temblor, I went about and took all the flowers out of the vases to save the water that was left; and that I went longer without washing my face than I ever expect to again.

I recall the red flare of a potted geranium undisturbed on a window ledge in a wall of which the brickwork dropped outward, while the roof had gone through the flooring; and the cross-section of a lodging house parted cleanly with all the little rooms unaltered, and the halls like burrows, as if it were the home of some superior sort of insect laid open to the microscope.

South of Market, in the district known as the Mission, there were cheap man-traps folded in like pasteboard, and from these, before the rip of the flames blotted out the sound, arose the thin, long scream of mortal agony.

Down on Market Street Wednesday morning, when the smoke from the burning blocks behind began to pour through the windows we saw an Italian woman kneeling on the street corner praying quietly. Her cheap belongings were scattered beside her on the ground and the crowd trampled them; a child lay on a heap of clothes and bedding beside her, covered and very quiet. The woman opened her eyes now and then, looked at the reddening smoke and addressed herself to prayer as one sure of the stroke of fate. It was not until several days later that it occurred to me why the baby lay so quiet, and why the woman prayed instead of flying.

Not far from there, a day-old bride waited while her husband went back to the ruined hotel for some papers he had left, and the cornice fell on him; then a man who had known him, but not that he was married, came by and carried away the body and shipped it out of the city, so that for four days the bride knew not what had become of him.

There was a young man who, seeing a broken and dismantled grocery, meant no more than to save some food, for already the certainty of famine was upon the city – and was shot for looting. Then his women came and carried the body away, mother and bethrothed, and laid it on the grass until space could be found for burial. They drew a handkerchief over its face, and sat quietly beside it without bitterness or weeping. It was all like this, broken bits of human tragedy, curiously unrelated, inconsequential, disrupted by the temblor, impossible to this day to gather up and compose into a proper picture.

The largeness of the event had the effect of reducing private sorrow to a mere pin prick and a point of time. Everybody tells you tales like this with more or less detail. It was reported that two blocks from us a man lay all day with a placard on his breast that he was shot for looting, and no one denied the aptness of the warning. The will of the people was toward authority, and everywhere the tread of soldiery brought a relieved sense of things orderly and secure. It was not as if the city had waited for martial law to be declared, but as if it precipitated itself into that state by instinct at its best refuge.

In the parks were the refugees huddled on the damp sod with insufficient bedding and less food and no water. They laughed. They had come out of their homes with scant possessions, often the least serviceable. They had lost business and clientage and tools, and they did not know if their friends had fared worse. Hot, stifling smoke billowed down upon them, cinders pattered like hail – and they laughed – not hysteria, but the laughter of unbroken courage.

That exodus to the park did not begin in our neighborhood until the second day; all the first day was spent in seeing such things as I relate, while confidently expecting the wind to blow the fire another way. Safe to say one-half the loss of household goods might have been averted, had not the residents been too sure of such exemption. It happened not infrequently that when a man had seen his women safe he went out to relief work and returning found smoking ashes – and the family had left no address. We were told of those who had died in their households who took them up and fled with them to the likeliest place in the hope of burial, but before it had been accomplished were pushed forward by the flames. Yet to have taken part in that agonized race for the open was worth all it cost in goods.

Before the red night paled into murky dawn thousands of people were vomited out of the angry throat of the street far down toward Market. Even the smallest child carried something, or pushed it before him on a rocking chair, or dragged it behind him in a trunk, and the thing he carried was the index of the refugee's strongest bent. All the women saved their best hats and their babies, and, if there were no babies, some of them pushed pianos up the cement pavements.

All the faces were smutched and pallid, all the figures sloped steadily forward toward the cleared places. Behind them the expelling fire bent out over the lines of flight, the writhing smoke stooped and waved, a fine rain of cinders pattered and rustled over all the folks, and charred bits of the burning fled in the heated air and dropped among the goods. There was a strange, hot, sickish smell in the street as if it had become the hollow slot of some fiery breathing snake. I came out and stood in the pale pinkish glow and saw a man I knew hurrying down toward the gutted district, the badge of a relief committee fluttering on his coat. "Bob," I said, "it looks like the day of judgment!" He cast back at me over his shoulder unveiled disgust at the inadequacy of my terms. "Aw!" he said, "it looks like hell!"...

No matter how the insurance totals foot up, what landmarks, what treasures of art are evanished, San Francisco, *our* San Francisco is all there yet. Fast as the tall banners of smoke rose up and the flames reddened them, rose up with it something impalpable, like an exhalation. We saw it breaking up in the movements of the refugees, heard it in the tones of their voices, felt it as they wrestled in the teeth of destruction. The sharp sentences by which men called to each other to note the behavior of brick and stone dwellings contained a hint of a warning already accepted for the new building before the old had crumbled. When the heat of conflagration outran the flames and reaching over wide avenues caught high gables and crosses of church steeples, men watching them smoke and blister and crackle into flame, said shortly, "No more wooden towers for San Francisco!" and saved their breath to run with the hose.

What distinguishes the personal experience of the destruction of the great city from all like disasters of record, is the keen appreciation of the deathlessness of the spirit of living.

1-7 The Alaskan Good Friday Earthquake – Jonathan Weiner

Jonathan Weiner, born in 1953, is an independent science writer and editor for The Sciences. He won the 1995 Pulitzer Prize for his book The Beak of the Finch – a study of rapid adaptation in Darwin's finches in the Galapagos Islands.

The present extract, taken from his book Planet Earth (1986), relates the terrible effects of the 1964 Alaskan Good Friday earthquake and links it to the subsequent development of the theory of plate tectonics.

Late on the afternoon of Good Friday, 1964, in Anchorage, Alaska, without warning, the land fluttered and snapped like a flag in a sudden wind. It shook for four to seven minutes. Some 200,000 megatons of energy were released, 400 times more than the combined force of all the nuclear bombs ever exploded in peace and war; it was the worst quake ever measured in North America, 8.5 on the Richter scale.

In the business district, shops and sidewalks sank twenty feet into the earth; parts of Fourth Avenue and L Street slid downhill and mingled at the corner; 130 acres of a suburb, Turnagain Heights, slid into the sea. The Alaskan town of Valdez was utterly destroyed, first by landslides, then by floods, and then by giant waves called tsunami. Tsunami sheared trees off cliffs and bluffs eighty feet above sea level. The quake was felt 800 miles north, at Point Hope and Point Barrow.

Amid such widespread destruction, the number of deaths was miraculously low. There were no tall buildings in Anchorage, and there was only one person per square mile in all Alaska. A hundred people died. If shocks of such power ever strike California, where there are several hundred people per square mile, the loss of life will be disastrous. The towering skyscrapers of San Francisco are engineering experiments that have never been tested.

At the time of the Good Friday Earthquake, geologists knew that quakes occurred along faults – cracks in the crust of the Earth. But why the cracks were there, and what moved the Earth along them, the geologists could not say.

Indeed, until that date geology had a bad press. The British physicist Lord Kelvin, in the heat of a dispute over the Earth's age, said geology was about as intellectually respectable as collecting postage stamps. The physicist was eventually proved wrong on both counts – wrong about the planet's age, and wrong about the earth sciences, But it was true that many geologists spent careers simply gathering rocks, and comparing the sandstone of Madagascar with the sandstone of India. They had no way to explain the endless comparisons and contrasts they catalogued. They could not say why those sandstones, separated by the Indian Ocean, were so similar. No single hypothesis explained mild tremors in western Africa and sudden and deadly ones in Alaska; or explained Hekla in Iceland and Kilauea in Hawaii. There was a globe, but no global theory. As one early student of the Earth, Alexander von Humboldt, lamented, "We examine the stones, but not the mountains, we have the materials but ignore how they fit together."

What are the forces that make and shape the Earth? What accounts for the torments of Mexico, Iceland, Hawaii, Zaire, New Zealand? Why is the pinnacle of Mount Everest made of rock and fossils that once lay in the bottom of the ocean – and, with some irony, now called "the roof of the world"? What accounts for the vestiges in the Sahara of massive glaciers; and in Alaska of tropical jungles? Have parcels of land wandered from one address to another?

Twenty years ago, geologists began a breathtaking succession of intellectual leaps that transformed their science. The years since the Great Alaskan Earthquake have brought a revolution in the study of the globe. Like other revolutions in the history of science, this one was a long time stirring. Indeed, certain keen minds anticipated parts of it decades, or even centuries, ago, long before they had tools with which to test their hunches. At last, in our time, the right tools, the sophisticated tests, and the decisive evidence arrived all in a rush.

The revolution brings together hundreds of thousands of observations, many of which were gained at some hazard in the field. They were won by scientists rappelling down ice cliffs in Antarctica, diving deep into sea floor rifts in tiny submarines, drilling the basement of the continent, helicoptering over the wasteland of Mount St. Helens's effluvia, lowering themselves into the Nyiragongo crater. The results have changed forever the way we see our world.

This new view is not easy to take in. It is hard to imagine that the continents are adrift, ferried about on great fragments of the Earth's shell, which incessantly grind against each other all around the Earth. The plates shudder and quake more than a million times a year. What we have been pleased to call "solid Earth" is not as solid as we thought. It is energetic, dynamic, and fundamentally restless. Its poles wander, and its magnetic field flickers and wavers like a fire in a grate. It is continually creating its own surface, destroying it, repairing it, renewing it like a skin. We have dreamed for years of life on other planets, but we have hardly noticed the life of our own. This Earth is Terra Nova – a land new to us.

1-8 Tsunami – Francis P. Shepard

Tsunami are giant waves, produced in the oceans by submarine earthquakes, and sometimes by explosive submarine volcanic eruptions. These "tidal waves," which may reach heights in excess of 100 feet, cause great destruction when they strike land. The account of the 1946 tsunami in Hawaii is based upon an eyewitness description by Francis P. Shepard, who had to escape from the violent effects of the wave that reached the Hawaiian Islands after crossing 2,300 miles of ocean. The tsunami originated around an earthquake near the Aleutian Island of Unimak, and the waves spread out from that area. They were imperceptible to ships moving across their path, because on the open ocean they had a height of only about a foot, although moving at an average speed of over 500 miles an hour. On coastlines, the waves reached heights of over 30 feet, and in a few restricted areas flung debris over 50 feet above normal beach level. Francis Shepard (1897–1985), who was Professor of Submarine Geology at Scripps Institute of Oceanography at La Jolla, California, and was one of the world's leading oceanographers, was living in a rented cottage on the waterfront near Kahuku Point, on the north shore of Ioahu, at the time of the 1946 tsunami, which he describes below. The extract is from Our Changing Coastlines *(1971).*

I was awakened at 6:30 in the morning by a hissing noise that sounded as if hundreds of locomotives were blowing off steam. I looked out in time to see the water lapping up around the edge of our house, rising 14 feet above its normal level. Grabbing my camera instead of clothes, I rushed out to take photographs. My wife and I watched the rapid recession of the water exposing the narrow reef in front of the house and stranding large numbers of fish. A few minutes later, we saw the water build up at the edge of the reef and move shoreward as a great breaking wave. Because the wave looked very threatening, we dashed in back of the house, getting there just in time to hear the water reach the front porch and smash in all of the glass at the front. As the water swept around the house and into the cane field, we saw our refrigerator carried past us and deposited right side up, eggs unbroken.

Water swept down the escape road to our right, leaving us no chance to get out by car. The neighbors' house, which was vacant at the time, had been completely torn apart, but the back portion of our house still stood, thanks probably to the casuarina (ironwood) trees growing along the edge of the berm in front.

As soon as the second wave had started to retreat, we ran along the beach ridge to our left, and then through the cane field by path to the main road, arriving there just ahead of the third wave.

We found quite a group on the road. The house of one family living at the edge of Kawela Bay had been carried bodily into the cane field and dropped without the wave having done much damage. In fact, their breakfast was still cooking when they were set down. Other unoccupied houses did not fare so well. Some of them were swept into a small pond, and others out into the bay by the retreat of the second wave.

We watched three or four more waves come into Kawela Bay at intervals (shown later by tide gauge records) of fifteen minutes. They had steep fronts, looking very much like the tidal bore that comes up the Bay of Fundy. Just before the eighth wave, I decided that the excitement was abating and ran back to the house to try to rescue some effects, particularly necessary as we were in pajamas and raincoats. Just as

I arrived, a wave that must have been the largest of the set came roaring in, and I had to run for a tree and climb it as the water surged beneath me. I hung on swaying back and forth as the water roared by into the cane field.

That was the end of the adventure, and we gradually got what we could rescue of our possessions, but had to leave behind quantities of notes and the beginning of a new book, *Submarine Geology*, which had been scattered in the cane field. We were given wonderfully hospitable treatment by new and old friends whom we shall never forget. After a few days, I got in touch with Gordon Macdonald, of the U.S. Geological Survey, and Doak Cox, the geologist of Hawaiian Planters Association, and together we started a program of investigating the tsunami effects around the different islands.

1-9 Not a Very Sensible Place for a Stroll – Haroun Tazieff

Haroun Tazieff (1914–1998), a volcanologist and filmmaker, was born in Warsaw, but moved to France. Educated in Belgium, he fought with the French resistance during World War II. He had a rich career, working as a geologist in the Congo, teaching at the University of Brussels, directing the French National Center for Scientific Research, and serving as Secretary of State for the Prevention of Natural and Technological Disasters.

Volcanoes provide the most dramatic of all illustrations of the cauldron-like energy of the Earth's interior. From his book, Craters of Fire *(1952), Tazieff contributes two descriptions of volcanoes, the "very substance of the Earth itself," as he describes them. Both these accounts convey something of the noise, the fumes, the molten lava, and the dangers of*

the volcanic eruptions that, on the pages of our textbooks, seem so ordinary. Tazieff, awakened one morning by a strange noise "like a herd of antelope galloping through the bush," provides not only a superb account of a volcano, but also describes the curious mixture of delight and desperation, of determination and despair, that represents the tension between personal discovery and personal danger. His account of his climb around the rim of an active crater of a "growling cone" is as good a piece of adventure writing as it is of scientific description. In the process, Tazieff, like the rest of us, learned something about himself. "The calm of this fiery pool. ... spoke to me in enigmatic terms of a mighty and mysterious power. I was spellbound, and literally had to wrench myself out of the fear-laden ecstasy. ..."

Standing on the summit of the growling cone, even before I got my breath back after the stiff climb, I peered down into the crater.

I was astonished. Two days previously the red lava had been boiling up to the level of the gigantic lip; now the funnel seemed to be empty. All that incandescent magma had disappeared, drawn back into the depths by the reflux of some mysterious ebb and flow, a sort of breathing. But there, about fifty feet below where I was standing, was the glow and the almost animate fury of the great throat which volcanologists call the conduit or chimney. It was quite a while before I could tear my eyes away from that lurid, fiery center, that weird palpitation of the abyss. At intervals of about a minute, heralded each time by a dry clacking, bursts of projectiles were flung up, running away up into the air, spreading out fan-wise, all aglare, and then falling back, whistling, on the outer sides of the cone. I was rather tense, ready to leap aside at any moment, as I watched these showers, with their menacing trajectories.

Each outburst of rage was followed by a short lull. Then heavy rolls of brown and bluish fumes came puffing out, while a muffled grumbling, rather like that of some monstrous watch-dog, set the whole bulk of the volcano quivering. There was not much chance for one's nerves to relax, so swiftly did each follow on the other – the sudden tremor, the burst, the momentary intensification of the incandescence, and the outbreak of a fresh salvo. The bombs went roaring up, the cone of fire opening out overhead, while I hung in suspense. Then came the hissing and sizzling, increased in speed and intensity, each 'whoosh' ending up in a muffled thud as the bomb fell. On their black bed of scoriae, the clots of molten magma lay with the fire slowly dying out of them, one after the other growing dark and cold.

Some minutes of observation were all I needed. I noted that today, apart from three narrow zones to the west, north and north-east, the edges of the crater had scarcely been damaged at all by the barrage from underground. The southern point where I stood was a mound rising some twelve or fifteen feet above the general level of the rim, that narrow, crumbling lip of scoriae nearer to the fire, where I had never risked setting foot. I looked at this rather alarming ledge all round the crater, and gradually felt an increasing desire to do something about it . . . It became irresistible. After all, as the level of the column of lava had dropped to such an exceptional degree, was this not the moment to try what I was so tempted to do and go right round the crater?

Still, I hesitated. This great maw, these jaws sending out heat that was like the heavy breathing of some living creature, thoroughly frightened me. Leaning forward over that hideous glow, I was no longer a geologist in search of information, but a terrified savage.

'If I lose my grip,' I said aloud, 'I shall simply run for it.'

The sound of my own voice restored me to normal awareness of myself. I got back my critical sense and began to think about what I could reasonably risk trying. 'De l'audace, encore de l'audace. . . .' That was all very well, of course, but one must also be careful. Past experience whispered a warning not to rush into anything blindly. Getting the upper hand of both anxiety and impatience, I spent several minutes considering, with the greatest of care, the monster's manner of behaving. Solitude has got me into the habit of talking to myself, and so it was more or less aloud that I gave myself permission to go ahead.

'Right, then. It can be done.'

I turned up my collar and buttoned my canvas jacket tight at the throat – I didn't want a sly cinder down the back of my neck! Then I tucked what was left of my hair under an old felt hat that did service for a helmet. And now for it!

Very cautiously indeed, I approach the few yards of pretty steep slope separating the peak from the rim I am going to explore. I cross, in a gingerly manner, a first incandescent crevasse. It is intense orange in colour and quivering with heat, as though opening straight into a mass of glowing embers. The fraction of a second it takes me to cross it is just long enough for it to scorch the thick cord of my breeches. I get a strong whiff of burnt wool.

A promising start, I must say!

Here comes a second break in the ground. Damn it, a wide one, too! I can't just stride across this one: I'll have to jump it. The incline makes me thoughtful. Standing there, I consider the unstable slope of scoriae that will have to serve me for a landing-ground. If I don't manage to pull up . . . if I go rolling along down this funnel with the flames lurking at the bottom of it. . . . My little expedition all at once strikes me as thoroughly rash, and I stay where I am, hesitating. But the heat under my feet is becoming unbearable. I can't endure it except by shifting around. It only needs ten seconds of standing still on this enemy territory, with the burning gases slowly and steadily seeping through it, and the soles of my feet are already baking hot. From second to second the alternative becomes increasingly urgent: I must jump for it or retreat.

Here I am! I have landed some way down the fissure. The ashes slide underfoot, but I stop without too much trouble. As so often happens, the anxiety caused by the obstacle made me over-estimate its importance.

Step by step, I set out on my way along the wide wall of slag-like debris that forms a sort of fortification all round the precipice. The explosions are still going on at regular intervals of between sixty and eighty seconds. So far no projectile had come down on this side, and this cheered me up considerably. With marked satisfaction I note that it is pretty rare for two bombs of the same salvo to fall less than three yards apart: the average distance between them seems to be one of several paces. This is encouraging. One of the great advantages of this sort of bombardment, compared with one by artillery, lies in the relative slowness with which the projectiles fall, the eye being able to follow them quite easily. Furthermore, these shells don't burst. But what an uproar, what an enormous, prolonged bellowing accompanies their being hurled out of the bowels of the earth!

I make use of a brief respite in order to get quickly across the ticklish north-eastern sector. Then I stop for a few seconds, just long enough to see yet another burst gush up and come showering down a little ahead of me, after which I start out for the conquest of the northern sector. Here the crest narrows down so much that it becomes a mere ridge, where walking is so difficult and balancing so precarious that I find myself forced to go on along the outer slope, very slightly lower down. Little by little, as I advance through all this tumult, a feeling of enthusiasm is overtaking me. The immediate imperative necessity for action has driven panic far into the background. And under the hot, dry skin of my face, taut on forehead and cheekbones, I can feel my compressed lips parting, of their own accord, in a smile of sheer delight. But look out!

A sudden intensification of the light warns me that I am approaching a point right in the prolongation of the fiery chimney. In fact, the chimney is not vertical, but slightly inclined in a north-westerly direction, and from here one can look straight down into it. These tellurian entrails, brilliantly yellow, seem to be surging with heat. The sight is so utterly amazing that I stand there, transfixed.

Suddenly, before I can make any move, the dazzling yellow changes to white, and in the same instant I feel a muffled tremor all through my body and there is a thunderous uproar in my ears. The burst of incandescent blocks is already in full swing. My throat tightens as, motionless, I follow with my gaze the clusters of red lumps rising in slow, perfect curves. There is an instant of uncertainty. And then down comes the hail of fire.

Suddenly I hurl myself backwards. The flight of projectiles has whizzed past my face. Hunched up again, instinctively trying to make as small a target of myself as I can, I once more go through the horrors that I am beginning to know. I am in the thick of this hair's-breadth game of anticipation and dodging.

And now it's all over; I take a last glance into the marvelous and terrible abyss, and am just getting ready to start off on the last stage of this burning circumnavigation, all two hundred yards of it, when I get a sudden sharp blow in the back. A delayed-action bomb! With all the breath knocked out of me, I stand rigid.

A moment passes. I wonder why I am not dead. But nothing seems to have happened to me – no pain, no change of any sort. Slowly I risk turning my head, and at my feet I see a sort of huge red loaf with the glow dying out of it.

I stretch my arms and wriggle my back. Nothing hurts. Everything seems to be in its proper place. Later on, examining my jacket, I discovered a brownish scorchmark with slightly charred edges, about the size of my hand, and I drew from it a conclusion of immense value to me in future explorations: so long as one is not straight in the line of fire, volcanic bombs, which fall in a still pasty state, but already covered with a kind of very thin elastic skin, graze one without having time to cause a deep burn.

I set off at a run, as lightly as my 165 pounds allow, for I must be as quick as I can in crossing this part of the crater-edge, which is one of the most heavily bombarded. But I am assailed by an unexpected blast of suffocating fumes. My eyes close, full of smarting tears. I am caught in a cloud of gas forced down by the wind. I fight for breath. It feels as if I were swallowing lumps of dry, corrosive cotton-wool. My head swims, but I urge myself at all costs to get the upper hand. The main thing is not to breathe this poisoned air. Groping, I fumble in a pocket. Damn, not this one. How about this other one, then? No. At last I get a handkerchief out and, still with my eyes shut, cover my mouth with it. Then, stumbling along, I try to get through the loathsome cloud. I no longer even bother to pay any attention to the series of bursts from the volcano, being too anxious to get out of this hell before I lose grip entirely. I am getting pretty exhausted, staggering. . . . The air filtered through the handkerchief just about keeps me going, but it is still too poisonous, and there is too little of it for the effort involved in making this agonising journey across rough and dangerous terrain. The gases are too concentrated, and the great maw that is belching them forth is too near.

A few steps ahead of me I catch a glimpse of the steep wall of the peak, or promontory, from the other side of which I started about a century ago, it seems to me now. The noxious mists are licking round the peak, which is almost vertical and twice the height of a man. It's so near! But I realize at once that I shall never have the strength to clamber up it.

In less than a second, the few possible solutions to this life-and-death problem race through my mind. Shall I turn my back to the crater and rush away down the outer slope, which is bombarded by the thickest barrages? No. About face and back along the ledge? Whatever I do, I must turn back. And then make my escape. By sliding down the northern slope? That is also under too heavy bombardment. And the worst of it would be that in making a descent of that sort there would be no time to keep a watch for blocks of lava coming down on one.

Only one possibility is left: to make my way back all along the circular ridge, more than a hundred yards of it, till I reach the eastern rim, where neither gas nor projectiles are so concentrated as to be necessarily fatal.

I swing round. I stumble and collapse on all fours, uncovering my mouth for an instant. The gulp of gas that I swallow hurts my lungs, and leaves me gasping. Red-hot scoriae are embedded in the palms of my hands. I shall never get out of this!

The first fifteen or twenty steps of this journey back through the acrid fumes of sulphur and chlorine are a slow nightmare; no step means any progress and no breath brings any oxygen into the lungs. The threat of bombs no longer counts. Only these gases exist now. Air! Air!

I came to myself again on the eastern rim, gasping down the clean air borne by the wind, washing out my lungs with deep fresh gulps of it, as though I could never get enough. How wide and comfortable this ledge is! What a paradise compared with the suffocating, torrid hell from which I have at last escaped! And yet this is where I was so anxious and so tense less than a quarter of an hour ago.

Several draughts of the prevailing breeze have relieved my agony. All at once, life is again worth living! I no longer feel that desire to escape from here as swiftly as possible. On the contrary, I feel a new upsurge of explorer's curiosity. Once more my gaze turns towards the mouth, out of which sporadic bursts of grape-shot are still spurting forth. Now and then there are bigger explosions and I have to keep a look-out for what may come down on my head, which momentarily interrupts the dance I keep up from one foot to the other, that *tresca* of which Dante speaks – the dance of the damned, harried by fire. True, I have come to the conclusion that the impact of these bombs is not necessarily fatal, but I am in no hurry to verify the observation.

The inner walls of the crater do not all incline at the same angle. To the north, west and south, they are practically vertical, not to say overhanging, but here on the east the slope drops away at an angle of no more than fifty degrees. So long as one moved along in a gingerly way, this might be an incline one could negotiate. It would mean going down into the very heart of the volcano. For an instant I am astounded by my own foolhardiness. Still, it's really too tempting . . .

Cautiously, I take a step forward . . . then another . . . and another . . . seems all right . . . it *is* all right. I begin the climb down, digging my heels as deep as I can into the red-hot scoriae. Gradually below me, the oval of the enormous maw comes nearer, growing bigger, and the terrifying uproar becomes more deafening. My eyes, open as wide as they will go, are drunken with its monstrous glory. Here are those ponderous draperies of molten gold and copper, so near – so near that I feel as if I, human being that I am, had entered right into their fabulous world. The air is stifling hot. I am right in the fiery furnace.

I linger before this fascinating spectacle. But then, by sheer effort, I tear myself away. It's time to get back to being 'scientific' and measure the temperatures, of the ground, and of the atmosphere. I plunge the long spike of the thermometer into the shifting scoriae, and the steel of it glitters among these brownish and grey screes with their dull shimmer. At a depth of six inches the temperature is two hundred and twenty degrees centigrade. It's amusing to think that when I used to dream it was always about polar exploration!

Suddenly, the monster vomits out another burst; so close that the noise deafens me. I bury my face in my arms. Fortunately almost every one of the projectiles comes down outside the crater. And now all at once I realize that it is I who am here – *alive* in this crater, surrounded by scorching walls, face to face with the very mouth of the fire. Why have I got myself into this trap, alone and without the slightest chance of help? Nobody in the world has any suspicion of the strange adventure on which I have embarked, and nobody, for that matter, could now do the slightest thing about it. Better not think about it . . .

Without a break the grim, steady growling continues to rise from the depths of that throat, only out-roared at intervals by the bellowing and belching of lava. It's too much; I can feel myself giving up. I turn my back on it, and try, on all fours, to scramble up the slope, which has now become incredibly steep and crumbles and gives way under my weight, which is dragging me down, down . . . 'Steady, now,' I say to myself. 'Keep calm for a moment. Let's work it out. Let's work it out properly. Or else, my boy, this is the end of *you*.'

Little by little, by immense exertions, I regain control of my movements, as well as the mental steadiness I need. I persuade myself to climb *calmly* up this short slope, which keeps crumbling away under my feet. When I reach the top, I stand upright for just a moment. Then, crossing the two glowing fissures that still intersect my course, I reach the part of the rim from where there is a way down into the world of ordinary peaceful things.

1-10   Last Days of St Pierre – Fairfax Downey

Most volcanoes lie in areas remote from centers of dense population. Those that do occur in inhabited areas are generally well studied and well known, their moods and tempers being, if not predictable, at least recognizable. One terrible exception is Mont Pelée, in the West Indies, which, after more than 50 years of dormancy, provided signs of renewed activity in the Spring of 1902. On the morning of the 8th of May, a cloud of incandescent gas and glowing volcanic fragments poured down the sides of the cone at speeds as high as 60 meters per second, destroying buildings in its path, and bringing instant death to some 40,000 people in and around St. Pierre, the capital of the island of Martinique. The account of this disaster by Fairfax Downey provides a vivid and terrible illustration of its effects on the lives of the members of the community. Fairfax Davis Downey (1893–1990) was born in Utah, educated at Yale, and served as an army officer in two world wars. He began his career on the staff of the Kansas City Star *and later the* New York Tribune *and* Herald Tribune, *eventually becoming a successful freelance writer and poet whose many books dealt often with military themes. The extract below is taken from* Disaster Fighters *(1938).*

How graciously had fortune smiled on Fernand Clerc. Little past the age of forty, in this year of 1902, he was the leading planter on the fair island of Martinique. Sugar from his broad cane fields, molasses, and mellow rum had made him a man of wealth, a millionaire. All his enterprises prospered.

Were the West Indies, for all their beauty and their bounty, sometimes powerless to prevent a sense of exile, an ache of homesickness in the heart of a citizen of the Republic? Then there again fate had been kind to Fernand Clerc. Elected a member of the Chamber of Deputies, it was periodically his duty and his pleasure to embark and sail home to attend its sessions – home to France, to Paris.

Able, respected, good-looking, blessed with a charming wife and children, M. Clerc found life good indeed. With energy undepleted by the tropics, he rode through the island visiting his properties. Tall and thick grew the cane stalks of his plantation at Vivé on the slopes of Mont Pelée. Mont Pelée – Naked Mountain – well named when lava erupting from its cone had stripped it bare of its verdure. But that was long ago. Not since 1851 had its subterranean fires flared up and then but insignificantly. Peaceful now, its crater held the lovely Lake of Palms, whose wooded shores were a favorite picnic spot for parties from St. Pierre and Fort-de-France. Who need fear towering Mont Pelée, once mighty, now mild, an extinct volcano?

Yet this spring M. Clerc and all Martinique received a rude shock. The mountain was not dead, it seemed. White vapors veiled her summit, and by May 2nd she had overlaid her green mantle with a gown of gray cinders. Pelée muttered and fumed like an angry woman told her day was long past. Black smoke poured forth, illumined by night of jets of flame and flashes of lightning. The grayish snow of cinders covered the countryside, and the milky waters of the Rivière Blanche altered into a muddy and menacing torrent.

Nor was Pelée uttering only empty threats. On May 5th, M. Clerc at Vivé beheld a cloud rolling from the mountain down the valley. Sparing his own acres, the cloud and the stream of smoking lava which it masked, enveloped the Guerin sugar factory, burying its owner, his wife, overseer, and twenty-five workmen and domestics.

Dismayed by this tragedy, M. Clerc and many others moved from the slopes into St. Pierre. The city was crowded, its population of 25,000 swollen to 40,000, and the throngs that filled the market and the cafés or strolled through the gorgeously luxuriant Jardin des Plantes lent an air of added animation, of almost hectic gaiety. When M. Clerc professed alarm at the behavior of Pelée to his friends, he was answered with shrugs of shoulders. Danger? On the slopes perhaps, but scarcely here in St. Pierre down by the sea.

Thunderous, scintillant, Mont Pelée staged a magnificent display of natural fireworks on the night of May 7th. [Everyone] stared up at it, fascinated. Some were frightened but more took a child-like joy in the vivid spectacle. It was as if the old volcano were celebrating the advent of tomorrow's fête day.

M. Fernand Clerc did not sleep well that night. He breakfasted early in the household where he and his family were guests and again expressed his apprehension to the large group of friends and relatives gathered at the table. Politely and deferentially – for one does not jeer a personage and man of proven courage – they heard him out, hiding their scepticism.

The voice of the planter halted in mid-sentence; he half rose, his eyes fixed on the barometer. Its needle was actually fluttering!

M. Clerc pushed back his chair abruptly and commanded his carriage at once. A meaning look to his wife and four children, and they hastened to make ready. Their hosts and the rest followed them to the door. *Non, merci*, none would join their exodus. *Au revoir. A demain.*

From the balcony of their home, the American Consul, Thomas Prentis, and his wife waved to the Clerc family driving by. "Stop," the planter ordered and the carriage pulled up. Best come along, the planter urged. His American friends thanked him. There was no danger, they laughed, and waved again to the carriage disappearing in gray dust as racing hoofs and wheels sped it out of the city of St. Pierre.

Governor Mouttet, ruling Martinique for the Republic of France, glared up at rebellious Mont Pelée. This *peste* of a volcano was deranging the island. There had been no such crisis since its captures by the English, who always relinquished it again to France, or the days when the slaves revolted. A great pity that circumstances beyond his control should damage the prosperous record of his administration, the Governor reflected.

That miserable mountain was disrupting commerce. Its rumblings drowned out the band concerts in the Savane. Its pyrotechnics distracted glances which might far better have dwelt admiringly on the proverbial beauty of the women of Martinique. . . . Now attention was diverted to a cruder work of Nature, a sputtering volcano. *Parbleu!* It was enough to scandalize any true Frenchman.

Governor Mouttet sighed and pored over the reports laid before him. He had appointed a commission to study the eruption and get at the bottom of *l'affaire Pelée*, but meanwhile alarm was spreading. People were fleeing the countryside and thronging into St. Pierre, deserting that city for Fort-de-France, planning even to leave the island. Steamship passage was in heavy demand. The *Roraima*, due May 8th, was booked solid out of St. Pierre, one said. This would never do. Steps must be taken to prevent a panic which would scatter fugitives through Martinique or drain a colony of France of its inhabitants.

A detachment of troops was despatched by the Governor to St. Pierre to preserve order and halt the exodus. His Excellency, no man to send others where he himself would not venture, followed with Mme. Mouttet and took up residence in that city. Certainly his presence must serve to calm these unreasoning, exaggerated fears.

He circulated among the populace, speaking soothing words. *Mes enfants*, the Governor avowed, Monte Pelée rumbling away there is only snoring soundly in deep slumber. Be tranquil.

Yet, on the ominous night of May 7th, as spurts of flame painted the heavens, the Governor privately confessed to inward qualms. What if the mountain should really rouse? Might it not then cast the mortals at its feet into a sleep deeper than its own had been, a sleep from which they would never awaken?

It was dark in the underground dungeon of the St. Pierre prison, but thin rays of light filtered through the grated opening in the upper part of the cell door. Enough so that August Ciparis could tell when it was night and when it was day.

Not that it mattered much unless a man desired to count the days until he should be free. What good was that? One could not hurry them by. Therefore Auguste stolidly endured them with the long patience of Africa. The judge had declared him a criminal and caused him to be locked up here. Thus it was settled and nothing was to be done. Yet it was hard, this being shut out of life up there in the gay city – hard when one was only twenty-five and strong and lusty.

Auguste slept and dozed all he could. Pelée was rumbling away in the distance – each day the jailer bringing him food and water seemed more excited about it – but the noise, reaching the subterranean cell only as faint thunder, failed to keep him awake . . .

Glimmerings of the dawn of May 8th filtered through the grating into the cell, and Auguste stirred into wakefulness. This being a fête day, imprisonment was less tolerable. What merriment his friends would be making up there in the squares of St. Pierre! He could imagine the sidelong glances and the swaying hips of the girls he might have been meeting today. Auguste stared sullenly at the cell door. At least the jailer might have been on time with his breakfast.

The patch of light in the grating winked out into blackness. *Ai! Ai!* All of a sudden it was night again.

On the morning of May 8th, 1902, the clocks of St. Pierre ticked on toward ten minutes of 8 when they would stop forever. Against a background of bright sunshine, a huge column of vapor rose from the cone of Mont Pelée.

A salvo of reports as from heavy artillery. Then, choked by lava boiled to a white heat by fires in the depths of the earth, Pelée with a terrific explosion blew its head off.

Like a colossal Roman candle it shot out streaks of flame and fiery globes. A pall of black smoke rose thousands of feet in the air, darkening the heavens. Silhouetted by a red, infernal glare, Pelée flung aloft viscid masses which rained incandescent ashes on land and sea.

Then, jagged and brilliant as the lighting flashes, a fissure opened in the flank of the mountain toward St. Pierre. Out of it issued an immense cloud which rushed with unbelievable rapidity down on the doomed city and the villages of Carbet and Le Precheur.

In three minutes that searing, suffocating cloud enveloped them, and 40,000 people died!

Fernand Clerc, the planter, watched from Mont Parnasse, one mile east of St. Pierre, where he had so recently breakfasted. Shrouded in such darkness as only the inmost depths of a cavern afford, he reached out for the wife and children he could not see and gathered them in blessed safety into his arms. But the relatives, the many friends he had left a short while ago, the American consul and his wife, who had waved him a gay good-by – them he would never see alive again. . . .

In that vast brazier which was St. Pierre, Governor Mouttet may have lived the instant long enough to realize that Pelée had in truth awakened and that eternal sleep was his lot and his wife's and that of all those whose flight he had discouraged. . . .

Down in that deep dungeon cell of his Auguste Ciparis blinked in the swift-fallen night. Through the grating blew a current of burning air, scorching his flesh. He leaped, writhing in agony and screaming for help. No one answered. . . .

Not until the afternoon of May 8th did the devastation of St. Pierre cool sufficiently to allow rescuers from Fort-de-France to enter. They could find none to rescue except one woman who died soon after she was taken from a cellar. . . .

Indeed St. Pierre might have been an ancient town, destroyed in some half-forgotten cataclysm and recently partly excavated – another Pompeii and Hercula-neum. Cinders, which had buried its streets six feet deep in a few minutes, were as the dust of centuries. Here was the same swift extinction Vesuvius had wrought.

Here was no slow flow of lava. That cloud disgorged by Pelée was a superheated hurricane issuing from the depths of the earth at a speed of ninety miles an hour. Such was the strength of the blast, it killed by concussion and by toppling walls on its victims. The fall of the fourteen-foot metal statue of Notre Dame de la Garde – Our Lady of Safety – symbolized the dreadful fact that tens of thousands never had a fighting chance for their lives. . . .

Then four days after the catastrophe, two [men] walking through the wreckage turned gray as they heard faint cries for help issuing from the depths of the earth.

"Who's that?" they shouted when they could speak. "Where are you?"

Up floated the feeble voice: "I'm down here in the dungeon of the jail. Help! Save me! Get me out!"

They dug down through the debris, broke open the dungeon door, and released Auguste Ciparis, the criminal.

Some days later, George Kennan and August F. Jaccaci, American journalists arriving to cover the disaster, located Ciparis in a village in the country. They secured medical attention for his severe burns, poorly cared for as yet, and obtained and authenticated his story. When the scorching air penetrated his cell that day, he smelled his own body burning but breathed as little as possible during the moment the intense heat lasted. Ignorant of what had occurred, not realizing that he was buried alive, he slowly starved for four days in his tomb of a cell. His scant supply of water was soon gone. Only echoes answered his shouts for help. When at last he was heard and freed, Ciparis, given a drink of water, managed with some assistance to walk six kilometers to Morne Rouge.

One who lived where 40,000 died! History records no escape more marvelous.

1-11 Beacons on the Passage Out – Hans Cloos

Hans Cloos (1885–1951) was born in Magdaburg, Germany, and received his geological training at the University of Freiburg. His early studies concerned the tectonics of the Jura Mountains, and, after receiving his doctorate, he worked as a mining geologist in South-western Africa, and then in petroleum exploration in Indonesia. Cloos later returned to Germany, and joined the faculty of the University of Marburg. He subsequently occupied the Chair of Geology at the University of Breslau and later the University of Bonn. An outstanding field worker, and a major contributor to our understanding of intrusion and tectonic processes, Cloos was especially well known for his work in experimental modeling, by which he sought to simulate geologic structures and their formation. He was awarded the Penrose Medal of the Geological Society of America in 1948.

The following passage "Beacons on the Passage Out" is from the book Conversation with the Earth *(1953) – a translation of the German* Gespräch mit der Erde *(1947) – and describes his voyage from Europe to North Africa, just after finishing his Freiburg studies. It was on the way that he witnessed Vesuvius for the first time, and gave the account which we include in this section. This is not an account of a geologist at work. Cloos made no scientific readings and recorded no new observations as he gazed on Vesuvius for the first time. He already knew all the technical explanations of vulcanism and was a well-informed exponent of present Earth processes as an adequate explanation for the former history of the Earth. And yet, as he gazed on that volcano, he knew that in reality, he had learned nothing, "because that concept of the physical world has not yet become a true possession of [his] own." This is a beautiful description of the difference between the acquisition of information and the possession of knowledge and understanding. Moments of real self-discovery, when the truth is revealed and assimilated on such a personal basis, are rare in our daily experience. But their very rarity makes them precious, and their significance is great, for they are major landmarks on our journey of personal intellectual inspiration. It is one such landmark that Cloos describes.*

When I arrived at Naples, it was a warm turmoil of noise and lights under a starless sky. From the hotel window I heard nothing but the soft lapping of the sea on the mole.

The next morning, however, I experienced the great moment which made me a real geologist. As I threw the shutters open, the whole splendor of the famous picture before me was unveiled like a vast triptych. From right to left; the sparkling bay hemmed in by mountainous shores; then, close by, the gloomy little fort on the beach; and finally, the colorful city rising landward in a thousand steps.

Upward the vision was cut off by a low cloud. Somewhat disappointed, I was about to turn back into the room when I saw a bright sheen above the clouds.

There the clear-cut triangular silhouette of the summit of Vesuvius seemed to be floating in the air, gleaming white with the new winter's snow, and from its sunken crater a little cloud of smoke idly detached itself. . . .

So it was really true!

Year after year I had read and learned little else but this: that our old earth had changed in countless ways during its endless history, and that the whole variegated mass of strata and rocks of the primeval world and of the present mountain ranges was but the result or relic of such changes; that the earth is still active today, living and working on its old material, adding new matter and energy to its old stores; that it is but an optical illusion to assume that the earth has reached a stability that provides an unshakeable foundation for human planning, and that recent changes are only superficial.

These teachings I had heard and believed. I had defended them against the incredulous and recited them before stern judges in searching examinations. But now I had to realize in an unguarded moment that in reality I had learned nothing at all because that concept of the physical world had not yet become a true possession of my own – not till this unique and unforgettable moment when I became a geologist forever by seeing with my own eyes: *the earth is alive!*

Up there in those noble isolated heights above seas and crowds the monstrous happens: the earth, this permanent and time-honored stage of our growing, being, and dying, of our digging and building, thunders and bursts and blows acrid fumes from dark caverns into the pure air we breathe. It spits red fire and flings up hot rocks to let them crash to the ground and smash to bits. There, too, glowing waves rise and flow, burning all life on their way, and freeze into black, crusty rock which adds to the height of the mountain and builds land, thereby adding another day to the geological past.

So it is that the earth itself growing, grows and renews itself.

And as we look, the bright cloud fades into the blue sky and a second one gushes up from the peak. It swells and rises, parting easily, like a soap bubble, from the stony crater rim, which remains bound by terrestrial gravity.

A third cloud follows. Hundreds more; some larger, and more powerful, wrecking the peak; and others, more gentle ones, quietly covering the mountain with dust and ashes, building it step by step to new heights.

Countless eruptions have preceded them, building and destroying, piling up and shattering again. In such fashion, this mountain, softly and smoothly contoured, this cone within a cone, has become for people all over the world the symbol of another, hotter, and more active realm under the cooler one we know, an ever-growing product of a cosmic process of construction patiently carried on through the ages.

The student of the earth thus sees for the very first time the significant relationship between the shape, structure, and history of the terrestrial formation. Deeply moved, he senses the inseparable bonds of matter and space with time and history. And he learns that mountain- and land-forms are the handwriting of the earth, and that he who wanders attentively over the earth's surface can follow the traces of their history.

Naples has a famous museum, but far richer and more impressive to the student of geology is the natural museum of the city's volcanic ground. Landward, to the east, Mount Vesuvius rises solitary and dangerous. To the west huddle low-lying cones and craters, a miniature lunar landscape. Some of these cones appear in the suburbs of the city itself. None of them is large or dangerous. But their casual proximity to roads and buildings makes them exciting. Here the earth is alive everywhere! Like the temperamental people who live on it, the ground itself is noisy and moves, changes, and adds to itself constantly. The present terrain is no longer what it was in Roman times. . . .

Wandering back toward Pozzuoli, the geologist looks for the Temple of Serapis between the road and the sea. Three pillars are still standing. The Temple is not famous in the history of art, but is sensational in the history of the earth. Each of these pillars is ringed to a height of about twenty-one feet by holes similar to those bored by marine mollusks in wood or stone. Between the time it was built and the present, the Temple must have been under water. Did the surface of the sea rise so high and retreat again in so short a period? Or did the little strip of land where the Temple stands rise after it had settled? A careful examination of the site has led to the latter conclusion. Not the ocean, but a rather small section of land has moved.

The height of sea level depends on the total amount of water, which could be increased by the melting of polar ice; on the capacity of the ocean basins; and on the rapidly changing elevation of the continents. Once in two thousand years the earth's crust has taken a deep breath. The eternal sea has spilled over and then slid back, as over the breathing chest of a man in a bath. Did the volcanic furnace first lower and then raise its plastic lid? Did it exhale the ashes of the volcanoes of the Phlegrean fields?

Or was it the pulsation of the whole earth itself? Did the lava yield to the movement of the earth's armor-plated crust itself?

Could it have been that in this little landscape the planet's two major forms of geological expression merged in a single, momentous rhythm? Which was the stronger, the more active force? Which the cause, and which the effect? . . .

The sun has disappeared into Homer's "Okeanos." A cool breeze blows and night is falling fast. The craters below are black and menacing. Will the Temple at Pozzuoli again subside into the sea, and the Solfatara again awaken to life?

The next morning I heard that the ship for Africa would be two days late, which made a trip to Pompeii possible.

Pompeii is a southern city with straight streets and light, rectangular walls. One can enter any of the buildings and find the sun pouring into the roofless rooms. I had a beautiful day. Even Vesuvius, the great villain, acted innocent. Leisurely I strolled into a *vineta* (an inn), whose floor is almost two thousand years old. All the vestiges of the living are still intact; there are even some traces of human form and gesture. I realized that these people felt and thought much as we do, yet I was also aware of the time-span separating their language and way of life from ours.

During the long period from Pliny the Elder to the days of Eleonora Duse, Vesuvius has remained always the same, has always been active and has always produced the same kind of lava and ashes. These two millennia have been but a day in the volcano's life, which had already begun millions of years before Pliny and Pompeii. And our own life is but the wink of an eye in the two-thousand-year existence of this fossil city.

Thoughtful and somewhat subdued, I rode back from the city of the dead to the city of the living on the beautiful bay.

The next afternoon the *Gertrud Woermann* arrived, and at eleven that night I embarked on my first long journey.

I stood at the stern and saw the city lights disappear. To the left and right beacons and channel-markers on cliffs and bars flashed their lights silently. Each performed its task punctually regardless of wind and weather.

Nature's lighthouses, however, rested. One I had visited three days before flared up only once. Epomeo and Vesuvius were somewhere, shadowy in the blackness of the night. When will they flare up again? The rhythm of their flashes is counted in centuries, perhaps millennia. As yet we do not know this rhythm, because our lives are too short to measure it. But some day it will be known. The beacons of the earth will be so scrupulously observed that they will yield the secrets of their construction and fuel supplies, so that periods of their eruptions and extinction will be known. And whoever solves the mystery of the earth's fiery breath will have caught and fettered the living earth itself. To do this was the hope and aim of the high science to which I had dedicated my life.

Silently the last lights ashore disappeared. The ship of my life bored its way into the black wall of the future.

1-12 Eruption of the Öraefajökull, 1727 – Jon Thorlakson

Our final description of a volcanic eruption concerns one which took place in the area of Öraefajökull, Iceland in 1727. The account which we include is a translation from a letter by Jon Thorlakson, who was minister of the church at Sandfell. The opening sentence sets the scene for a balanced, sober and touching account of the impact of a catastrophe on the life of a local community. The account is taken from Ebenezer Henderson's Iceland; or the Journal of a Residence in that Island during the Years 1814 and 1815 (1819). The book contains extensive accounts of Iceland's natural history, including its volcanic phenomena. Henderson credits the first publication of the account to Ólafur Olavius, the original recipient of the Thorlakson letter, in his book Economical Travels in Iceland, *published in Copenhagen in 1780. The account is taken from the book* Travels in Iceland *(1811) by Sir George Steuart Mackenzie (1780–1848), written soon after a scientific expedition to Iceland in 1810 by Mackenzi and two students, Henry Holland (1788–1873) and Richard Bright (1789–1858),* both of whom later achieved great prominence as physicians and travellers. At the time little was known about this area of the world. The three spent several months exploring and studying – along with many elaborate sketches which appear in their book – the botany, zoology, geology, history, literature, customs, government, music, weather, diseases, and flora of Iceland.

In the year 1727, on the 7th August, after the commencement of divine service in the church of Sandfell, as I stood before the altar, I was sensible of a gentle concussion under my feet, which I did not mind at first; but, during the delivery of the sermon, the rocking continued to increase, so as to alarm the whole congregation; yet they remarked that the like had often happened before. One of them, a very aged man, repaired to a spring, a little below the house, where he prostrated himself on the ground, and was laughed at by the rest for his pains; but, on his return, I asked him what it was he wished to ascertain, to which he replied, "Be on your guard, Sir; the earth is on fire!" Turning, at the same moment, towards the church door, it appeared to me, and all who were present, as if the house contracted and drew itself together. I now left the church, necessarily ruminating on what the old man had said; and as I came opposite to Mount Flega, and looked up towards the summit, it appeared alternately to expand and be heaved up, and fall again to its former state. Nor was I mistaken in this, as the event shewed; for on the morning of the 8th, we not only felt frequent and violent earthquakes, but also heard dreadful reports, in no respect inferior to thunder. Everything that was standing in the houses was thrown down by these shocks; and there was reason to apprehend, that mountains as well as houses would be overturned in the catastrophe. What most augmented the terror of the people was, that nobody could divine in what place the disaster would originate, or where it would end.

After nine o'clock, three particularly loud reports were heard, which were almost instantaneously followed by several eruptions of water that gushed out, the last of which was the greatest, and completely carried away the horses and other animals that it overtook in its course. When these exudations were over, the ice mountain itself ran down into the plain, just like melted metal poured out of a crucible; and on settling, filled it to such a height, that I could not discover more of the well-known mountain Lounagrupr than about the size of a bird. The water now rushed down the east side without intermission, and totally destroyed what little of the pasture-grounds remained. It was a most pitiable sight to behold the females crying, and my neighbours destitute both of counsel and courage: however, as I observed that the current directed its course towards my house, I removed my family up to the top of a high rock, on the side of the mountain, called Dalskardstorfa, where I caused a tent to be pitched, and all the church utensils, together with our food, clothes and other things that were most necessary, to be conveyed thither; drawing the conclusion that should the eruption break forth at some other place, this height would escape the longest, if it were the will of God, to whom we committed ourselves, and remained there.

Things now assumed quite a different appearance. The Jökull itself exploded, and precipitated masses of ice, many of which were hurled out to the sea; but the thickest remained on the plain, at a short distance from the foot of the mountain. The noise and reports continuing, the atmosphere was so completely filled with fire and ashes, that day could scarcely be distinguished from night, by reason of the darkness which followed, and which was barely rendered visible by the light of the fire that had broken through five or six cracks in the mountain. In this manner the parish of Oraefa was tormented for three days together; yet it is not easy to describe the disaster as it was in reality; for the surface of the ground was entirely covered with pumice-sand, and it was impossible to go out in the open air with safety, on account of the red-hot stones that fell from the atmosphere. Any who did venture out, had to cover their heads with buckets, and such other wooden utensils as could afford them some protection.

On the 11th it cleared up a little in the neighbourhood; but the ice-mountain still continued to send forth smoke and flames. The same day I rode, in company with three others, to see how matters stood with the parsonage, as it was most exposed, but we could only proceed with the utmost danger, as there was no other way except between the ice-mountain and the Jökull which had been precipitated into the plain, where the water was so hot that the horses almost got unmanageable: and, just as we entertained the hope of getting through by this passage, I happened to look behind me, when I descried a fresh deluge of hot water directly above me, which, had it reached us, must inevitably have swept us before it. Contriving, of a sudden, to get on the ice, I called to my companions to make the utmost expedition in following me; and by this means, we reached Sandfell in safety. The whole of the farm, together with the cottages of two tenants, had been destroyed; only the dwelling houses remained, and a few spots of the tuns. The people stood crying in the church. The cows which, contrary to all expectation, both here and elsewhere, had escaped the disaster, were lowing beside a few haystacks that had been damaged during the eruption. At the time the exudation of the Jökull broke forth, the half of the people belonging to the parsonage were in four nearly-constructed sheep-cotes, where two women and a boy took refuge on the roof of the highest; but they had hardly reached it when, being unable to resist the force of the thick mud that was borne against it, it was carried away by the deluge of hot water and, as far as the eye could reach, the three unfortunate persons were seen clinging to the roof. One of the women was afterwards found among the substances that had proceeded from the Jökull, but burnt and, as it were, parboiled; her body was so soft that it could scarcely be touched. Everything was in the most deplorable condition. The sheep were lost; some of which were washed up dead from the sea in the third parish from Oraefa. The hay that was saved was found insufficient for the cows so that a fifth part of them had to be killed; and most of the horses which had not been swept into the ocean were afterwards found completely mangled. The eastern part of the parish of Sida was also destroyed by the pumice and sand; and the inhabitants were on that account obliged to kill many of their cattle.

The mountain continued to burn night and day from the 8th of August, as already mentioned, till the beginning of Summer in the month of April the following year, at which time the stones were still so hot that they could not be touched; and it did not cease to emit smoke till near the end of the Summer. Some of them had been completely calcined; some were black and full of holes; and others were so loose in their contexture that one could blow through them. On the first day of Summer 1728, I went in company with a person of quality to examine the cracks in the mountain, most of which were so large that we could creep into them. I found here a quantity of saltpeter and could have collected it, but did not choose to stay long in the excessive heat. At one place a heavy calcined stone lay across a large aperture; and as it rested on a small basis, we easily dislodged it into the chasm but could not observe the least sign of its having reached the bottom. These are the more remarkable particulars that have occurred to me with respect to this mountain; and thus God hath led me through fire and water, and brought me through much trouble and adversity to my eightieth year. To Him be the honour, the praise, and the glory for ever.

2. Exploration

To explore is to investigate, to seek, to search, to journey to the untraveled, to address the unknown. It requires not only curiosity, but also imagination, fortitude and skill. And, though it may, and does, involve investigating unknown corners and hidden places of the Earth, it can also involve the investigation of the familiar and the intensive study of Earth processes, mechanisms and theories.

Our selection includes a first-hand record of a journey to Antarctica, John Wesley Powell's narrative of the exploration of the Colorado River, an account of a paleonto-logical collecting expedition to Patagonia, the discovery of early cave art, the hazards of a volcanologist, and a dive to explore the mid-ocean ridge.

2-1   The Voyage of the Beagle – Charles Darwin

Charles Darwin (1809–1882) was the son and grandson of physicians. Educated at Edinburgh and Cambridge, he showed little promise as a student, but the voyage here described set him on a new path. Though he is remembered chiefly as a biologist and the founder of modern evolutionary theory, his first love was geology and he made major contributions to the subject. Our extract is taken from The Voyage of the Beagle, *first published as* Journal of Researches Into the Geology and Natural History of the Various Countries Visited by H.M.S. Beagle ... *(1839). The following account concerns Darwin's voyage with the Beagle and describes his visit in October, 1835, to the Galapagos Islands, a group of ten volcanic islands lying on the equator some 500 miles west of Ecuador. It was here that Darwin was brought into contact with "that mystery of mysteries – the first appearance of new beings on Earth." The extract indicates his meticulous observations, and the astonishing gift he had of comparison and synthesis, by which he was able from isolated phenomena to draw wider conclusions. This capacity was, itself, shaped by the voyage. Darwin later wrote "The voyage of the 'Beagle' has been by far the most important event in my life and has deter-mined my whole career. ... I have always felt that I owe to the voyage the first real training or education of my mind."*

October 8th. – We arrived at James Island: this island, as well as Charles Island, were long since thus named after our kings of the Stuart line. Mr. Bynoe, myself, and our servants were left here for a week, with provisions and a tent, whilst the Beagle went for water. We found here a party of Spaniards, who had been sent from

Charles Island to dry fish, and to salt tortoise-meat. About six miles inland, and at the height of nearly 2000 feet, a hovel had been built in which two men lived, who were employed in catching tortoises, whilst the others were fishing on the coast. I paid this party two visits, and slept there one night. As in the other islands, the lower region was covered by nearly leafless bushes, but the trees were here of a larger growth than elsewhere, several being two feet and some even two feet nine inches in diameter. The upper region being kept damp by the clouds, supports a green and flourishing vegetation. So damp was the ground, that there were large beds of a coarse cyperus, in which great numbers of a very small water-rail lived and bred. While staying in this upper region, we lived entirely upon tortoise-meat: the breast-plate roasted (as the Gauchos do *carne con cuero*), with the flesh on it, is very good; and the young tortoises make excellent soup; but otherwise the meat to my taste is indifferent.

One day we accompanied a party of the Spaniards in their whaleboat to a salina, or lake from which salt is procured. After landing, we had a very rough walk over a rugged field of recent lava, which has almost surrounded a tuff-crater, at the bottom of which the salt-lake lies. The water is only three or four inches deep, and rests on a layer of beautifully crystallized, white salt. The lake is quite circular, and is fringed with a border of bright green succulent plants; the almost precipitous walls of the crater were clothed with wood, so that the scene was altogether both picturesque and curious. A few years since, the sailors belonging to a sealing-vessel murdered their captain in this quiet spot; and we saw his skull lying among the bushes.

During the greater part of our stay of a week, the sky was cloudless, and if the trade-wind failed for an hour, the heat became very oppressive. On two days, the thermometer within the tent stood for some hours at 93°; but in the open air, in the wind and sun, at only 85°. The sand was extremely hot; the thermometer placed in some of a brown colour immediately rose to 137°, and how much above that it would have risen, I do not know, for it was not graduated any higher. The black sand felt much hotter, so that even in thick boots it was quite disagreeable to walk over it.

The natural history of these islands is eminently curious, and well deserves attention. Most of the organic productions are aboriginal creations, found nowhere else; there is even a difference between the inhabitants of the different islands; yet all show a marked relationship with those of America, though separated from that continent by an open space of ocean, between 500 and 600 miles in width. The archipelago is a little world within itself, or rather a satellite attached to America, whence it has derived a few stray colonists, and has received the general character of its indigenous productions. Considering the small size of the islands, we feel the more astonished at the number of their aboriginal beings, and at their confined range. Seeing every height crowned with its crater, and the boundaries of most of the lava-streams still distinct, we are led to believe that within a period geologically recent the unbroken ocean was here spread out. Hence, both in space and time, we seem to be brought somewhat near to that great fact-that mystery of mysteries-the first appearance of new beings on this earth.

Of land-birds I obtained twenty-six kinds, all peculiar to the group and found nowhere else, with the exception of one lark-like finch from North America (Dolichonyx oryzivorus), which ranges on that continent as far north as 54°, and generally frequents marshes. The other twenty-five birds consist, firstly, of a hawk, curiously intermediate in structure between a buzzard and the American group of carrion-feeding Polybori; and with these latter birds it agrees most closely in every habit and even tone of voice. Secondly, there are two owls, representing the short-eared and white barn-owls of Europe. Thirdly, a wren, three tyrant-flycatchers (two of them species of Pyrocephalus, one or both of which would be ranked by some ornithologists as only varieties), and a dove-all analogous to, but distinct from, American species. Fourthly, a swallow, which though differing from the Progne purpurea of both Americas, only in being rather duller colored, smaller, and slenderer, is considered by Mr. Gould as specifically distinct. Fifthly, there are three species of mocking thrush-a form highly characteristic of America. The remaining land-birds form a most singular group of finches, related to each other in the structure of their beaks, short tails, form of body and plumage: there are thirteen species, which Mr. Gould has divided into four sub-groups. All these species are peculiar to this archipelago; and so is the whole group, with the exception of one species of the sub-group Cactornis, lately brought from Bow Island, in the Low Archipelago. Of Cactornis, the two species may be often seen climbing about the flowers of the great cactus-trees; but all the other species of this group of finches, mingled together in flocks, feed on the dry and sterile ground of the lower districts. The males of all, or certainly of the greater number, are jet black; and the females (with perhaps one or two exceptions) are brown. The most curious fact is the perfect gradation in the size of the beaks in the different species of Geospiza, from one as large as that of a hawfinch to that of a chaffinch, and (if Mr. Gould is right in including his sub-group, Certhidea, in the main group) even to that of a warbler. Seeing this gradation and diversity of structure in one small, intimately related group of birds, one might really fancy that from an original paucity of birds in this archipelago, one species had been taken and modified for different ends. In a like manner it might be fancied that a bird originally a buzzard, had been induced here to undertake the office of the carrion-feeding Polybori of the American continent.

2-2 The Map that Changed the World – Simon Winchester

In 1815, William Smith (1769–1839), a surveyor and canal digger, published a "map that changed the world." For the previous twenty years, he had labored, single-handed, mapping every corner of England and Wales to produce this remarkable map. Four years after its publication, "with his young wife going steadily mad. … Smith ended up in debtors' prison, a victim of plagiarism, swindled out of his recognition and his profits." This is the story that Simon Winchester recounts in his book The Map That Changed the World (2001). Winchester, born in 1944, an Oxford geology graduate, is a journalist and author who has written other books on the British peerage, Northern Ireland, Korea, Japan and China, and the best-selling work The Professor and the Madman (1998).

Above one of the many grand marble staircases within the east wing of Burlington House, the great Palladian mansion on the north side of London's Piccadilly, hangs a pair of huge sky blue velvet curtains, twisted and tasseled silk ropes beside them. Although many may wonder in passing, rarely does any one of the scores of people who climb and descend the stairs inquire as to what lies behind the drapes. A blocked-off window, perhaps? A painting too grotesque to show? A rare Continental tapestry, faded by the sunlight?

Once in a while someone curious and bold will demand a look, whereupon a functionary will emerge from behind a door marked Private, and with practiced hand will tug gently on the silk ropes. The curtains will slowly part, revealing an enormous and magnificent map of England and Wales, engraved and colored – in sea blue, green, bright yellow, orange, umber – in a beguiling and unfamiliar mixture of lines, patches, and stippled shapes.

"The German Ocean," it says to the east of the English coast, instead of today's "North Sea." There is, in an inset, a small cross-section of what is said to be the underside of the country from Wales to the river Thames. Otherwise all is readily familiar, comfortingly recognizable. The document is exquisitely beautiful – a beauty set off by its great size, more than eight feet by six – and by the fact that it towers – looms, indeed – above those who stand on the staircase to see it. The care and attention to its detail is clear: This is the work of a craftsman, lovingly done, the culmination of years of study, months of careful labor.

At the top right is its description, engraved in copperplate flourishes: "A Delineation of The Strata of England and Wales with part of Scotland; exhibiting the Collieries and Mines; the Marshes and Fen Lands originally Overflowed by the Sea; and the Varieties of Soil according to the Variations in the Sub Strata; illustrated by the Most Descriptive Names." There is a signature: "By W. Smith." There is a date: "August 1, 1815."

This, the official will explain, is the first true geological map of anywhere in the world. It is a map that heralded the beginnings of a new science. It is a document that laid the groundwork for the making of great fortunes – in oil, in iron, in coal, and in other countries in diamonds, tin, platinum, and silver – that were won by explorers who used such maps. It is a map that laid the foundations of a field of study that culminated in the work of Charles Darwin. It is a map whose making signified the beginnings of an era not yet over, that has been marked ever since by the excitement and astonishment of scientific discoveries that allowed human beings to start at last to stagger out from the fogs of religious dogma, and to come to understand something certain about their own origins – and those of the planet they inhabit. It is a map that had an importance, symbolic and real, for the development of one of the great fundamental fields of study – geology – which, arguably like physics and mathematics, is a field of learning and endeavor that underpins all knowledge, all understanding.

The map is in many ways a classic representation of the ambitions of its day. It was, like so many other grand projects that survive as a testament to their times – the *Oxford English Dictionary*, the Grand Triangulation of India, the Manhattan Project, the Concorde, the Human Genome – a project of almost unimaginably vast scope that required great vision, energy, patience, and commitment to complete.

But a signal difference sets the map apart. Each of the other projects, grand in scale, formidable in execution, and unassailable in historical importance, required the labor of thousands. The *OED* needed entire armies of volunteers. To build the Concorde demanded the participation of two entire governments. More men died during the Indian triangulation than in scores of modest wars. The offices at Los Alamos may have housed behind their chain-link fences shadowy figures who would turn out to be Nobel laureates or spies, but they were all hemmed in by immense battalions of physicists. And to attend to all their various needs – be they bomb makers, plane builders, lexicographers, codifiers of chemistry, or measurers of the land – were legions upon legions of minions, runners, amanuenses, and drones.

The incomparably beautiful geological map of 1815, however, required none of these. As vital as it turned out to be for the future of humankind, it stands apart – because it was conceived, imagined, begun, undertaken and continued and completed against all odds by just one man. All the Herculean labors involved in the mapping of the imagined underside of an entire country were accomplished not by an army or a legion or a committee or a team, but by the single individual who finally put his signature to the completed document – William Smith, then forty-six years old, the orphaned son of the village blacksmith from the unsung hamlet of Churchill, in Oxfordshire.

And yet William Smith, who created this great map in solitary endeavor, and from whose work all manner of benefits – commercial, intellectual, and nationalistic – then flowed, was truly at first a prophet without honor. Smith had little enough going for him: He was of simple yeoman stock, more or less self-taught, stubborn and visionary, highly motivated, and single-minded. Although he had to suffer the most horrendous frustrations during the long making of the map, he never once gave up or even thought of doing so. And yet very soon after the map was made, he became ruined, completely.

He was forced to leave London, where he had drawn and finished the map and which he considered home. All that he owned was confiscated. He was compelled to live as a homeless man for years, utterly without recognition. His life was wretched: His wife went mad – nymphomania being but one of her recorded symptoms – he fell ill, he had few friends, and his work seemed to him to have been without point, without merit.

Ironically and cruelly, part of the reason for his humiliation lies behind another set of faded velvet curtains that hang nearby, on another of Burlington House's many elaborate staircases. There, it turns out, is quite another map, made and published shortly after William Smith's. It was in all essentials a copy, made by rivals, and it was made – if not expressly then at least in part – with the intention of ruining the reputation of this great and unsung pioneer from Oxfordshire: a man who was not gently born, and who was therefore compelled, like so many others in those times, to bear the ungenerous consequences of his class.

But in the very long run William Smith was fortunate. A long while after the map had been published, a kindly and liberal-minded nobleman for whom Smith had been performing tasks on his estate in a small village in Yorkshire, recognized him – knew, somehow, that this was the man who had created the extraordinary and beautiful map which, it was said, all learned England and all the world of science outside was talking.

This aristocratic figure let people – influential and connected people – know about the man he had discovered. He reported that he was hidden, incognito, in the depths of the English countryside. He supposedly had no expectation that anyone would now ever remember, or would ever recognize, the solitary masterpiece that he once had made. He imagined he was doomed to suffer an undeserved oblivion.

But on this occasion his pessimism was misplaced: The messages that had been sent *did* get through – with the consequence that, eventually, William Smith was persuaded to return to London, to receive at last the honors and rewards that were due him, and to be acknowledged as the founding father of the whole new science of English geology, a science that remains at the core of intellectual endeavor to this day.

2-3 The Exploration of the Colorado River – John Wesley Powell

John Wesley Powell (1834–1902) was one of the nineteenth century's greatest leaders. He was born in Mount Morris, New York, the son of a Wesleyan minister. When the family moved to Illinois, he became fascinated with the natural history of the Ohio and Mississippi

Rivers. After study at Oberlin and Wheaton Colleges, he served in the Union Army during the Civil War, rising from private to major and losing his right arm at the Battle of Shiloh. He became Professor of Geology at Illinois Wesleyan College in 1865 and afterwards at Illinois Normal College. In 1867 he began his western explorations and in 1869 he was sponsored by the Smithsonian Institution to lead a party from Green River, Wyoming down the Green and Colorado rivers and through the Grand Canyon. It is this journey, described in The Exploration of the Colorado River of the West, first published in 1875, that our extract records.

Powell continued his western surveys for another decade, combining his geological interest with a study of the Native American peoples of the region. He became the first Director of the United States Bureau of Ethnology in the Smithsonian in 1879, publishing the first complete classification of the 58 native language stocks in 1891. He also served as the second head of the United States Geological Survey from 1881–1894, where he provided outstanding leadership.

August 9 [1869]. – And now, the scenery is on a grand scale. The walls of the canyon, 2,500 feet high, are of marble, of many beautiful colors, and often polished below by the waves, or far up the sides, where showers have washed the sands over the cliffs.

At one place I have a walk, for more than a mile, on a marble pavement, all polished and fretted with strange devices, and embossed in a thousand fantastic patterns. Through a cleft in the wall the sun shines on this pavement, which gleams in iridescent beauty.

I pass up into the cleft. It is very narrow, with a succession of pools standing at higher levels as I go back. The water in these pools is clear and cool, coming down from springs. Then I return to the pavement, which is but a terrace or bench, over which the river runs at its flood, but left bare at present. Along the pavement, in many places, are basins of clear water, in strange contrast to the red mud of the river. At length I come to the end of this marble terrace, and take again to the boat.

Riding down a short distance, a beautiful view is presented. The river turns sharply to the east, and seems inclosed by a wall, set with a million brilliant gems. What can it mean? Every eye is engaged, every one wonders. On coming nearer, we find fountains bursting from the rock, high overhead, and the spray in the sunshine forms the gems which bedeck the wall. The rocks below the fountain are covered with mosses, and ferns, and many beautiful flowering plants. We name it Vasey's Paradise, in honor of the botanist who traveled with us last year.

We pass many side canyons today, that are dark, gloomy passages, back into the heart of the rocks that form the plateau through which this canyon is cut.

It rains again this afternoon. Scarcely do the first drops fall, when little rills run down the walls. As the storm comes on, the little rills increase in size, until great streams are formed. Although the walls of the canyon are chiefly limestone, the adjacent country is of red sandstone; and now the waters, loaded with these sands, come down in rivers of bright red mud, leaping over the walls in innumerable cascades. It is plain now how these walls are polished in many places.

At last, the storm ceases, and we go on. We have cut through the sandstones and limestones met in the upper part of the canyon, and through one great bed of marble a thousand feet in thickness. In this, great numbers of caves are hollowed out, and carvings are seen, which suggest architectural forms, though on a scale so grand that architectural terms belittle them. As this great bed forms a distinctive feature of the canyon, we call it Marble Canyon.

It is a peculiar feature of these walls, that many projections are set out into the river, as if the wall was buttressed for support. The walls themselves are half a mile high, and these buttresses are on a corresponding scale, jutting into the river scores of feet. In the recesses between these projections there are quiet bays, except at the foot of a rapid, when they are dancing eddies or whirlpools. Sometimes these alcoves have caves at the back, giving them the appearance of great depth. Then other caves are seen above, forming vast, dome-shaped chambers. The walls, and buttresses, and chambers are all of marble.

The river is now quiet; the canyon wider. Above, when the river is at its flood, the waters gorge up, so that the difference between high- and low-water marks is often fifty or even seventy feet; but here, high-water mark is not more than twenty feet above the present stage of the river. Sometimes there is a narrow flood plain between the water and the wall.

Here we first discover mesquite shrubs, or small trees, with finely divided leaves and pods, somewhat like the locust.

2-4 Mono Lake–Aurora–Sonora Pass – William H. Brewer

Between 1860 and 1864, William H. Brewer (1828–1910) traveled extensively in California as the first assistant to Josiah D. Whitney, Chief of the California Geological Survey. A Yale graduate and a member of the first class at the Sheffield Scientific School, Brewer was by training an agricultural chemist. He later taught for many years at Yale.

Brewer's account of his explorations and geological investigations of the western frontier area was later included in an engaging book, Up and Down California, 1860–1864, which was not published until 1930. Our sample from Brewer is of a visit to Mono Lake, a saline body of water in eastern California. For another view of the same area, and for a contrast in style and interpretation, compare the description of Mark Twain (see page 209).

We descended safely, and camped in the high grass and weeds by a stream a short distance south of Lake Mono. This camp had none of the picturesque beauty of our mountain camps, and a pack of coyotes barked and howled around us all night.

July 8 Hoffmann and I visited a chain of extinct volcanoes which stretches south of Lake Mono. They are remarkable hills, a series of truncated cones, which rise about 9,700 feet above the sea. Rock peeps out in places, but most of the surface is of dry, loose, volcanic ashes, lying as steep as the material will allow. The rocks of these volcanoes are a gray lava, pumice stone so light that it will float on water, obsidian or volcanic glass, and similar volcanic products. It was a laborious climb to get to the summit. We sank to the ankles or deeper at every step, and slid back most of each step. But it was easy enough getting down – one slope that took three hours to ascend we came down leisurely in forty-five minutes. The scene from the top is desolate enough – barren volcanic mountains standing in a desert cannot form a cheering picture. Lake Mono, that American "Dead Sea," lies at the foot. Between these hills and our camp lie about six miles of desert, which is very tedious to ride over – dry sand, with pebbles of pumice, supporting a growth of crabbed, dry sagebrushes, whose yellow-gray foliage does not enliven the scene.

July 9 we came on about ten miles north over the plain and camped at the northwest corner of Lake Mono. This is the most remarkable lake I have ever seen. It lies in a basin at the height of 6,800 feet above the sea. Like the Dead Sea, it is without an outlet. The streams running into it all evaporate from the surface, so of course it is very salt – not common salt. There are hot springs in it, which feed it with peculiar mineral salts. It is said that it contains borax, also boracic acid, in addition to the materials generally found in saline lakes. I have bottled water for analysis and hope to know some time. . . .

July 11 we were up at dawn, a clear, calm morning. Clouds of gulls screamed around us. An early breakfast, then a tour of examination. These islands are entirely volcanic, and in one place the action can hardly be said to have ceased, for there are hundreds of hot springs over a surface of many acres. Steam and hot gases issue from fissures in the rocks, and one can hear the boiling and gurgling far beneath. Some of the springs are very copious, discharging large quantities of hot water with a very peculiar odor. Some boil up in the lake, near the shore, so large that the lake is warmed for many rods – no wonder that the waters hold such strange mineral ingredients. The rock is all lava, pumice, and cinders. At the northeast corner of the island are two old craters with water in them. The smaller, or north island, has no fresh water – it looks scathed and withered by fire. One volcanic cone, three hundred or four hundred feet high, looked more recent than any other I have seen in the state.

2-5   Thrills in Fossil Hunting – George F. Sternberg

George F. Sternberg (1883–1969) was the son of Charles H. Sternberg, one of the great fossil hunters of the American west. It was Charles who pioneered the plaster jacket technique for removing and transporting fossil vertebrates and it was also Charles who headed the family fossil hunting business of father and three sons, whose discoveries made a major contribution to some of the nation's leading museum collections. George Sternberg, who later became curator of the Fort Hays Kansas Teachers' College Museum, here describes the intense excitement at the discovery of a "mummified" duck-billed dinosaur in 1909. The extract is from an article originally published in The Aerend *in 1930.*

Will you go with me for a few moments into Converse County, Wyoming, where I was fortunate enough to discover a remarkable skeleton of a duckbilled dinosaur wrapped in its skin impression, which at the time of discovery was said to be the finest example of a dinosaur known. It was in July, 1908, that I drove our outfit into the rough sheep and cattle country north of Lusk, Wyoming, along the breaks of the Cheyenne River to search for dinosaur remains. My father, C.H. Sternberg, was in charge of the party. We were assisted by my two younger brothers, Charles and Levi. Father was the only one of the party who had ever collected a dinosaur bone. But we boys were looking forward to doing great things. We trudged over the rough exposures day after day and week after week only to return to our humble camp, pitched some sixty miles from Lusk where we must go for our supplies. As time went on we began to realize that finding dinosaur specimens was not as easy as we had expected. We wished that we were back in Kansas where the sunflowers grew. Camp supplies were getting low, and a trip to Lusk for more food must be made. We had gathered together a few unimportant specimens which could be hauled in, and it was decided that my younger brother, Levi, and I should remain in camp while Charles, who was the official camp cook, should accompany father on the trip.

Charles informed me that if I would ride to a sheep wagon some seven miles away and get some baking powder, there would be enough food in camp to last the two of us until they could return three days later. This I did; and upon arriving on the spot where the sheep wagon had been located for some time, we found it just ready to move some eight or ten miles farther away. We were informed that the dry weather had caused the water to give out. The herder also told us that he was running short of grub. However, I secured the necessary baking powder and returned to camp thinking all was well.

The evening before, I had accidently found some bones of a duckbilled dinosaur protruding from the base of a high sandstone ledge. Although it looked as if something worthwhile might be buried there, still the amount of work necessary to uncover the prospect did not lend any encouragement to us boys. Besides, the quarry was a mile and a half from our camp, and we had only one saddle horse between us to ride back and forth. Nevertheless, we went to work in earnest, and after two days of hard digging we laid bare a floor large enough to trace out most of the skeleton. Then with tools we began to follow the bones into the bank starting in the region of the hips. Slowly and carefully we worked. The removal of each bit of dirt revealed more and more of the skeleton. We were beginning to get hopeful. "But what if the head were gone?", we began to think. For it so often happens that after finding an articulated skeleton one finds the head has been severed from the body before burial and is nowhere to be found. This practically destroys the skeleton for museum purposes.

About this time our food gave out except for a sack of potatoes. I began to realize the baking powder for which I had ridden fourteen miles was of little use to us as there was only enough flour for two days. Had I been told that there was practically no salt in camp I might have borrowed some from the sheep herder. But the wagon should be back almost any time now, so rather than leave our interesting specimen and ride to the nearest ranch for food we decided to "go it on spuds." And "spuds" it was for three whole days. It would not have been so bad if we had had some salt to season them.

Finally by the evening of the third day, I had traced the skeleton to the breast bone, for it lay on its back with the ends of the ribs sticking up. There was nothing unusual about that. But when I removed a rather large piece of sandstone rock from over the breast I found, much to my surprise, a perfect cast of the skin impression beautifully preserved. Imagine the feeling that crept over me when I realized that here for the first time a skeleton of a dinosaur had been discovered wrapped in its skin. That was a sleepless night for me. Had I missed my regular cup of coffee or eaten too many potatoes for supper?

We were loathe to leave our treasure and go for food though we knew there was not a human being for miles to disturb it. Few cattle men ever rode this way as this was a high and dry region where cattle were seldom found ranging. The sheep men came here with their flocks only when there was melting snow to furnish water.

It was about dusk on the evening of the fifth day when we saw the wagon loaded with provisions roll into camp. "What luck?" was my father's first question. And before he could leave his seat I had given him a vivid sketch of my find; for we had found every bone in place except the tail and hind feet which had protruded from the rock and had washed away. The head lay bent back under the body. Traces of the skin were to be seen everywhere. "Let's go and see it," he shouted as he jumped from his seat on the wagon. I grabbed some food from the boxes on the wagon and away the two of us went, leaving the others to prepare a meal for us. Darkness was nearly upon us when we reached the quarry and there laid out before us was the specimen. One glance was enough for my father to realize what I had found and what it meant to science. Will I ever forget his first remark as we stood there in the fast approaching twilight? It thrills me now as I repeat it. "George, this is a finer fossil than I have ever found. And the only thing which would have given me greater pleasure would have been to have discovered it myself."

I do not remember what we had for supper that night. I do not remember what news was brought from home and loved ones. I do remember, however, that is was another restless night for not only myself but for my father as well. I could hear him roll in his bed and cough or make some noise which told me he too was spending a sleepless night. The day was beginning to break before I finally dozed off, only to awake as the sun came up over the hill. There was much to be done to collect the specimen. But by the time it was boxed and ready for shipment, Professor Osborn of the American Museum of Natural History, New York City, had sent a man to see it and had arranged to secure it for his great institution.

It was not until 1911 that I again had the privilege of seeing my prize specimen. There it lay in that great institution just as it was when I found it resting on its back with the head bent down under its body and its fore limbs stretched out on either side. It had been so skillfully prepared that the impression of the skin had been preserved over nearly all parts of the body. In Professor Osborn's description of the specimen he has called it the "Mummy Dinosaur." Could there be a greater thrill in the life of a man than to know he has been able to discover one of these buried treasures of bygone days and to have it placed before the world as an everlasting memorial? Is a fossil hunter's life worthwhile? Are there any thrills in it?

2-6   The Creative Explosion – John E. Pfeiffer

The discovery by a young girl – Maria – of large animal paintings on the ceiling of a cave at Altamira, in the Cantabrian Mountains of northern Spain, opened up a new chapter in the study of human prehistory. The number of paintings was immense – bison, deer, horses, some up to six feet long – full of action and movement; many were polychrome, others were black outlines, still others were engraved. And the cave was huge: 300 yards long. John Edward Pfeiffer, born in 1914, describes that remarkable moment in his book The Creative Explosion: an inquiry into the origins of art and religion *(1982).*

Maria's father – Don Marcelino de Sautuola, a Spanish nobleman and amateur archeologist – interpreted these paintings as prehistoric, the work of long vanished human ancestors. Though the public was impressed – King Alfonso XII visited the cave – the professionals, alas, were not. The International Congress of Anthropology and Prehistoric Archeology, meeting in Lisbon in 1880, a year after the discovery, invited to view the paintings, declined to do so, accepting instead a report which denounced them as a fraud. Today they are celebrated as one of humankind's greatest works of art, created over 15,000 years ago.

The first person to recognize the cave paintings for what they were, the works of people who lived many thousands of years ago, rested his case on the art of Altamira, a cave located in the foothills of the Cantabrian Mountains of northern Spain – and died labeled as a charlatan. A century ago Don Marcelino de Sautuola, a Spanish nobleman and amateur archeologist, found himself holding views that conflicted with those of individuals who knew far more about times past than he did. He was ahead of his time, about a generation ahead. He saw something plain and clear, naively if you will, lacking the education to doubt himself or to appreciate the reasons for the doubts of the experts, to realize how startling his conclusion would be to them. So he ran headlong and innocently into a barrage of personal criticism severe even by the standards of a profession noted for such attacks.

In any case, his is one of the great tales in the annals of prehistory. He had dug near the entrance to Altamira on a number of occasions without noticing anything special, until one summer day in 1879 when, according to a story which has become legend, his daughter Maria, aged five, seven, or twelve depending on the storyteller, wandered into one of the cave's side chambers. Suddenly by the light of her lantern she saw paintings on the ceiling, large paintings of animals in vivid blacks and reds, pinks and browns, and ran out to tell her father.

Nothing had prepared Sautuola for the shock of such a discovery. He had explored that chamber, and thought he knew what was in it. He had been there several times before, stooping or crawling, since the chamber was only 3 to 5 feet high in most places, using his lantern to avoid bumping his head against rounded rocky protuber-ances on the ceiling. And yet he never noticed that there were paintings on those protuberances. They were covered with paintings. Artists had taken advantage of them some 15,000 years before to represent the bodies of animals in relief, to give a living, three-dimensional effect.

Archeological records include many cases of art overlooked. The eye never comes innocent to its subject. Everything seen is a blend of what actually exists out there, the "real" object, and the viewer's expectations, upbringing, and current state of mind. It is amazing what you can miss when you do not expect to see anything or, given a strong enough motive, what you can see that is not there. Unless the mind is properly adjusted or set, anticipating revelation of a particular sort, nothing happens. Things may remain invisible – like the figures at Altamira.

Actually, Sautuola had no real interest in the walls or ceilings of the cave. He was an excavator interested above all in what he could find at his feet, on the floor, such things as flint artifacts and bones and the remains of hearths. The low ceiling of the side chamber, now known as the Great Hall, was only a hazard to him, something to avoid. Bumping your head against rock is one of the dangers of exploring caves. Maria, being shorter, had rather less to fear on that score. Beyond that, she was too young to have acquired a bias against looking up rather than looking down.

Father and daughter spent the rest of the day, and many subsequent days, walking through the decorated galleries of Altamira. Theirs was the high excitement of being the first to experience the full impact of the totally unexpected. Finding a single polychrome painting in such a setting would have been enough of an adventure. Imagine the cumulative effect of moving from place to place, deeper and deeper into the galleries, and coming across figure after figure in a chain reaction of discovery. The experience was overwhelming even for visitors, entering a century later and knowing what to expect, having seen photographs and reproductions of the paintings in this cave and in many others.

2-7 Attending Marvels: A Patagonian Journal – George Gaylord Simpson

George Gaylord Simpson (1902–1984) was one of the giants of twentieth century paleontology. Simpson was born in Chicago and grew up in Denver; he graduated from the University of Colorado and Yale, and served at the American Museum of Natural History for almost 30 years. He held professorships at Columbia, Harvard, and at the University of Arizona. His writings were prolific, and his 500 books and articles dealt with topics ranging from primitive Mesozoic mammals of the American west, to Tertiary faunas of North and South America, penguins, horses, statistics, taxonomy, and evolution. He was one of the founders of the Synthetic Theory of Evolution. Simpson, though a small and rather frail man, was a tireless field geologist. It is from an account of his field work – Attending Marvels: a Patagonian journal (1934) – that this extract is taken.

"The señor doctor is crazy, isn't he?" asked Manuel, not realizing that I could not avoid overhearing him.

"I don't think so. Why?" put in loyal Justino.

"He came down to this desert for no reason! No one watches him and he still works hard! And what sort of work is that? Climbing around the barranca and getting all tired out, just to pick up scraps of rock. As if there were not rocks everywhere, even in that America of the North!"

"But those are not rocks. They are bones."

"Then why doesn't he stay in Buenos Aires and get bones from a slaughter-house, if anyone is fool enough to pay him for them?"

"They are not common bones. They are not sheep or guanaco. They are very old, so antique that they have turned to stone. They are bones of animals that now do not exist, and they are not found anywhere but here at Colhué–Huapí."

"The señor doctor is fooling you. Even if they were old bones they would not be any good. What would you do with them?"

"He says that some of them are the ancestors of our own animals and he wants to learn where they came from and what they were like millions of years ago. And others are strange beasts that are not like those of today. He will put them in a museum and people will come to see them because they are so queer."

"Oh, well, then maybe it does make sense. If many of those North American millionaires will pay to see them, then he can sell them at big prices. But then he should pay us more, because we are helping him and he is making his fortune."

"He says that no one pays to see the bones. And he does not sell them. He is just paid wages like us to come and find them."

"Why, then, with no one to watch him, he could stay in Buenos Aires and then go back to Nueva York and say he could not find any bones. No, friend Justino, it doesn't make sense. If the señor doctor is not crazy, he is too clever for you, and you are swallowing his lies."

And so the argument ended, as they always did, with each thoroughly convinced that he had been right all along. Manuel, like almost all the local inhabitants, was never able to see any logical reason or excuse for our activities. Rare Justino, on the other hand, took naturally to bone digging. He had tireless energy and a keen eye, as so many fossils now in New York testify. He also had that consuming curiosity which is the real reason for any scientific research, whatever higher motives we scientists sometimes like to claim for ourselves.

2-8   Explorations – Robert D. Ballard

Robert Ballard, president of the Sea Research Institute's Institute for Exploration in Mystic, Connecticut, is, perhaps, best known for his discovery of the wreck of the Titanic, *deep under the waters of the North Atlantic. In the course of more than 100 deep sea dives, he has not only explored other wrecks (the* Bismarck, *the* Yorktown, *and ancient Phoenician vessels among them) but has established a major reputation in marine geology and geophysics.*

Born in Wichita, Kansas, in 1942, he grew up in San Diego, graduated with a PhD in marine geology from the University of Rhode Island, and has most recently been involved in studies in the Black Sea, showing great submergence, perhaps linked to the flood of Noah. In this extract, taken from his book Explorations *(1995), he describes a dive that opened the exploration of the rift along the mid-ocean ridge, and the unexpected discoveries it produced.*

The first *Alvin* dive on the rift discovered not one but two hydrothermal vents along the linear mound of the central axis. When *Alvin* returned to the surface, the specimen tray contained five live mussels with chocolate-colored shells, a clam, and several chunks of lava encased with mats of living matter that could have been either colonial microorganisms or coral. Unfortunately, the normally reliable seawater sampler had malfunctioned, and they retrieved only half a liter.

On the next day's dive, Jerry van Andel and his Oregon State colleague, Jack Dymond, were not so lucky. Dudley Foster took them to a likely looking mound near the first clam colonies. But they found no hydrothermal vents or any unusual signs of life. They did discover some bizarre pillarlike structures of basalt streaked with bright smears of purple, green, and yellow, but the ambient sea temperature never rose above 3 degrees Celsius.

On the third dive in as many days, Corliss and MIT chemist John Edmond returned to the first clam colony. The milky blue shimmering clouds were in much greater evidence than they had been earlier. Scores of white crabs clambered over the clam shells, scavenging morsels among the occasional dead bivalve. Today, pilot Jack Donnelly managed to retrieve a full 12 liters of water samples before returning to the surface.

It was after dark by the time the Boston Whaler returned from *Lulu* with the water samples, and most of us were at dinner when John Edmond depressurized the sample flasks to begin his chemical assays. But the overpowering, unmistakable, rotten-egg stench of hydrogen sulfide was quickly spread throughout the ship by *Knorr*'s efficient air conditioning. We rushed to the lab but didn't stay long. Even John, used to playing with sulfur-rich volcanic minerals, was having a hard time breathing.

That night we discovered that the water collected around the clam colony contained an incredible amount of dissolved hydrogen sulfide, enough to poison most land creatures. But obviously Clambake I, as we now called this site, was rich with life. We sat around the dayroom, sipping beers like so many sophomores in over their intellectual heads, speculating on the nature of this discovery. If only there was a qualified biologist among the fifteen-odd Earth scientists out here on the rift.

But as geologists, we recognized that the chemistry of this hydrothermal system had to be complex. Seawater seeping down through fissures in the lava floor toward the magma chambers was heated to the point where its chemical composition was altered drastically. At high temperatures and pressures, the seawater must have lost some of its suspended minerals to the surrounding rock while drawing other mineral compounds. On percolating back toward the sea floor, these sulfates were probably converted to hydrogen sulfide, which precipitated out as the heated water rapidly cooled, and produced the shimmering milky clouds the dive teams had observed and photographed.

"Hydrogen sulfide," Edmond told us, "can be metabolized by certain anaerobic bacteria."

That made sense. Most of us had smelled the unforgettable bad-egg stench of the mud in swamps and bogs. And we all had a vague, high-school biology recollection of paleo-microorganisms that had evolved in the early epochs before the planet's atmosphere contained free molecular oxygen. But was such a primitive food chain somehow functioning around those warm water vents on the floor of the Galapagos Rift?

For answers we turned to the nearest cooperative biologists we could find, Holger Jannasch and Fred Grassle at Woods Hole. I arranged the single-sideband radio equivalent of a conference call the next morning. After Corliss, van Andel, and Jack Edmond carefully described the variety of the creatures and details of their habitat, we asked Holger and Fred for advice on how to proceed, in effect, trying to compress four years of undergraduate biology fieldwork into one scratchy radio-telephone call.

"First off," Holger advised, "take core samples in a careful grid pattern every couple of meters."

"We'll need to carefully analyze the organic qualities of the mud," Fred added.

The four of us grouped around the big radio console in the *Knorr*'s communications room exchanged quizzical glances.

"There's no mud, Fred," I explained, looking at the color prints from the latest dive.

Jack Corliss took the microphone. "It's all bare lava, I'm afraid, just naked stone."

The speaker hissed with static, then Holger's voice returned. "I don't see how that's possible. There must be some mistake."

But there was no mistake.

Over the next five weeks, we completed twenty-one successful dives on the rift. In the process, we discovered and thoroughly explored four more unique colonies of living creatures in the near-sterile desolation of the deep-sea bottom.

At one site, Clambake II, we found a virtual graveyard: hundreds of empty clamshells, picked clean by scavenging crabs. Significantly, there were no warm-water vents at this once-thriving colony. Some obscure change in the substrata plumbing had shut down the life-giving hydrothermal vent, breaking the unique food chain and dooming the immobile clams to extinction.

As we explored further down the axis of the rift, we discovered a distinctive feature of these oases. Some were dominated by clams. Another – like the mistakenly named "Oyster Bed" – was the domain of brown-shelled mussels, while the most prevalent large organism at the Dandelion Patch was the hitherto unknown stalklike creature with the bulbous multipetaled head. The colony around the teeming Garden of Eden vent was made up of distinctive rings of dominant animals: first the dandelion creatures, then white crabs, then a ring of white-stalked, red-fleshed worms with shimmering crimson crowns.

Because we had no way of transmitting pictures to biologists ashore, it was up to Earth scientists to venture away from their own field to speculate about the underlying causes of this dominant species pattern.

Tanya Atwater, a geophysicist from MIT, came up with as likely an explanation as any other. "Maybe," she told us one night, as *Knorr* cruised slowly beneath the glittering dome of equatorial stars, "all these animals started out life as free-swimming larvae. Then," she twirled her fingers to mimic falling marine snow, "they drifted down and down until they became lost on the dark ocean bottom. Then they discovered the plumes of hot water, grew up, and reproduced to start the colonies."

Jack Corliss added a variation to this hypothesis, which he called the "Founder Principle." Under this theory, the dominant species at any given hydrothermal vent oasis was the first to drift there in larval form, be it clams, mussels, worms, or the strange orangish pink dandelions, which the biologists had been unable to classify by radio conference.

The more we considered these hypotheses, the more logical they became. Obviously, complex creatures such as clams and mussels did not spontaneously generate from inanimate lava, as in some medieval pseudoscience. And the newly formed sea floor along the rift's central axis had equally obviously never been a coastal shallow that had sunk to these depths by geological subsidence. Still, the complex metabolic linkage we call the food chain operating down there without the benefit of photosynthesis remained mysterious for much of the expedition.

The only common feature to all these vents was the prevalence of rich hydrogen sulfide solution in the water.

This unusual evidence, coupled with our impromptu self-tutorials in biology and a good deal of radio-telephone coaching from Woods Hole biologists, led us toward a startling conclusion: there *was* a hydrogen sulfide-based food chain operating almost 2 miles beneath the surface along the central axis of the Galapagos Rift.

In this "chemosynthesis," which was completely independent of the sunlight-fueled photosynthesis food chain that prevailed elsewhere on the planet, primitive bacteria metabolized the sulfides and were, in turn, eaten by an ascending order of microorganisms that, in their turn, supported even higher life forms all the way up to the clams and crabs.

Halfway into the cruise, we realized that we had stumbled onto a major scientific discovery. I saw normally staid colleagues actually skipping with pleasure down the ship's corridors to and from the lab. All of us relished the joy of discovery, even of nature's more mundane secrets. But the implications of the bizarre and fragile chemosynthesis oases clustered around those warm-water vents on the Galapagos Rift was anything but mundane.

The fact that this chain of life existed in the black cold of the deep ocean and was utterly independent of sunlight – previously thought to be the font of *all* Earth's life – had startling ramifications. If life could flourish, nurtured by a complex chemical process based on geothermal heat, then life could exist under similar conditions on planets far removed from the nurturing light of our parent star, the Sun.

I could picture conditions similar to those of the Galapagos Rift in sea-floor lava tubes hidden beneath the frozen surface of the Jovian moons.

In effect, we had discovered what amounted to an entire separate branch of evolution down in those dark lava mounds, a virtual new planet for biologists to explore.

One night, drinking a cold beer on the fantail as ANGUS slalomed unseen below, back and forth across the rift, searching for more vents, John Edmond summed up what we all felt.

"You know, Bob," he said, this is what it must have been like sailing with Columbus."

2-9   The Blue Planet – Louise B. Young

Louise Young was born in Ohio in 1919, graduated from Vassar and is a geophysicist, environmentalist, traveler, and writer. Confronted by a life-threatening illness, she responded by devoting herself to explore and describe those places which illuminate our relationship with nature. The result over 35 years has been a succession of highly praised books, one of which is The Blue Planet (1983), from which our extract is taken and which won the Carl Sandburg Literary Arts Award. In it, Young describes the exhilarating effects of a sailplane flight above the island of Ohau.

I was born too soon to ever see with my own eyes the whole earth silhouetted against the dark abyss of space, or coming in closer to watch the continents emerge from their veil of clouds, revealing the sculptured profile of their coasts against the shining darkness of the deep blue sea. But today I have seen a small piece of the earth – a tiny microcosm – from the perspective of space, as I soared on the tradewinds above the island of Oahu.

On this jewel-like Hawaiian morning I was towed aloft in a sailplane from a field on the northern shore of the island where the Waianae Range meets the sea. The breeze came clean and steady out of the northeast after traveling uninterrupted over thousands of miles of open water. Encountering the ridge of mountains, it swings upward in a wave of rising current; so when I released from tow I was borne aloft as a gull rides the ascending winds. Smooth as silk the air rose, carrying me up over glistening coral beaches, over velvet green mountain slopes, over cobalt sea – higher and higher until I seemed suspended in a sunlit bubble of space and a great quietness – the voice of the wind muted to a whisper as it flowed gently over the wings. ...

As I followed the ridge westward, I could catch glimpses between the jagged peaks of the land beyond the hills. And then where the mountains descended suddenly to the shore the whole leeward coast came into view: valleys bathed in sunshine, mile upon mile of pineapple plantations with young plants set in neat rows against the mahogany-red Hawaiian soil, and vast seas of silvery green sugarcane fields ripe for harvest. ...

Soon I could see the far coastlines of the island: the south shore with the three deeply indented lobes of Pearl Harbor; the east shore, soft and misty, veiled in clouds hanging low upon the dark hulking shapes of the Koolau Range. The whole island was there below me – complete and self-contained – a tiny model of the earth in time as well as in space.

3. Geologists are also Human

Because knowledge is rooted in the knower, science is rooted in the scientist just as surely as art is rooted in the artist. Scientific knowledge, therefore, represents not only an enclosing edifice, in which each of the component parts blends into a more or less harmonious whole, but an edifice which is the product of individual insight. It is these individual insights, hard-won from patient reflection, and prolonged encounter with the natural world, which are the real elements of scientific knowledge. Yet the insights are achieved, not by white-coated technocrats, the high priests of some particular terrestrial vision, nor by committees, but by individual Earth scientists who are also very human. It is in that sense that scientific knowledge is both private and public, for when a geologist is not practicing his or her craft, he or she engages in the same activities and shares the same hopes and prejudices as his or her neighbors, digging the garden, listening to Mozart, voting in elections, and walking the dog. For this reason, the character of the individual observer is inevitably involved in subtle ways in the character of the observations which he or she makes and the conclusions which he or she draws. The individual's frame of reference, the commitment to devote weeks or years to the study of some isolated and perhaps obscure phenomenon, the determination to journey to remote and often inhospitable regions to gain firsthand access to certain phenomena, all these reflect the relationship between individual character and the scientific work which the individual produces.

Nor is the history of geology confined to an account of the data accumulated by professional geologists during the last two hundred years. It stretches back as far as early humans, who, in collecting flints for weapons, and using fossils to decorate the graves of the dead, asserted their belief in the dignity and persistence of life. It stretches back to Pliny the Elder, Roman naturalist, philosopher and sailor, who lost his life during the great eruption of Vesuvius. His nephew, Gaius Plinius (Pliny the Younger), records in a letter to his friend Tacitus that Pliny was at Misenum, in command of the Roman fleet, when at one o'clock on the 24th of August 79 AD his attention was drawn to a great cloud which "seemed of importance and worthy of nearer investigation to the philosopher." He ordered a light boat to be got ready "but as he was about to embark, Pliny received a plea from Rectina, wife of Caesius Bassus, living on the other side of the bay, begging him to rescue her from what she thought was certain death." This converted his plan of observation into a more serious purpose. "He steered directly for the dangerous spot whence others were flying, watching it so fearlessly as to be able to dictate a description and be able to take notes of all the movements and appearances of this catastrophe as he observed them." After landing, but being unable, because of violent seas, to embark, he encouraged and reassured the community. When day came, they found his body perfect and uninjured, seeming by his attitude to be asleep, rather than dead.

In various ways, the world view of particular individuals had profound effects upon their commitment to science. In the case of Nicholas Steno, for example, the seventeenth century Danish priest and philosopher, and the first to formulate the essential principles of historical geology, his conversion to Catholicism led to the end of his scientific career, for his efforts thereafter were devoted wholly to the service of the Catholic Church. In other cases, such as that of James Hutton, religious commitment was a catalytic influence in scientific thought, as shown by Hutton's commitment to the idea of providence in Nature, which was a major influence in his determination to develop his theory of the Earth, even though the work was bitterly attacked by some of his theological critics. In still other cases, chance encounters with the Earth while following some other career have been all important, as in the case of William Smith, the self-taught surveyor and engineer, whose occupation in constructing canals in southern England involved trenching through stratified rocks and, thus, allowed him the opportunity to formulate fundamental principles of stratigraphy.

There is also an interaction in the other direction, however, because our sense of the Earth on which we live may, in turn, influence our world view. At least one Hindu tribe, for example, living in an area subject to earthquakes, regarded the Earth as supported by elephants standing on a giant turtle (an incarnation of the God Vishnu) which, in turn, rested on a cobra, the symbol of water. This floating concept of the Earth fitted well with their experience of earthquakes, which were believed to be caused by movements of the elephants, adjusting their stance in response to the load of the Earth.

The selections that follow range from the nineteenth century to the present, reflecting the different ways in which working geologists and others respond to the challenge of unraveling the history of the Earth. Several accounts describe the ways in which individuals were first attracted to geology and the course of events which led them to devote their lives to it.

3-1 Stepping Stones – Stephen Drury

One of the most lively recent debates has been that between scientists – arguing for the objectivity, and even the truth, of their conclusions – and the post-modernist critics, claiming that science, like all other endeavors, is culturally bound and is thus but one description of reality among many, either competing with it or complementary to others. It was this debate that led to the now infamous publication of Alan D. Sokal's paper, "Transgressing the Boundaries: towards a transformative hermeneutics of quantum gravity" in 1996 in Social Text, *a fashionable literary publication. The paper was a parody, a spoof, and brought forth both fury and amusement. But the questions raised by this public practical joke remain.*

Stephen Drury, born in 1946, is at the Department of Earth Sciences in the UK's Open University. He reflects on the objectivity of science in his book Stepping Stones: the making of our home world *(1999).*

While Salvador Allende's social democratic regime was being overthrown by the Chilean military in 1973, a cameraman was filming the drama on the street. In his footage we see an ordinary soldier on a truck turn to glower, to aim his rifle towards us and then to fire. The film jerks but continues. A second shot and the scene falls to the sidewalk. The unfortunate journalist felt himself isolated from events because he was peering through a lens. Many scientists consider themselves to be part of such a 'Fourth Estate', dispassionate observers as the world unfolds, though few suffer deadly physical damage.

Being 'objective' supposes that any part of the material world, indeed the whole of it, can be ring-fenced and observed without bias from beyond its perimeter. It demands not only excluding the observer from the observations, but in many cases ignoring everything except a particular phenomenon. Any social scientist, anthropologist, psychologist or even a student of other animals' behavior knows that this approach is, for them, an unattainable delusion. Is it possible to be 'objective' in the study of things that do not move perceptively, such as rocks, or that are immensely remote, such as planets, stars and galaxies, and the time spanned by their history? Earth's history goes back for almost five million millennia, nearly five billion years, so surely Earth scientists who attempt to grapple with it can be detached from its course?

This is what the science of geology amounts to: people with special skills, experience and knowledge examining rocks and time's record in them. But they are people, above all else. Their objective is not rocks, the information locked within them, nor even the history of the world and the way it works, but a general enrichment of human knowledge through each individual's self-enrichment. Rocks, like anything else, are pored over for entirely human motives, not for their own sakes: so that canals might safely transport coal; to find gold and, more importantly, water; to warn of impending disaster; to understand which soils are fertile or not; to know why landscapes are the way they are; even for curiosity or a cussed feeling that everyone else is wrong! Studying the Earth can never be 'objective'. Scientists can end up as nervous wrecks striving to attain the god-like standard of 'objectivity'. Others can sit smugly on the sidelines flinging the ultimate epithet of 'subjectivity', of allowing the human element in. But that is what people do, in every walk of life, simply because they are human; not 'only human', but conscious, self-creating beings whose only true wealth is shared knowledge of their world and the culture that it supports.

3-2 William Buckland – Elizabeth O.B. Gordon

In Harper's Magazine some years ago an article was published entitled, "Oxford's Magnificent Oddballs," in which William Buckland (1784–1856), a geologist, was honored by being named the most lurid and fabled of the Oxford eccentrics. Yet Buckland was a geologist of ability, author of 53 articles, twice President of the Geological Society, Fellow of the Royal Society, first President of the British Association for the Advancement of Science, and the recipient of the Wollaston Medal, the highest geological honor that could be bestowed. Buckland was a truly eloquent and enthusiastic lecturer and highly regarded in contemporary circles. In fact, a readership at Oxford was created specifically for him and was financed by a treasury stipend advanced by the Prince Regent. His daughter, Elizabeth Oke (Buckland) Gordon, prepared

The Life and Correspondence of William Buckland, D.D., F.R.S., sometime Dean of Westminster, twice president of the Geological Society and first President of the British Association *(1894)* which shows William Buckland's well-developed sense of humor *and considerable stage-presence. Extracted from this book is an account of a student's reaction to a Buckland lecture and a poem about Bucklandian antics in the classroom by a fellow professor. Humor, eccentricity, and ability all characterize this unusual Oxford don.*

"He lectured on the Cavern of Torquay, the now famous Kent's Cavern. He paced like a Franciscan Preacher up and down behind a long show-case, up two steps, in a room in the old Clarendon. He had in his hand a huge hyena's skull. He suddenly dashed down the steps – rushed, skull in hand, at the first undergraduate on the front bench – and shouted, 'What rules the world?' The youth, terrified, threw himself against the next back seat, and answered not a word. He rushed then on me, pointing the hyena full in my face – 'What rules the world?' 'Haven't an idea,' I said. 'The stomach, sir,' he cried (again mounting his rostrum), 'rules the world. The great ones eat the less, and the less the lesser still.' "

The Professor's *forte* as a lecturer in these early days excited the rhyming propensities of his College friend Shuttleworth, afterwards Warden of New College, and subsequently Bishop of Chichester. The lecture which suggested the following lines was probably delivered early in 1822.

> In Ashmole's ample dome, with look sedate,
> 'Midst heads of mammoths, heads of houses sate;
> And tutors, close with undergraduates jammed,
> Released from cramming, waiting to be crammed.
> Above, around, in order due displayed,
> The garniture of former worlds was laid:
> Sponges and shells in lias moulds immersed,
> From Deluge fiftieth, back to Deluge first;
> And wedged by boys in artificial stones,
> Huge bones of horses, now called mammoths' bones;
> Lichens and ferns which schistose beds enwrap,
> And – understood by most professors – trap.
> Before the rest, in contemplative mood,
> With sidelong glance, the inventive Master stood,
> And numbering o'er his class with still delight,
> Longed to possess them cased in stalactite.
> Then thus with smile suppressed: "In days of yore
> One dreary face Earth's infant planet bore;
> Nor land was there, nor ocean's lucid flood,
> But, mixed of both, one dark abyss of mud;
> Till each repelled, repelling by degrees,
> This shrunk to rock, that filtered into seas;
> Then slow upheaved by subterranean fires,
> Earth's ponderous crystals shot their prismy spires;
> Then granite rose from out the trackless sea,
> And slate, for boys to scrawl – when boys should be –
> But earth, as yet, lay desolate and bare;

Man was not then, – but Paramoudras were.
'Twas silence all, and solitude; the sun,
If sun there were, yet rose and set to none,
Till fiercer grown the elemental strife,
Astonished tadpoles wriggled into life;
Young encrini their quivering tendrils spread,
And tails of lizards felt the sprouting head.
(The specimen I hand about is rare.
And very brittle; bless me, sir, take care!)
And high upraised from ocean's inmost caves,
Protruded corals broke the indignant waves.
These tribes extinct, a nobler race succeeds:
Now sea-fowl scream amid the plashing reeds;
Now mammoths range, where yet in silence deep
Unborn Ohio's hoarded waters sleep,
Now ponerous whales. . . .
                    (Here, by the way, a tale
I'll tell of something, very like a whale.
An odd experiment of late I tried,
Placing a snake and hedgehog side by side;
Awhile the snake his neighbor tried t' assail,
When the sly hedgehog caught him by the tail,
And gravely munched him upwards joint by joint, –
The story's somewhat shocking, but in point.)
Now to proceed: –
The earth, what is it? Mark its scanty bound –
'Tis but a larger football's narrow round;
Its mightiest tracts of ocean – what are these?
At best but breakfast tea-cups, full of seas.
O'er these a thousand deluges have burst,
And quasi-deluges have done their worst."

The lecture ends with a couplet which the facetious writer observes of its own accord "slides into verse, and hitches in a rhyme": –

Of this enough. On Secondary Rock,
To-morrow, gentlemen, at two o'clock.

Buckland was greatly pleased, on his return from this long sojourn on the Continent, to be greeted with the following epitaph written by his friend Whately, afterwards the famous Archbishop of Dublin. He had the verses lithographed, and gave copies to his friends, so that they are more known than many of the clever verses written by Dr. Shuttleworth and Mr. Duncan.

ELEGY
Intended for Professor Buckland. December 1st, 1820.
By Richard Whately

Mourn, Ammonites, mourn o'er his funeral urn,
  Whose neck ye must grace no more.
Gneiss, granite, and slate, he settled your date,

And his ye must now deplore.
Weep, caverns, weep with unfiltering drip,
   Your recesses he'll cease to explore;
For mineral veins and organic remains
   No stratum again will he bore.

Oh, his wit shone like crystal; his knowledge profound
   From gravel to granite descended,
No trap could deceive him, no slip could confound,
   Nor specimen, true or pretended;
He knew the birth-rock of each pebble so round,
   And how far its tour had extended.

His eloquence rolled like the Deluge retiring,
   Where mastodon carcases floated;
To a subject obscure he gave charms so inspiring,
   Young and old on geology doated.
He stood out like an Outlier; his hearers, admiring,
   In pencil each anecdote noted.

Where shall we our great Professor inter,
   That in peace may rest his bones?
If we hew him a rocky sepulchre,
   He'll rise and break the stones,
And examine each stratum that lies around,
For he's quite in his element underground.

If with mattock and spade his body we lay
   In the common alluvial soil,
He'll start up and snatch those tools away
   Of his own geological toil;
In a stratum so young the Professor disdains
That embedded should lie his organic remains.

Then exposed to the drip of some case-hardening spring
   His carcase let stalactite cover,
And to Oxford the petrified sage let us bring
   When he is encrusted all over;
There 'mid mammoths and crocodiles, high on a shelf,
Let him stand as a monument raised to himself."

Buckland's championship of the glacial theory was the subject of a poetic "Dialogue between Dr. Buckland and a Rocky Boulder," written by his friend Philip Duncan. The following are the lines:—

Buckland, *loquitur.*

Say when, and whence, and how, huge Mister Boulder,
And by what wondrous force hast thou been rolled here?
Has some strong torrent driven thee from afar,
Or hast thou ridden on an icy car?
Which, from its native rock once torn like thee,
Has floundered many a mile throughout the sea,
And stranded thee at last upon this earth,

So distant from thy primal place of birth;
And having done its office with due care,
Was changed to vapour, and was mixed in air.

Boulder, *respondit*.

Thou great idolater of stocks and stones,
Of fossil shells and plants and buried bones;
Thou wise Professor, who wert ever curious
To learn the true, and to reject the spurious,
Know that in ancient days an icy band
Encompassed around the frozen land,
Until a red-hot comet, wandering near
The strong attraction of this rolling sphere,
Struck on the mountain summit, from whence torn
Was many a vast and massive iceberg borne,
And many a rock, indented with sharp force
And still-seen striae, shows my ancient course;
And if you doubt it, go with friend Agassiz
And view the signs in Scotland and Swiss' passes.

3-3   The Old Red Sandstone – Hugh Miller

Hugh Miller (1802–1856) was a Scottish author, philosopher and geologist. After a rather wild youth – including forming his companions into a gang of rovers and orchard robbers – he was apprenticed to a stonemason at 17 and became a journeyman mason, following his trade all over Scotland. He later became an accountant in a local bank, and embarked on a writing career. This included some early poetry, and antiquarian and natural history studies of northern Scotland, and his skills subsequently brought him the editorship of a biweekly political newspaper, Witness. *A man of deep religious conviction, Miller pursued geology as a hobby. His best-known work,* The Old Red Sandstone, *from which our extract is taken, first appeared serially during 1840 in* Witness. *His self-taught technical skill in describing the fossil fishes of the Old Red Sandstone was highly praised by Huxley. Amongst his dozen or so other books,* Footprints of the Creator *was a vigorous attack on Chamber's evolutionary* Vestiges of Creation. *His later years were marred by lung complications, induced by his early quarrying work.*

It was twenty years last February since I set out, a little before sunrise, to make my first acquaintance with a life of labour and restraint; and I have rarely had a heavier heart than on that morning. I was but a slim, loose-jointed boy at the time, fond of the pretty intangibilities of romance, and of dreaming when broad awake; and, woeful change! I was now going to work at what Burns has instanced, in his "Twa Dogs," as one of the most disagreeable of all employments, – to work in a quarry. Bating [sic] the passing uneasiness occasioned by a few gloomy anticipations, the portion of my life which had already gone by had been happy beyond the common lot. I had been a wanderer among rocks and woods, a reader of curious books when I could get them, a gleaner of old traditionary stories; and now I was going to exchange all my day-dreams, and all my amusements, for the kind of life in which men toil every day that they may be enabled to eat, and eat every day that they may be enabled to toil!

The quarry in which I wrought lay on the southern shore of a noble inland bay, or frith rather, with a little clear stream on one side, and a thick fir wood on the other. It had been opened in the Old Red Sandstone of the district, and was overtopped by a huge bank of diluvial clay, which rose over it in some places to the height of nearly thirty feet, and which at this time was rent and shivered, wherever it presented an open front to the weather, by a recent frost. A heap of loose fragments, which had fallen from above, blocked up the face of the quarry, and my first employment was to clear them away. The friction of the shovel soon blistered my hands, but the pain was by no means very severe, and I wrought hard and willingly, that I might see how the huge strata below, which presented so firm and unbroken a frontage, were to be torn up and removed. Picks, and wedges, and levers, were applied by my brother-workmen; and, simple and rude as I had been accustomed to regard these imple-ments, I found I had much to learn in the way of using them. They all proved inefficient, however, and the workmen had to bore into one of the inferior strata, and employ gunpowder. The process was new to me, and I deemed it a highly amusing one: it had the merit, too, of being attended with some such degree of danger as a boating or rock excursion, and had thus an interest independent of its novelty. We had a few capital shots: the fragments flew in every direction; and an immense mass of the diluvium came toppling down, bearing with it two dead birds, that in a recent storm had crept into one of the deeper fissures, to die in the shelter. I felt a new interest in examining them. The one was a pretty cock goldfinch, with its hood of vermilion, and its wings inlaid with the gold to which it owes its name, as unsoiled and smooth as if it had been preserved for a museum. The other, a somewhat rarer bird, of the woodpecker tribe, was variegated with light blue and a grayish yellow. I was engaged in admiring the poor little things, more disposed to be sentimental, perhaps, than if I had been ten years older, and thinking of the contrast between the warmth and jollity of their green summer haunts, and the cold and darkness of their last retreat, when I heard our employer bidding the workmen lay by their tools. I looked up and saw the sun sinking behind the thick fir wood beside us, and the long dark shadows of the trees stretching downwards towards the shore.

This was no very formidable beginning of the course of life I had so much dreaded. To be sure, my hands were a little sore, and I felt nearly as much fatigued as if I had been climbing among the rocks; but I had wrought and been useful, and had yet enjoyed the day fully as much as usual. It was no small matter, too, that the evening, converted, by a rare transmutation, into the delicious "blink of rest" which Burns so truthfully describes, was all my own. I was as light of heart next morning as any of my brother-workmen. There had been a smart frost during the night, and the rime lay white on the grass as we passed onwards through the fields; but the sun rose in a clear atmosphere, and the day mellowed, as it advanced, into one of those delightful days of early spring which give so pleasing an earnest of whatever is mild and genial in the better half of the year. All the workmen rested at mid-day, and I went to enjoy my half-hour alone on a mossy knoll in the neighbouring wood, which commands through the trees a wide prospect of the bay and the opposite shore. There was not a wrinkle on the water, nor a cloud in the sky, and the branches were as moveless in the calm as if they had been traced on canvas. From a wooded promontory that stretched half-way across the frith there ascended a thin column of smoke. It rose straight as the line of a plummet for more than a thousand yards, and then, on reaching a thinner stratum of air, spread out equally on every side, like the foliage of a stately tree. Ben Wyvis rose to the west, white with the yet unwasted snows of winter, and as sharply defined in the clear atmosphere as if all its sunny slopes and blue retiring hollows had been chiseled in marble. A line of snow ran along the opposite hills: all above was white, and all below was purple. I returned to the quarry, convinced that a very exquisite pleasure may be a very cheap one, and that the busiest employments may afford leisure enough to enjoy it.

The gunpowder had loosened a large mass in one of the inferior strata, and our first employment, on resuming our labours, was to raise it from its bed. I assisted the other workmen in placing it on edge, and was much struck by the appearance of the platform on which it had rested. The entire surface was ridged and furrowed like a bank of sand that had been left by the tide an hour before. I could trace every bend and curvature, every cross hollow and counter ridge, of the corresponding phenomena; for the resemblance was no half resemblance, – it was the thing itself; and I had observed it a hundred and a hundred times, when sailing my little schooner in the shallows left by the ebb. But what had become of the waves that had thus fretted the solid rock, or of what element had they been composed? I felt as completely at fault as Robinson Crusoe did on his discovering the print of the man's foot in the sand. The evening furnished me with still further cause of wonder. We raised another block in a different part of the quarry, and found that the area of a circular depression in the stratum below was broken and flawed in every direction, as if it had been the bottom of a pool recently dried up, which had shrunk and split in the hardening. Several large stones came rolling down from the diluvium in the course of the afternoon. They were of different qualities from the sandstone below, and from one another; and, what was more wonderful still, they were all rounded and water-worn, as if they had been tossed about in the sea or the bed of the river for hundreds of years. There could not, surely, be a more conclusive proof that the bank which had enclosed them so long could not have been created on the rock on which it rested. No workman ever manufactures a half-worn article, and the stones were all half-worn! And if not the bank, why then the sandstone underneath? I was lost in conjecture, and found I had food enough for thought that evening, without once thinking of the unhappiness of a life of labour.

The immense masses of diluvium which we had to clear away rendered the working of the quarry laborious and expensive, and all the party quitted it in a few days, to make trial of another that seemed to promise better. The one we left is situated, as I have said, on the southern shore of an island bay, – the Bay of Cromarty; the one to which we removed has been opened in a lofty wall of cliffs that overhangs the northern shore of the Moray Frith. I soon found I was to be no loser by the change. Not the united labours of a thousand men for more than a thousand years could have furnished a better section of the geology of the district than this range of cliffs. It may be regarded as a sort of chance dissection on the earth's crust. We see in one place the primary rock, with its veins of granite and quartz, its dizzy precipices of gneiss, and its huge masses of hornblende; we find the secondary rock in another, with its beds of sandstone and shale, its spars, its clays, and its nodular limestones. We discover the still little-known but highly interesting fossils of the Old Red Sandstone in one deposition, we find the beautifully preserved shells and lignites of the Lias in another. There are the remains of two several creations at once before us. The shore, too, is heaped with rolled fragments of almost every variety of rock, – basalts, ironstones, hyperstenes, porphyries, bituminous shales, and micaceous schists. In short, the young geologist, had he all Europe before him, could hardly choose for himself a better field. I had, however, no one to tell me so at the time, for Geology had not yet traveled so far north; and so, without guide or vocabulary, I had to grope my way as I best might, and find out all its wonders for myself. But so slow was the process, and so much was I a seeker in the dark, that the facts contained in these few sentences were the patient gatherings of years.

In the course of the first day's employment I picked up a nodular mass of blue limestone, and laid it open by a stroke of the hammer. Wonderful to relate, it contained inside a beautifully finished piece of sculpture, one of the volutes, apparently, of an Ionic capital; and not the far-famed walnut of the fairy tale, had I broken the shell and found the little dog lying within, could have surprised me more. Was there another such curiosity in the whole world? I broke open a few other nodules of similar appearance, – for they lay pretty thickly on the shore, – and found that there might. In one of these there were what seemed to be the scales of fishes, and the impressions of a few minute bivalves, prettily striated; in the centre of another there was actually a piece of decayed wood. Of all Nature's riddles, these seemed to me to be at once the most interesting and the most difficult to expound. I treasured them carefully up, and was told by one of the workmen to whom I showed them, that there was a part of the shore about two miles farther to the west where curiously-shaped stones, somewhat like the heads of boarding-pikes, were occasionally picked up; and that in his father's days the country people called them thunderbolts, and deemed them of sovereign efficacy in curing bewitched cattle. Our employer, on quitting the quarry for the building on which we were to be engaged, gave all the workmen a half-holiday. I employed it in visiting the place where the thunderbolts had fallen so thickly, and found it a richer scene of wonder than I could have fancied in even my dreams.

What first attracted my notice was a detached group of low-lying skerries, wholly different in form and colour from the sandstone cliffs above or the primary rocks a little farther to the west. I found them composed of thin strata of limestone, alternating with thicker beds of a black slaty substance, which, as I ascertained in the course of the evening, burns with a powerful flame, and emits a strong bituminous odour. The layers into which the beds readily separate are hardly an eighth part of an inch in thickness, and yet on every layer there are the impressions of thousands and tens of thousands of the various fossils peculiar to the Lias. We may turn over these wonderful leaves one after one, like the leaves of a herbarium, and find the pictorial records of a former creation in every page: scallops, and gryphites, and ammonites, of almost every variety peculiar to the formation, and at least some eight or ten varieties of belemnite; twigs of wood, leaves of plants, cones of an extinct species of pine, bits of charcoal, and the scales of fishes; and, as if to render their pictorial appearance more striking, though the leaves of this interesting volume are of a deep black, most of the impressions are of a chalky whiteness. I was lost in admiration and astonishment, and found my very imagination paralysed by an assemblage of wonders that seemed to outrival in the fantastic and the extravagant even its wildest conceptions. I passed on from ledge to ledge, like the traveler of the tale through the city of statues, and at length found one of the supposed aerolites I had come in quest of firmly imbedded in a mass of shale. But I had skill enough to determine that it was other than what it had been deemed. A very near relative, who had been a sailor in his time on almost every ocean, and had visited almost every quarter of the globe, had brought home one of these meteoric stones with him from the coast of Java. It was of a cylindrical shape and vitreous texture, and it seemed to have parted in the middle when in a half-molten state, and to have united again, somewhat awry, ere it had cooled enough to have lost the adhesive quality. But there was nothing organic in its structure; whereas the stone I had now found was organized very curiously indeed. It was of a conical form and filamentary texture, the filaments radiating in straight lines from the centre to the circumference.

Finely-marked veins like white threads ran transversely through these in its upper half to the point; while the space below was occupied by an internal cone, formed of plates that lay parallel to the base, and which, like watchglasses, were concave on the under side and convex on the upper. I learned in time to call this stone a belemnite, and became acquainted with enough of its history to know that it once formed part of a variety of cuttle-fish, long since extinct.

My first year of labour came to a close, and I found that the amount of my happiness had not been less than in the last of my boyhood. My knowledge, too, had increased in more than the ratio of former seasons; and as I had acquired the skill of at least the common mechanic, I had fitted myself for independence. The traditional experience of twenty years has not shown me that there is any necessary connection between a life of toil and a life of wretchedness; and when I have found good men anticipating a better and a happier time than either the present or the past, the conviction that in every period of the world's history the great bulk of mankind must pass their days in labour, has not in the least inclined me to scepticism. ....

One important truth I would fain press on the attention of my lowlier readers: there are few professions, however humble, that do not present their peculiar advantages of observation; there are none, I repeat, in which the exercise of the faculties does not lead to enjoyment. I advise the stonemason, for instance, to acquaint himself with Geology. Much of his time must be spent amid the rocks and quarries of widely-separated localities. The bridge or harbour is no sooner completed in one district than he has to remove to where the gentleman's seat or farm-steading is to be erected in another; and so, in the course of a few years, he may pass over the whole geological scale. ....

Should the working man be encouraged by my modicum of success to improve his opportunities by observation, I shall have accomplished the whole of it. It cannot be too extensively known, that nature is vast and knowledge limited, and that no individual, however humble in place or acquirement, need despair of adding to the general fund.

3-4   A Long Life's Work – Sir Archibald Geikie

Sir Archibald Geikie (1835–1924) was one of the great scientist–statesmen of the late nineteenth and early twentieth centuries. He had a distinguished career as a geologist with what was then the Geological Survey of the United Kingdom, ultimately becoming its director. He also served as the first Murchison Professor of Geology at the University of Edinburgh. Some of his early work was carried out jointly with his fellow-Scot, Sir Roderic Murchison, with whom in 1862 he published one of the earliest geologic maps of Scotland. An outstanding field geologist, he made major contributions to the study of geomorphology, vulcanism, and the geology of Scotland. His textbooks were widely read, but he was also a perceptive and literate author of more general works, as the following account illustrates. In the present extract taken from his autobiography A Long Life's Work, published in 1924, Geikie described the discovery of fossils made early in his career, and reflected upon his own personal commitment to geology. The passage also includes a fascinating account of a meeting with Hugh Miller. The events described took place in the autumn of 1856.

A new interest in the valleys and ravines of the Pentland Hills arose when I had the good fortune to discover among them, underneath conglomerates and massive sheets of lava of Old Red Sandstone age, a group of shales crowded with fossils. Some twenty years earlier Charles Maclaren had found a shell (*Orthoceras*) there, but nothing further had been done in searching the strata for more organic remains which would determine their geological age. I disinterred a good many species of brachiopods, lamellibranchs and other fossils, well preserved and identical with familiar organisms in the Upper Silurian part of the Geological Record. When subsequently the Survey Collector was put on the ground he obtained a number of additional species.

Such a "find" as this gave special emphasis to a thought which from the beginning had often been in my mind. The work on which I was now engaged, and to which I had dedicated my life, was not merely an industrial employment; the means of gaining a livelihood; a pleasant occupation for mind and body. It often wore to me an aspect infinitely higher and nobler. It was in reality a methodical study of the works of the Creator of the Universe, a deciphering of His legibly-written record of some of the stages through which this part of our planet passed in His hands before it was shaped into its present form. A few of the broader outlines of this terrestrial history had been noted by previous observers in the Lothians, but many other features still remained to be recognised and interpreted, while the mass of complicated detail, so needful for an adequate comprehension of the chronicle as a whole, was practically still unknown. I felt like an explorer entering an untrodden land. Every day the rocks were yielding to me the story of their birth, and thus making fresh, and often deeply interesting, additions to my stock of knowledge. Lucretius, in a well-known passage of his great poem on Nature, has given expression to the enthusiasm wherewith a mind may be filled by the contemplation of a region of philosophical thought which no human foot has yet entered, and he warmly pictures the joy of the first comer who approaches and drinks from springs which no human lips have yet tasted. It is true that the poet confessed that he felt goaded onward by the sharp spur of fame, and by the hope that from the new flowers of the unexplored realm he would gather a splendid coronal, such as had never yet graced the head of any mortal. I had no such ambitious dreams, though I looked forward to rising in my profession to something higher than the "surveyor's drudge or Scottish quarryman" of Macculloch's sarcasm. Meanwhile the discovery of new facts in the ancient history of the land gave me in itself an ample store of pleasure, augmented by the delight of fitting them into each other and extracting from them the connected narrative which they had to reveal. I remember the indignation with which, about this time, in reading Cowper's poems, as I loved to do, I came upon his ignorant and contemptuous denunciation of astronomers, geologists and men of science in general whose work in life was described as

> the toil
> Of dropping buckets into empty wells
> And growing old in drawing nothing up.

More tolerable was Wordsworth's condescending description of the geologist, going about with a pocket-hammer, smiting the "luckless rock or prominent stone," classing the splinter by "some barbarous name" and hurrying on, thinking himself enriched, "wealthier, and doubtless wiser than before."

Samuel Johnson is recorded by his biographer to have expressed his belief that "there is no profession to which a man gives a very great proportion of his time. It is wonderful, when a calculation is made, how little the mind is actually employed in the discharge of any profession." Whether the active clergyman, lawyer or medical man would subscribe to this statement may be questioned. But I found that the profession which I had chosen demanded all the time and thought which I could give to it, both with body and mind.

The discovery of a large group of organic remains, previously unknown to exist where I found them, might have been made by anybody, and conferred no particular credit on the finder. But it made a profound impression on my mind, as its full meaning in the unravelling of the geological history of the district gradually revealed itself. I lost no time in including it, with an account of my recent doings in Skye, in a letter to my kind friend Hugh Miller. His note in reply, from its interest in reference to his own field-observations, which he never lived to publish, may be inserted here.

Wednesday evening 9th October, 1856

My dear Sir

Could you drop in upon me on the evening of Saturday first, and share in a quiet cup of tea. I am delighted to hear that you have succeeded in reading off the Liassic deposits of Skye, and shall have much pleasure in being made acquainted with the result of your labours. Like you, I want greatly a good work on the English Lias, but know not that there is any such. My explorations this season have been chiefly in the Pleistocene and the Old Red. I have now got Boreal shells in the very middle of Scotland, about equally removed from the eastern and western seas. But the details of our respective explorations we shall discuss at our meeting.

Yours ever

Hugh Miller.

Having been fortunately able to accept this invitation, I spent a couple of hours with him in his home at Portobello. He was interested in the discovery of so large a series of well-preserved fossils in the basement rocks of the Pentland Hills, and not less pleased with the details of my work in Skye. He had spread out on the table his trophy of northern shells, which enabled him to affirm that, at a late date in geological time, Scotland was cut in two by a sea-strait that connected the firths of Forth and Clyde. He had found marine shells at Buchlyvie in Stirlingshire, on the low ground between the two estuaries. Finding me not quite clear as to the precise location of Buchlyvie, he burst out triumphantly with the lines which Scott puts at the head of the immortal chapter of *Rob Roy* wherein are told the adventures at the Clachan of Aberfoyle:

> Baron o' Buchlyvie,
> May the foul fiend drive ye,
> And a' to pieces rive ye,
> For building sic a toun,
> Where there's neither horse-meat, nor man's meat, nor a chair
> to sit down.

3-5   Life, Time, and Darwin – Frank H.T. Rhodes

In this article, taken from an inaugural lecture published by the University of Wales, Swansea in 1958, Frank H.T. Rhodes discusses the *background to Charles Darwin's voyage on HMS Beagle (1831–1836) and the events which followed from the voyage.*

At a meeting of the Linnean Society of London on 1 July 1858, Charles Darwin and Alfred Russell Wallace presented a joint paper entitled "On the Tendency of Species to form Varieties; and on the Perpetuation of Varieties and Species by Natural Means of Selection." The association of the two authors was remarkable. In February 1858, Wallace lay stricken with fever at Ternate, in the jungles of the Moluccas. As his mind wandered over the problem of the development of species, a subject that had exercised his attention for a number of years, he suddenly recalled an *Essay on Population* by the Rev. Robert Malthus, which he had read twelve years before. Malthus argued that the human race would increase in a geometric progression were it not for the fact that many of its members failed to survive and to reproduce. In a sudden flash of insight, Wallace realized the applicability of this to the organic world as a whole and conceived the idea of natural selection in the development of species. Within a week he sent to Darwin a summary of his conclusions under the title "On the Tendency of Varieties to depart indefinitely from the Original Type." Wallace wrote that the idea expressed seemed to him to be new, and asked Darwin, if he also thought it new, to show it to Sir Charles Lyell.

Darwin received the essay with astonishment and dismay, for Wallace's hypothesis was identical with that which he himself had formulated. Darwin, who curiously enough had also been much influenced by Malthus's essay, had devoted the previous twenty years to the patient accumulation of evidence which he proposed to publish as a book. Year after year Darwin accumulated more and more data, and slowly the enormous treatise took form, although he had, in fact, prepared an outline of his theory as far back as 1842, and a more lengthy account two years later. Of these, he wrote to Lyell, "I never saw a more striking coincidence; if Wallace had my MS. Sketch written out in 1842, he could not have made a better short abstract! Even his terms now stand as heads of my chapters." It was under such circumstances that Sir Charles Lyell and Sir Joseph Hooker suggested a joint presentation of papers announcing the theory. Darwin and Wallace readily agreed, and the joint publication included Wallace's essay and an extract from Darwin's manuscript of 1844, together with an extract from one of his letters to Asa Grey written in October 1857 "in which," (as Lyell and Hooker noted in their accompanying letter of presentation) "he repeats his views, and which shows that these remained unaltered from 1839 to 1857." The paper was calmly received and few of those who heard it could have predicted the way in which this new theory of evolution was soon to shatter the tranquillity of Victorian thought.

The following year Darwin completed a brief abstract (as he called it) of his researches and on 24 November 1859, there was published the most important book of the century – *The Origin of Species*. On the day of its publication, the first edition of 1,250 copies was sold out.

The effect of *The Origin* upon public opinion was immediate and profound, but its effect upon the natural sciences was revolutionary. It provided the key that not only integrated and interpreted the maze of biological data but also gave new impetus and urgency to every avenue of research.

To Darwin, as to Wallace, the ultimate solution of the problem of the origin of species came suddenly: "In October 1838," he wrote "... I happened to read for amusement Malthus on *Population*, and being well prepared to appreciate the struggle for existence which everywhere goes on, from long-continued observation of the habits of animals and plants, it at once struck me that under these circumstances favourable variations would tend to be preserved, and unfavourable ones to be destroyed. ... I can remember the very spot in the road whilst in my carriage, when to my joy the solution came to me." Yet the path by which Darwin reached this conclusion was a long and laborious one, and certainly his early years gave little hint of the development of genius. He was born at Shrewsbury in 1809, the son and grandson of physicians. His grandfather, Erasmus Darwin, achieved considerable recognition for his poetical exposition of evolutionary views on the origin of species, similar to, but achieved independently of those of Lamarck. Darwin's school days made little impression upon him, although he became an avid collector of minerals and insects.

He himself has described these years in his autobiographical notes.

"In the summer of 1818 I went to Dr. Butler's great school in Shrewsbury, and remained there for seven years. ... Nothing could have been worse for the development of my mind than [this] school, as it was strictly classical, nothing else being taught, except a little ancient geography and history. The school as a means of education to me was simply a blank. ... Towards the close of my school life, my brother worked hard at chemistry, and made a fair laboratory with proper apparatus in the tool-house in the garden, and I was allowed to aid him as a servant in most of his experiments. ... The subject interested me greatly, and we often used to go on working till rather late at night. This was the best part of my education at school ... but I was once publicly rebuked by the headmaster ... for thus wasting my time on such useless subjects. ...

As I was doing no good at school, my father wisely took me away at a rather earlier age than usual, and sent me [October 1825] to Edinburgh University with my brother, where I stayed two years. ... My brother was completing his medical studies ... and I was sent there to commence mine. But soon after this period I became convinced ... that my father would leave me property enough to subsist on with some comfort ... [and] my belief was sufficient to check any strenuous effort to learn medicine. ... During my second year at Edinburgh I attended Jameson's lectures on Geology and Zoology but they were incredibly dull. The sole effect they produced on me was the determination never as long as I lived to read a book on Geology, or in any way to study the science. ...

After having spent two sessions in Edinburgh, my father perceived ... that I did not like the thought of being a physician, so he proposed that I should become a clergyman ... and [I] went to Cambridge after the Christmas vacation, early in 1828. ... During the three years which I spent at Cambridge my time was wasted, as far as the academical studies were concerned, as completely as at Edinburgh and at school ... [but] by answering well the examination questions in Paley, by doing Euclid well, and by not failing miserably in Classics, I gained a good place among the oi polloi, or crowd of men who do not go in for honours."

Yet in spite of his own somewhat melancholy assessment, Darwin's Cambridge years were to mark the turning point of his career, for there grew up a lasting friendship with J.S. Henslow, the Professor of Botany. It was Henslow who urged him to pursue the despised science of geology, which he did under Adam Sedgwick for an extra term at Cambridge after his graduation. It was also Henslow who arranged for Darwin to accompany H.M.S. Beagle (a 240-ton, ten-gun brig) on a survey voyage to South America and thence round the world. Henslow's parting gesture to the young Darwin was to suggest that he study carefully Lyell's newly published first volume of the *Principles of Geology*, "but on no account accept the views therein advocated." The five-year voyage was, for Darwin, a time of diligent observation and collecting, which gradually opened up a new world of study before him, a world in which most of the treasures were geological and a world which (in spite of Henslow's warning) he saw through the eyes and studied by the methods of Lyell. "I clambered over the mountains of Ascension with a bounding step", he wrote, "and made the volcanic rocks resound under my geological hammer."

3-6   King's Formative Years – R.A. Bartlett

Clarence Rivers King (1842–1901) graduated from Yale in 1862 and spent the following winter studying under Alexander Agassiz. In 1863 he followed the North Platte and the Humbolt rivers in the west, and, after losing all his belongings in a fire, was forced to cross the Sierra Nevada on foot. For the three following years he served as assistant geologist to the Geological Survey of California. Returning to Washington in 1866, it was largely his persistence that led to the appropriation by Congress of funds to finance a survey of the 40th parallel. He was only 25 years old at the time he made the expedition, but he devoted the next seven years of his life to developing the work. The survey, which began from California, collected over 5,000 rock samples, and published its results as a series of articles on systematic geology. It was King who later convinced Congress to establish the United States Geological Survey, and he became its first director. He occupied the position for only two years, however, before returning to work as a consulting mining geologist and engineer. His death in 1901 was hastened by pneumonia, which he contracted while working in mining properties in Butte, Montana. He remains one of the most colorful, if controversial, figures of North American geology. Richard Adams Bartlett was born in Colorado in 1920, educated at the Universities of Colorado and Chicago, and taught at Texas A & M and Florida State University. Our extract is from his 1962 book Great Surveys of the American West.

Clarence King was born at Newport, Rhode Island, on January 6, 1842, the newest addition to a long line of transplanted Englishmen. Tradition had it that his great grandfather, Samuel King, of Newport, had been an artist of some ability and a friend of Benjamin Franklin's, even helping the good doctor with some of his electrical experiments. And throughout the American part of their history, the Kings had enjoyed a generous sprinkling of gentlemen, scholars, and men of wealth.

Clarence King's father died in China, at the port of Amoy, June 2, 1848. He left his fortune, what there was of it, invested in a trading firm, and when it went bankrupt in 1857, the widow King and her young son were left nearly penniless.

But the widowed Mrs. King was a remarkably intelligent young woman who was determined to help her little son in every way possible. In the pre-Darwinian era, fossils and pretty rocks were strange things indeed for a small boy to collect, but it is related that she shared his enthusiasms. Once, when Clare was just seven, she let him take her by the hand and lead her over a mile of snow to show her what was perhaps his first geological discovery, a fossil fern in a stone wall. This so aroused the boy's interests that from then on the room he and his mother inhabited became "a veritable museum." The close mother-son relationship was to last throughout King's life. In manhood, he found in his mother his closest intellectual companion, and even on his death bed his closest thoughts were of her. . . .

In the summer of 1862, King was graduated with the first class to receive bachelor of science degrees from Sheffield Scientific School. For a summer vacation he rowed up Lake Champlain and down the St. Lawrence to Quebec, accompanied by three friends including Gardner. That winter of 1862–63 found him studying glacial geology with Alexander Agassiz and even dabbling in the study of the Pre-Raphaelite school of art. Not until the spring of 1863 did the emergency arise that determined the course of his life.

It concerned the health of a rather frail little fellow, of considerably smaller stature than King, who was nevertheless his best friend. James Terry Gardner had known King since they were both lads of fourteen, and, as Gardner once testified before a congressional committee, he and King had lived together "on terms of intimacy closer than those of most brothers." During their summers together they had learned the joys of hunting and fishing and botanizing, and when the time came for them to choose their professions, both had chosen science. Gardner had attended the Polytechnic School (Rensselaer) at Troy and then had gone on to the Sheffield School, where he and King had renewed their friendship. But the cold climate and the hard work had taxed Gardner's constitution, and the two close friends, King, handsome and robust, and Gardner, haggard and exhausted, decided to head for California. ...

One wonders at the sad partings from home. What did King's mother have to say about her favorite child leaving for the Far West? What did Jamie's parents say when their frail son, harboring an insistent cough, proudly announced that California was the place for *him*? But it was 1863. That summer witnessed Gettysburg and the fall of Vicksburg. The argument that going to California was safer then enlisting in the Union Army had a telling effect. The young men won out, and, with a third companion named Jim Hyde, they hit the road for Kansas, the great jumping-off place for the long trek to the land of the setting sun.

At St. Joseph, Missouri, they met a man named Speers who was leading a wagon train to California, and they joined up with him. They bought horses, gathered their frontiersman's gear together, and set out. Hardly had they hit the trail before their adventures began. ...

[O]ne day, still in June, the wagon train came to a place among the hills where the earth had been grooved by thousands of wagon wheels, and as they coasted through it, they realized that they had just crossed South Pass. It was quite a letdown, for they were not in mountains, and there was not a pine tree in sight. Southward lay high bluffs, however, and far to the northwest, rising abruptly in the summer haze, lay the ominous hulks of the Wind River Mountains. Straight west lay the cool, shallow waters of the Green River. The divide was crossed, and the true West lay beyond: Mormons, the desert, the Comstock Lode at Virginia City, the high Sierras, and California. With renewed energy they clattered on. ...

Westward they traveled, to Fort Bridger, crowded with nine hundred Indians held prisoners; to Echo Canyon and the Weber Valley; and finally to Salt Lake City, where they encamped on a square in the center of the city and feasted on lettuce and green onions.

But California still called. The wagon train worked its way around the northern shores of the Great Salt Lake, then westward across the parched sagebrush and alkali flats to the Humboldt River in Nevada. In this valley the three young men turned away from the main party and toward Virginia City. Until then, they had followed the approximate route of the future railroad and had included in their journey much of the area which would later be studied by the Fortieth Parallel Survey. The trip so far had taken them three months, and each of the three thought of himself a completely different man than he had been just three months before back in New York. ...

As they tramped up into the Sierras (they had sold their horses), they paused occasionally for breath and turned their eyes eastward, surveying the land over which they had passed. It was hot and dry out there, but a few clouds shared the hills, and the delicate blues of the distant mountains would have defied the painting abilities of a great artist. Both King and Gardner had fallen in love with the Great Basin, and when Jamie wrote home to his mother, he told her so. "But seriously," he said, "before we left the plains we had become so fascinated with the life and so interested in the vast loneliness in those deserts … that I would gladly have turned around and traveled right back over the same road."

They trekked across the divide and down through the Mother Lode country, analyzing the lay of the land as they went. Soon they were at Sacramento, where they took passage on a river steamer for San Francisco. The steamer was filled with men from the mines, rough, sunburned men in flannel shirts and high boots, wearing belts and revolvers. But one of them, though dressed much the same way, had a different face. Gardner walked past him, then sat down opposite him and pretended to be reading a newspaper. "An old felt hat, a quick eye, a sunburned face with different lines from the other mountaineers, a long weather-beaten neck protruding from a coarse gray flannel shirt and a rough coat, a heavy revolver belt and long legs made up the man," Gardner recalled. "And yet," he felt, "he is an intellectual man: I know it." The "intellectual" turned out to be William H. Brewer, assistant to Josiah Whitney on the California Geological Survey. Brewer had worn down his exploring party to a single packer, and he eagerly hired King to go northward with him to Lassen Peak and Mount Shasta. While Gardner went on to San Francisco and took work with the United States Topographical Engineers, King got in his first large scale geological work and had an opportunity for a first-hand study of volcanism-a branch of geology in which he later became the foremost authority in the United States.

For the next four years, King's and Gardner's stars were hitched to the California Geological Survey. Once during this period, King took time off to help in a geological survey of the huge Mariposa estate, owned by John C. Fremont. On another occasion, he and Gardner accompanied General McDowell on a reconnaissance of the desert regions of southern California and part of Arizona. Gardner had joined the survey in April, 1864. The bracing San Francisco climate had not improved his health, and he had resigned his job rather than work on the Sabbath. He had about decided to return home when Professor Whitney invited him to go into the southern Sierras with King, Brewer, and Charles F. Hoffman, the survey topographer. Gardner was to be an "assistant," furnishing his own horse and blankets, for a trip of about four months. As the party progressed, Gardner helped Hoffman and thus learned the secrets of the good topographical methodology which he put to good use during his service with the Fortieth Parallel Survey and with Hayden.

The importance of the association of King and Gardner with the California Geological Survey cannot be overestimated. Josiah Whitney was one of the most respected geologists then at work in the United States, and his appointment as head of the survey had been urged by such eminent scientists as the Sillimans, Dana, Marsh, Leidy, and Meek. His associate, Professor Brewer, like King and Gardner, was a product of the Sheffield Scientific School. Charles F. Hoffman, the topographer, was a well-trained young German who became one of the most valuable men on the survey.

The California Survey had been created by an act of the state legislature in 1860, and Whitney was appointed state geologist. He was supposed to make a complete geological survey of the state and furnish a complete scientific report. From the first, however, he failed to placate the state legislature, and that old bogey appropriations – or lack of them – brought the Survey to an end in 1872, although Whitney continued as state geologist until 1874.

Whitney was a scholar and a scientist, and he never understood the necessity of coming down from the clouds of scientific speculation and doing some earthly lobbying. American legislators tend to be pragmatic, and California legislators in the 1860s were no exception. They had to have an idea of a practical, industrial application for the results of a geological survey, or else they would reject the survey entirely. Whitney's overall plan embraced a preliminary topographical and geological survey as preparation for more specialized studies which ultimately would have practical significance. But the legislature could neither understand this nor wait for the transformation from lofty speculation to concrete and applicable data. When a lawmaker brought one of the survey publications, a fossil-like tome on paleontology, onto the floor of the legislature and proceeded to read from it, the California Geological Survey was as doomed as the dinosaur.

King learned both good and bad things from his experiences with the California Survey. To his benefit were nearly four years with some of the nation's leading geologists, four years of working with scientific field expeditions, and four years of roughing it in the unexplored mountains of the Sierras. Then there was Whitney's idealism. His men, wrote Brewster, "were about the only persons in California who were concerned with the earth, and were not trying to make money out of it." Nor could any member of the survey make a penny for himself.

Such insistence upon personal integrity was of course good. But the personal aspect embraced the policy of the survey as a whole: it reflected no interest in the mining industry. It was engaged in pure science, something very close to a scientist's heart. But its appropriations came from a less intellectual body, and so the survey failed for lack of funds.

3-7   With Shackleton in the Antarctic – M.E. David

Sir Tannatt William Edgeworth David (1858–1934) was born in Wales, and was educated at Oxford, intending to enter holy orders. His health broke down, however, and he emigrated to Australia, first joining the Geological Survey of New South Wales, and subsequently, in 1891, becoming Professor of Geology at the University of Sydney. He served as leader of the Royal Society's Funafuti expedition in 1897, and was chief of the scientific staff to Shackleton's Antarctic Expedition of 1907–1909. Though 50 years old when he reached the Antarctic, he led the climb of Mt. Erebus in 1908 and was one of the small party that reached the

South Magnetic Pole in 1909. Mary Edgeworth David (1888–1987) was Edgeworth David's second daughter, and was known as a conservationist and a writer, living an almost entirely self-sufficient life on a hectare of land in Hornsby until her 99th year. She wrote a biography of her father Professor David, the life of Sir Edgeworth David (1937) from which the present extract is taken. The events described in the present extract took place on the memorable journey to the top of Mt. Erebus. David's colleague, "Mawson," who rescued him, was Sir Douglas Mawson (1882–1958), British-born Australian geologist and polar explorer.

On January the 28th, the Professor's fiftieth birthday, they sighted the top of the cone of Mount Erebus. He said:

> The excitement grew on board as we drew nearer and nearer to the marvellous mass of land which formed such a dramatic termination to Ross' ever memorable voyage in 1841.
>
> Little by little the giant peak of Erebus and its smoke-cloud were shut out from our view by the sister peak of Mount Terror. ... At midnight of January 28th the sun was still shining brightly at the back of Mount Terror, and the

scene was one of exquisite beauty. … In the middle distance, across a strip of purplish-grey sea, were the ice-cliffs of the heavily crevassed glaciers creeping seawards down the slopes of Mount Terror, with here and there immense masses of black lava breaking the even line of white. To the left, the slopes of Terror showed the most delicate tints of bluish-purple and pale blue violet, while to the west, where the sunlight at the back of Mount Terror was reflected down by a dense cumulus cloud, the snow-slopes glowed with a soft golden radiance. … In the distance was the skyey peak of Erebus with its great steam-cloud flushed with pink below and all dark grey above. One understood then in some measure what must have been the feelings of Ross when he discovered his wonderful land.

… While working steadily at the daily tasks that lay about him, his eyes were often turned towards that majestic mountain, Erebus, which, rising rapidly from sea-level, rears itself aloft from near the western side of Ross Island to a height of over 13,000 feet. To the Professor this volcano,

> … its flanks and foothills clothed with spotless snow, patched with the pale blue of glacier ice, its active crater crowned with a spreading smoke-cloud and overlooking the vast white plain of the Barrier to the east and south, is one of the fairest and most majestic sights that Earth can show. … For us, living under its shadow, the longing to climb it and penetrate the mysteries beyond the veil, soon became irresistibly strong.

Though several expeditions had been in its neighbourhood, it had never been climbed, and when the suggestion to do so had been made various difficulties had to be considered by the leader before he gave his consent to the project.

> In the first place, the only party who had ascended the foothills of Erebus had found their path barred by heavily crevassed ice. … Then, too, the winter was fast approaching, bringing with it blizzards, and temperatures likely to be specially low at high altitudes. … After careful consideration Lieutenant Shackleton decided a reconnaissance in the direction of Mount Erebus might be made, and that, if the risk did not appear to be too great, an attempt might be made to reach the summit of the mountain.

The date for starting was fixed for the following morning, March the 5th, and immediately all was excitement and bustle at the winter-quarters, which 'literally rang with the clang of preparation.' It was past midnight before this ceased.

At nine the following morning a party of six, consisting of the summit party and a supporting party, set out with an eleven-foot sledge. The summit party consisted of David, Mawson, and Dr. Mackay, and was under the Professor's leadership. The supporting party was composed of Lieutenant Adams, Dr. Marshall, and Sir Philip Brocklehurst, and was led by Adams. The summit party was provisioned for eleven days, and the supporting party for six. …

All that day and the following night they were marooned in their sleeping-bags by the blizzard. As the tent-poles had been left in the sledge they merely doubled the tents over the top of the sleeping-bags. Adams, Marshall, and Brocklehurst, who shared a three-man sleeping-bag, had a narrow escape during the day, Brocklehurst being blown away down the gully when he emerged from the bag, and Adams, who had accompanied him, was also blown down. They managed, by crawling on their hands and knees, and with the utmost difficulty, to make their way back to the sleeping-bag, where they arrived so numb and exhausted that they could hardly struggle into it. The party managed to get some sleep that night, and when they awoke at 4 a.m. the following morning were devoutly thankful to find that the blizzard was over.

This experience upholds the old saying that 'God tempers the wind to the shorn lamb,' for to live in sleeping bags without the shelter of a tent for two nights and a day in a blizzard, at an altitude of 8,750 feet above sea-level, without any ill effects, is, to say the least, unusual in polar travel. It also speaks well for the endurance of the party, particularly for the fifty-year-old member of it, who provided his own warmth in a one-man sleeping-bag. As Shackleton afterwards said, they 'were winning their spurs not only on their first Antarctic campaign, but in their first attempts at serious mountaineering.'

They started the ascent again at 5:30 a.m. The gradient was now a rise of 1 in 1½.

Burdened as we were with our forty-pound loads, and more or less stiff after thirty continuous hours in our sleeping-bags, and beginning besides to find respiration difficult as the altitude increased, we felt exhausted while we were still 800 feet below the rim of the main crater. Accordingly we halted at noon, thawed some snow with the primus, and were soon reveling in cups of delicious tea, hot and strong, which at once invigorated us. Once more we tackled the ascent.

Mackay, underestimating the effects of altitude on a man carrying a heavy load, took a short cut up a dangerously steep snow slope, cutting steps with his ice-axe instead of following the more gradual rocky route taken by the others. He was soon heard calling for help, so the rest of the party hurried to the top of the ridge and Dr. Marshall and the Professor dropped down to his assistance. Mackay fainted on reaching safety, but soon recovered.

On arriving at the rim of the main crater they found themselves on the brink of a massive precipice, 80 to 100 feet high, of black rock forming the inner edge of the vast depression. Beyond this was an extensive snow-field, the floor of the old crater, which sloped up to the lip of the active cone and crater at its south end, the latter emitting great volumes of steam.

Having chosen for their camping-place a little rocky gully on the slope of the main cone, they prepared a meal, while Dr. Marshall examined Brocklehurst's feet, which had felt numb for some time. They were found to be so badly frost-bitten that the doctor decided that Brocklehurst must stay in camp in his sleeping-bag. The party lunched at 3 p.m. and all but the invalid set off to explore the snow slope leading up to the active crater. . . .

They arrived at the camp soon after 6 p.m. and as they sat on the rocks enjoying their tea, they had a glorious view to the west.

While the foothills of Erebus flushed rosy red in the sunset, a vast rolling sea of cumulus cloud covered all the land from Cape Bird to Cape Royds. . . . Far away the western mountains glowed with the purest tints of greenish-purple and amethyst. That night we had nothing but hard rock rubble under our sleeping-bags, and quite expected another blizzard; nevertheless, 'weariness can snore upon the flint,' and thus we slept soundly couched on Kenyte lava.

The following morning had two surprises for us; first, when we arose at 4 a.m. there was no sign of a blizzard, and next, while we were preparing breakfast, someone exclaimed, 'Look at the great shadow of Erebus!' and a truly wonderful sight it was. All the land below the base of the main cone, and for forty miles to the west of it, across MacMurdo Sound, was a rolling sea of dense cumulus cloud. Projected obliquely on this, as on a vast magic lantern screen, was the huge bulk of the giant volcano. The sun had just risen, and flung the shadow of Erebus right across the Sound and against the foothills of the Western Mountains. Every detail of the profile of Erebus, as outlined on the clouds, could be readily recognized. There to the right was the great black fang, the relic of the first crater; far above and beyond that was to be seen the rim of the main crater, near our camp; then further to the left and still higher rose the active crater with its canopy of steam faithfully portrayed on the cloud screen. Still further to the left the dark shadow dipped rapidly down into the shining fields of cloud below. All within the shadow of Erebus was a soft bluish grey; all without was warm, bright, and golden. Words fail to describe a scene of such transcendent majesty and beauty.

At 6 a.m. they left camp and made all speed to reach the crater summit. Roped together, they climbed several slopes formed by alternating beds of hard snow and vast quantities of large and perfect felspar crystals mixed with pumice.

A little further on we reached the foot of the recent cone of the active crater; here we unroped, as there was no possibility of any crevasses ahead of us.

Our progress was now painfully slow, as the altitude and cold combined to make respiration difficult. . . . A shout of joy and surprise broke from the leading files when, a little after 10 a.m., the edge of the active crater was at last reached. We had traveled only about two and a half miles from our camp, and had ascended just 2,000 feet, and yet this had taken us, with a few short halts, just four hours.

The scene that now suddenly burst upon us was magnificent and awe-inspiring. We stood on the edge of a vast abyss, and at first could neither see to the bottom nor across it, on account of the huge mass of steam filling the crater and soaring aloft in a column 500 to 1,000 feet high. After a continuous loud hissing sound, lasting for some minutes, there would come from below a big dull boom, and immediately afterwards a great globular mass of steam would rush upwards to swell the volume of the snow-white cloud which ever sways over the crater. These phenomena recurred at intervals of a few minutes during the whole of our stay at the crater. Meanwhile the air around us was redolent of burning sulphur.

Presently a gentle northerly breeze fanned away the steam cloud, and at once the whole crater stood revealed to us in all its vast extent and depth. Mawson's measurements made the depth 900 feet, and the greatest width about half a mile. . . . While at the crater's edge we made a boiling point determination with the hypsometer. . . . As the result of averaging aneroid levels, together with the hypsometer determination at our camp at the top of the old crater, calculations made by us showed that the summit of Erebus is probably about 13,370 feet above sea-level.

As soon as our measurements had been made, and some photographs had been taken by Mawson, we hurried back towards our camp, as it was imperatively necessary to get Brocklehurst down to the base of the main cone that day, and this meant a descent in all of nearly 8,000 feet.

On arrival in camp they had a hurried meal, shouldered their heavy packs, and started down the steep mountain slope. . . . As the Professor in his 'Narrative' does not relate all that followed, Mawson's account of this episode is now quoted:

> I was busy changing photographic plates in the only place where it could be done – inside the sleeping-bag. . . . Soon after I had done up the bag, having got safely inside, I heard a voice from outside – a gentle voice – calling:
> 'Mawson, Mawson.'
> 'Hullo!' said I.
> 'Oh, you're in the bag changing plates, are you?'
> 'Yes, Professor.'
> There was silence for some time. Then I heard the Professor calling in a louder tone:
> 'Mawson!'
> I answered again. Well, the Professor heard by the sound I was still in the bag, so he said:
> 'Oh, still changing plates, are you?'
> 'Yes.'
> More silence for some time. After a minute, in a rather loud and anxious tone:
> 'Mawson!'
> I thought there was something up, but could not tell what he was after. I was getting rather tired of it and called out:
> 'Hullo. What is it? What can I do?'
> 'Well, Mawson, I am in a rather dangerous position. I am really hanging on by my fingers to the edge of a crevasse, and I don't think I can hold on much longer. I shall have to trouble you to come out and assist me.'
> I came out rather quicker than I can say. There was the Professor, just his head showing, and hanging on to the edge of a dangerous crevasse.

Mawson had brought an ice-axe with him. He now inserted the chisel edge of this into a hole which he chipped in firm ice at the crevasse's edge, and holding on to the pick-point, swung the handle towards the Professor, who seized it and managed to clamber on to the solid ice, after which he went off to continue his sketching.

3-8 The Great Diamond Hoax – William H. Goetzmann

William H. Goetzmann, born in 1930, was educated at Yale, where he obtained a PhD. He went on to teach there and later became Stiles Professor of American Studies and History at the University of Texas, Austin. The book from which our extract is taken – Exploration and Empire *(1966) – won a Pulitzer Prize in History and the Francis Parkman Award. The selection that follows has all the elements of a first-class detective tale; a setting in flamboyant San Francisco in the immediate post-Civil War era, two mysterious prospectors, a bagful of uncut diamonds, involvement of a prominent mining consultant, a huge public stock offering, and a detective-geologist – Clarence King (see p. 69) – who was to become the first Director of the United States Geological Survey.*

In the spring and summer of 1872, a series of events were taking place that would come to a climax as the capstone of King's entire career. Early in 1872, two grizzled prospectors, Philip Arnold and John Slack, had appeared at a San Francisco bank and attempted to deposit a bagful of uncut diamonds. Having done so, they disappeared. But the diamonds came into instant prominence. They were shown to William Ralston, a director of the Bank of California, and then to a number of other capitalists in the city, who were all intrigued as to their origin. A search was instituted and Arnold and Slack were quickly located. When quizzed, they reported that they had found the diamonds at an undisclosed site. They consented to take a representative of the capitalists to the site, but insisted on blindfolding him first. After disembarking from the Union Pacific Railroad, they led General David Colton on a four-day trip through mountains and canyons to the site of their discovery, where he, too, found diamonds, and rubies, and a number of other precious stones.

Extremely cautious, the California businessmen then took Arnold and his diamonds to New York, where Charles Tiffany, upon examination, pronounced the gems "a rajah's ransom." They had, as one writer put it, discovered "the American Golconda" – a vast field of gems virtually untouched by the hand of man, just waiting to be exploited by sound business enterprise. If they succeeded, Amsterdam itself might well move West. Immediately they formed a nationwide syndicate, the New York and San Francisco Mining and Commercial Company, which bought the rights to the diamond fields from Arnold and Slack for about $600,000, and they prepared to float a public stock issue of $12,000,000. To insure that everything would be absolutely safe and sure, the syndicate hired the most cautious mining engineer in the West, Henry Janin, and he too examined and was dazzled by the diamond field, reporting in the public print that he had personally invested in it and that he considered "any investment at the rate of forty dollars per share or at the rate of four million dollars for the whole property, a safe and attractive one."

During this time, the syndicate had managed to keep the location of the diamond fields a strict secret, and as the news of the discovery leaked out, men with imaginations and the strong acquisitive instincts of the day went wild. Most people located Golconda in Arizona, others in Nevada, and almost every day prospectors wandered into Denver announcing they had found gems. Not since Coronado had followed Cibola halfway across North America had so many men so assiduously pursued windfall wealth. All along the transcontinental railroad, spies watched the trains for suspicious parties of prospectors. Even the telegraphers were not to be trusted.

King and the other members of the Fortieth Parallel Survey were of course fascinated by the rumors of the diamond discovery, particularly inasmuch as it very probably lay in the heart of country they had crossed many times. They were suspicious. Never in all the weary years of surveying and geologizing had they found gems or even a formation that looked as if it might contain such precious stones. When King visited Emmons and Wilson at camp near Fort Bridger in midsummer, they begged to be allowed to institute a search of their own for the stones, but King, with one eye on the War Department, put it off until the end of the season, when they had completed their routine work. Meanwhile he did some speculating on his own. The location of the diamond field could not be in Arizona, he knew, because the flooded condition of the Snake [Little Snake], Bear [Yampa], and Green rivers precluded any foray in that direction during the past summer. Piecing this fact together with Janin's description of the location of the field "upon a mesa near pine timber," King realized that "there was only one place in that country which answers to the description, and ... that place lay within the limits of the Fortieth Parallel Survey."

When King, Emmons, Gardner, and A.D. Wilson all met at the office in Montgomery Block on the eighteenth of October, they compared the results of their thinking and, King remembered: "Curiously enough Mr. Gardner and I had without formerly mentioning the subject reached an identical conclusion as to where the spot was."

Actually, some ingenious detective work had been done by Emmons and Wilson, as well as Gardner and King. On the fifth of October, while on the westbound train just out of Battle Mountain, Nevada, Emmons and Gardner had noticed a party of Eastern surveyors who were joined at Alta Station in the Sierras by Henry Janin. They were obviously returning from the diamond fields, so Gardner and Emmons struck up a conversation and gleaned what bits of information they could. They didn't get much out of Janin and his friends, but they did learn that Janin's earlier trip to the diamond fields had taken only three weeks, not enough time to go to Arizona. A.D. Wilson, who had been in the Green river country that summer, reported that the party had disembarked from the train somewhere between Green River and Rawlins. Putting this together with the data about the flooded Green, Yampa, and Snake rivers, they narrowed the field of possibilities considerably. Emmons had further learned that Janin had camped at the foot of a pine-covered mountain which had snow on its slopes as late as June, and Gardner pried from his surveyor the observation that their camp had been on the northeast side of a mountain from which they could see no other high peaks either to the north or to the east. Emmons, Gardner, Wilson, and Hague, who joined the game, concluded that there was only one place the diamonds could be – a peak just north of Brown's Hole, some forty miles east of Green River, right in the country where Utah, Wyoming, and Colorado came together. King agreed completely, and they made plans for an immediate departure for the site.

Everything had to be kept secret, of course. King did not even inform General Humphreys. Instead, just before they boarded the train – in separate parties, with sieves and shovels – he sent a dispatch from Fort Bridger to the General, on October 28, telling him he was off on a reconnaissance of Brown's Hole.

They departed from Fort Bridger on a bitter-cold day. It was October 29. For the next few days they worked their way north through the most unpleasant, freezing weather imaginable. Emmons reported that the horses "became encased in balls of ice." Four days out, they crossed the icy Green River and struck a deep gulley cutting into the side of the mountain they were aiming for. There, out of the biting wind, they made camp and the search began. A few hundred yards out of camp, they found a mark on a tree, then Henry Janin's official water claim, then more mining notices, and finally a small sandstone mesa stained with iron deposits. This must be the spot. Quickly they got down on hands and knees. They probed until dark, finding one diamond apiece on the first try. "That night," remembered Emmons, "we were full-believers in the verity of Janin's reports, and dreamed of untold wealth that might be gathered.

The next day, ignoring the sweeping wind and the shriveling cold, they surveyed the entire area, picking up quantities of stones. They learned, however, that the diamonds did not occur anywhere else but in the mesa where they first found them. By the afternoon they began to get suspicious. King discovered that the gems were always found in a ratio of twelve rubies to one diamond. Then they noticed that most of the stones were in anthills and places where the earth had been ever so slightly disturbed. Their suspicions heightened when one man found a diamond delicately perched atop an anthill where, had it been there long, it would surely have been washed away. Later they found evidence that the stones had been pushed into the anthills with a stick, and some disturbed places even had human footprints around them, though this was not conclusive since Janin and others had worked the site. To be sure, on the third day they checked the stream beds and other places where diamonds must surely have settled, and they dug a deep pit far into the sandstone formation. All this yielded nothing. Golconda was a fraud.

Just as they concluded this, a stranger appeared. He was an unscrupulous New York City diamond dealer who had followed their labors for days through field glasses. When he learned that the diamonds were a hoax, he declared: "What a chance to sell short." This reminded King of the grave responsibility they bore, and that night they made plans to slip out and beat the diamond dealer back to civilization. The next day King and Wilson left before dawn and rode all day, forty-five miles through the badlands, directly to Black Butte Station. They arrived just in time to catch the train for San Francisco.

In San Francisco, King first sought out his friend Janin and confronted him with the evidence. Then they went to the company directors, and King presented a letter with his findings. Ralston and Company stalled for time and begged King to keep quiet until they could sell out, but he refused. They did gain a few days, however, by persuading King to take Janin and the others to the diamond fields for a final check. When they returned, crowds had gathered outside the bank waiting for a dramatic announcement from Ralston. The San Francisco tycoons had been duped by Arnold and Slack and caught by King. There was nothing to do but reveal the whole story. Ralston and his partners published King's letter and the syndicate assumed full responsibility and repaid the stockholders. There were many who still continued to believe, however, that the scheme had originated with the syndicate in the first place. Slack disappeared, and Arnold, back in Kentucky, bought a safe and coolly stored his loot. In 1879 he was shot in the back and the loot disappeared.

Meanwhile King was the hero of the hour. The San Francisco *Bulletin* called him "a cool headed man of scientific education who esteemed it a duty to investigate the matter in the only right way, and who proceeded about his task with a degree of spirit and strong common sense as striking as his success." The Fortieth Parallel Survey as a scientific endeavor, according to the editor, had proven its practical value. Back East, Whitney, who was still fighting his own battle against the oil swindles, rejoiced at King's success. "Who's the King of Diamonds now?" he said. "And isn't he trumps?"

Unbelievably, however, General Humphreys was displeased. Upon receiving King's official report, he replied soberly: "As these Fields are situated within the limits of the Survey you have charge of, it was eminently proper that they should be included in your operations and an exhaustive examination of them be made." But, he added: "The manner of publicly announcing the results of the examination should I think have been different." In short, King should have gone through channels. But perhaps General Humphreys's stiffness can be excused, for by 1873 the Army was beginning to feel the competition from rival civilian surveys under the Interior Department. In view of increased Congressional opposition, it had to be jealous of whatever laurels fortune cast its way.

3-9   Sand, Wind, and War – Foreword by Luna B. Leopold, Paul D. Komar, and Vance Haynes

Ralph Alger Bagnold (1896–1990) was a professional soldier, who served in the British Army through two world wars. He was also a major contributor to our understanding of deserts and to the dynamics of sediment transport and deposition, with his books and articles frequently referenced in the scientific literature. His sister, the playwright and novelist Enid Bagnold (1889–1981), was the author of National Velvet (1935), a still popular children's book. At 94 years old, and a month before his death, Ralph Bagnold finished his auto-biography Sand, Wind and War: memoirs of a desert explorer (1990). In a foreword to this book, three well-known scientists – Luna Leopold, Paul Komar, and Vance Haynes – describe Bagnold's career. In an extract from this foreword, we see how a combination of Bagnold's understanding of how sand moves in the desert and his role as a professional soldier made a strong contribution to the desert battle that took place in North Africa during World War II. A much more complete account is given in his autobiography.

It is unusual to find a professional solider who is known primarily for his contributions to intellectual and scientific thought. This is true, however, of Brigadier Ralph A. Bagnold, whose research spanned several of the physical sciences, including hydrology, geophysics, oceanography, and geography. His earliest scientific publications appeared in the 1930s and dealt with the movement of desert sand; his two last papers were in 1983 and 1986, respectively examining the random distributions of word lengths and sediment transport in water. This productivity and originality seems astonishing when one considers that Bagnold was not formally trained as a scientist and had never been a professional academician.

... Commissioned as a regular officer in the Royal Engineers in 1915, Bagnold served in the trenches of France and Flanders during World War I. After the war, he took time out from the army to earn an honors degree in engineering at Cambridge and then rejoined the army, serving in Egypt, India, and China. While stationed in Egypt, he and other young officers began tinkering with Model-T and Model-A Fords, modifying them in ways similar to those that youngsters in California would use in the 1960s and later to design "dune buggies." Gradually this tinkering led to a more serious interest in exploring the dunes, and during periods of leave they organized trips into the surrounding desert country. Such trips developed into serious large-scale geographic exploration of the last remaining unmapped areas of the earth. They developed techniques for taking these simple cars deep into the dune country, inventing ingenious devices to keep themselves from becoming stranded in these remote seas of sand.

Bagnold learned from experience what part of a dune would support the car and where the wheels would sink in. This was the beginning of a growing knowledge about dunes and the movement of sand. As his young friends were gradually reassigned elsewhere, Bagnold turned his attention to examining the processes by which dunes are formed and maintained. His retirement from the [Army] in 1935 allowed him to devote full time to laboratory experiment designed to interpret his observations in the desert. The results were published in 1941 in a book which remains the classic in the field today, The Physics of Blown Sand and Desert Dunes. Meanwhile, he published a delightful chronicle of his explorations entitled Libyan Sands, a book filled with his zest for adventure and his keen eye for detail.

As Britain was plunged into World War II, Bagnold was recalled to the [Army]. In 1939 he was en route to an unexciting position in East Equatorial Africa when his ship was involved in a collision and had to put in to Port Said for repairs. His presence in Egypt came to the attention of the commanding general for that region, who knew of Bagnold's experience in the desert and quickly had his orders changed. Bagnold was given a free hand to develop the Long Range Desert Group, an unconventional "private army" that traveled undetected through remote regions of the desert to harass the enemy with unexpected attacks. So important and useful was the work of the Long Range Desert Group that Bagnold rose to the rank of brigadier and directed the operations from headquarters. He retired again in 1944, and after a few years as Director of Research for Shell Oil, he once more took up his principal hobby of studying the transport of sand. Now he extended his work beyond the movement of sand by wind to include grains moved in any fluid.

. . . The position of importance he has earned in science stems from his insistence on critical experimentation combined with the use of pure physics. His goal was to eliminate the empiricism that so long dominated the problem of transportation of debris in water. This led to a productive association with the United States Geological Survey, which published a series of his papers, the most important being a 1966 Professional Paper entitled "An Approach to the Sediment Transport Problem from General Physics." In the field of oceanography he studied the motion of sand on beaches and the hydraulics of submarine density currents.

Although Bagnold still tended to view himself as an amateur, he was accorded the highest honors of professional societies. In geography, his explorations in Libya won him the Founders' Gold Medal of the Royal Geographical Society; in hydraulics and related fields, he was given the G.K. Warren Prize of the National Academy of Sciences, the Penrose Medal of the Geological Society of America, and the Sorby Medal of the International Association of Sedimentologists. However, Bagnold must have taken greatest pleasure in being elected a Fellow of the Royal Society. He recounts that even early in his military career he expressed to his sister that he would rather be an FRS than a brigadier general. It is a measure of his accomplishments that he managed both.

Bagnold had many fascinating experiences, both as an explorer and in the military, and he had much to share concerning his views on scientific research. We are fortunate that he completed his autobiography shortly before his death in May 1990. Those who knew him only through his research papers can now come to appreciate the total wealth of his experience.

3-10 Ship's Wake – Hans Cloos

Our second extract by Hans Cloos (1885–1951) – see page 32 for the first – is also from his book Conversation With the Earth *(1953).*

This extract describes the impression made upon him by Stromboli and Aetna.

> The present is a distraction; the future a dream; only memory can unlock the meaning of life.
>
> Desmond MacCarthy

The next day was gloomy and rainy. The gray triangle of Stromboli, most punctual volcano on earth and Italy's weather prophet, slid past. Three times I saw the clouds of smoke which Stromboli has been exhaling at regular intervals since the days of antiquity. These intervals decrease slightly with increasing barometric pressure. That afternoon Sicily came into view, with Mt. Aetna dominating, its snowy head ringed with little clouds of steam, and Messina at its base. There had just been a severe earthquake, and through field glasses I saw the ruins of hotels on the quay. Again the earth's two faces showed themselves – the black of the tenebrous depths and the light one of the non-volcanic crust. For although Messina had not been destroyed by the volcano which towers over the city, the crust had been rent asunder by great tremors, and the volcano's action showed that the fissures must reach down into the still fluid depths of the earth.

A little later the ship turned left, and henceforth sailed straight through the open Mediterranean for the mouth of the Nile.

The next morning, like countless seafarers before and after me, I made a discovery: on ahead, where the steady bow of the ship cut through the blue sea as if it were molten glass, one looks into the future; astern, at the taffrail, above the noisy propellers, images of the past seem to emerge from the greenish-white wake. One peers into the restless Charybdian whirlpool boiling behind the ship; unfathomable masses of water and endless horizons recede before one's gaze, and dreams turn to the past.

While I stood there, looking back at the gulls as they swooped this way and that, becoming smaller and smaller, then racing after the ship from which they got their food, a traveling companion joined me. He too for a time stared silently at the water, and finally observed that geology must be a curious profession. How had I happened to take it up? How could anybody? Every geologist has been asked that question a few times. I told him about my own intellectual development, and about the evolution of a geologist. I told him that my earliest childhood memory – which goes back almost to the age of two – involved an attempt to catch a sparrow-hawk that had flown into the room. I told him how my relationship with the earth had developed out of many experiences, and I told him about this dramatic one:

The land had been drenched by a cloudburst. The gardener of the estate carried me – then aged ten – across the flooded village street which had become a stream. When the water had subsided, I followed its course. There was a fresh, deep washout in the ravine in the little wood, and a bridge had been covered with gravel. Beyond and below the bridge, at the mouth of the ravine, limestone debris and mud had spread out over the meadows, forming a new layer which raised the level of the ground. Yesterday's meadow was now covered and obliterated. Did the earth grow like a tree, adding a ring at a time? Could the land change its shape from one day to the next? Even at that early age I had been half-aware of these questions.

A little later I developed an interest in fossils and minerals. A deep gorge, or "Klamm," was cut into the "Muschelkalk" (shell-limestone) near my home. The stony creek bed and adjacent fields were full of fossils like ammonites, fluted bivalves, and smooth terebratula as large as nuts – all petrified into yellowish-white limestone and washed free of the surrounding rock. Today this region has been much picked over and it may take a long time before it is again a good spot for collectors. I inspected the limestone fields and the stone piles at their boundaries, and collected a rich booty. Small finds I kept in a cigar box; large slabs found a place on the garden wall. Since that time some selected samples have accompanied me wherever I have gone. A splendid ceratite that my father found, resembling a curled snake and weighing several pounds, now decorates my desk. At the beginning my unbridled imagination led me to fancy that I had found fossilized snakes, skulls, heads, and worms. But gradually I learned the limits of fossil preservation and at the same time began to understand the strange historical character of these extinct forms of life, which evolved under entirely different conditions from those of the present.

Never again did nature seem so permeated with wonder, and never again was my relationship with it so strong and direct as during these years of transition from my childhood when I was stimulated by fairy-tales and play, to the period of sober and objective scientific observation of the adult. The former taught me to see the richness of forms, the latter added color and life to diversity. Bluebells bloomed in the spring woods, the perfume of daphne was sweet on the air. Sand-yellow vipers sunned themselves in the summer's heat, ringed snakes glided across the path. Fat emerald lizards clung to the warm walls. . . .

During those years I understood the voices of animals and stones as did Siegfried after he had tasted dragon's blood. Later, when I was able to feel less strange in distant lands, deserts, and jungles, and when I learned something of the strange languages of the wide, wild world, I was thankful for those summers of communion with nature during my boyhood. The language of nature must be learned like the mother tongue. Anyone who begins only at the age of twenty has little prospect of mastering that language.

It was somewhat later when I became enchanted by the brightly colored minerals and sparkling crystals. Even now I can feel my astonishment when I first found gypsum in the gray wall of an abandoned quarry. It was clear as glass, yet soft as wax or wood. To think there should be anything like that in the rocks of the ground! When a well-house was hewn out of the limestone, little cavities were exposed; tiny caves evolved like doll house rooms. In the dumps of an ore mine I found multi-colored crystals, growing side by side and on top of one another like flowers. A summer vacation in the Hunsrück led me to the old agate cutting works of the Idar Valley. For days I would dig in the tailings, looking for purple amethysts, red ribbon agates, fibrous and yellow cat's-eyes. Even today a certain odor, apparently that of the oil used by the cutters, is ineradicably linked to my early impressions of this dead and yet so colorful and multiform world.

Then suddenly schooldays were over, and the choice was between architecture and geology. Was that not vacillating, indicative of a lack of character? At the time I reproached myself for this indecision, and I realized only many years later that architecture can be applied to the study of the earth's crust and that, indeed, during my entire life I have done little else.

And now my own choice, geology! For me, geology was far superior to all other fields, and that is as it should be. In geology, too, there were books, numbers and other ballast for my memories. Like the rest, I wore out books and filled blank papers with notes. But by far the most important books for geology students were the quarries and clay pits, the cliffs and creek beds, the road and railroad cuts in woods and fields. Our words and letters were the imprints of plants and animals in stone, the minerals and crystals, and our vast, inexhaustible, incorruptible, and infallible library was nature itself. . . .

And to this end we went out into the field, time and time again. Often at brief intervals we returned to the same outcrop where our Pythia, nature, opened her mouth from time to time to utter her equivocal oracles. Time and again we studied the same stratification, or the same interpenetration of rocks, and yet each time advanced one stop further, because of what we had learned on the last visit, because the previous impression had had time to settle, or because this time our eyes were a little keener and now observed what hitherto had escaped them. . . .

The day came when I was far enough along with my work to dare to ask the experienced teacher for a project of my own. Now real research began. Every good eological doctorate dissertation should be based on a first modest research expedition, and countless theses really were. . . .

The first independent piece of research had an educational effect: I stood alone, face to face with nature. In that chosen small portion of the earth's surface there were no reference books to depend on, and no professors to turn to for help. The student had to depend on his own eyes and head, and on his legs and hobnailed boots as well.

As a rule nature was very reserved about her revelations. She gave very little freely. Ask, and ask intelligently and persistently, ask again and again, and then listen carefully: perhaps nature will answer.

. . . Then the knapsack was packed a bit more carefully than usual and I got underway. For the first few weeks, I groped around uncertainly. Evenings I returned to the village inn, either in despair, or filled with hope or pride, but invariably in possession of many errors! Yet slowly I began to understand, and during ensuing days and weeks the future geologist commenced to evolve. . . .

Toward the end of the first summer in the field, the professor came out for two or three days. He looked over the exposures, the mapping, and the drawings. He listened to what we had to say, and drank some of the good wine that had grown on his student's layered rocks. He praised the wine and the stratified rocks, and returned to his comfortable Institute, where a thousand finished theses looked down from the shadowy bookshelves. Then I was on my own again, and had to figure out how to muddle through. . . .

After the final examination I was completely at a loss. The task I had worked on for two solid years was completed; what I had learned in four years had been divulged in two hours. Then I had some luck, without which the best man in the world cannot succeed. Even so, I had to open the door, and reach out for it. Four months later I was sitting in the Freiburg–Basel–Olten–Lucerne express, passing swiftly through my first modest field of geological enterprise, and on toward the second and vastly larger one. By evening of that day I was already in Milan on the south side of the Alps, the following day I was in Florence, at the western base of the Apennines, and a few days later in Rome and Naples. And now the mighty limestone walls of Crete were sliding by to the left over the smooth blue of the sunny Mediterranean. They are the extreme southerly margin of the Eurasian belt of mountain ranges, on whose northern border at Freiburg I had just passed my doctoral examination. Now it was a case of: "Hic Rhodus, hic salta!"; that is, "Let's see what you have learned, my boy!" The actual Rhodes, of course, after Crete the most easterly pier in the marginal wall of Europe, remained hidden somewhere to the left beyond the horizon. But already the new continent, the black continent with the big blank places on its map, was blowing its glowing desert air towards me. Africa. . . .!

4. Celebrities

If war, as it has been said, is too important to be left to the generals, how much more is the study of Earth too important to be left to professionals. This is so because, in large part, Earth science – as all science – is hugely reductionist, selecting only those few common features that are both quantifiable and "meaningful." But the very process of abstraction and analysis – at once both the heart of science and the secret of its power – means that much eludes the meshes of the scientific net, and much escapes the scientific description. If we seek a more comprehensive view of Earth and a fuller understanding of its nature, we need other approaches, other views, and other voices. For what is meaningful for particular scientific purposes may be less meaningful for others and our grasp of some subject or enjoyment of some place may be enriched by features that have no place in science. A geologic map of Mount St. Victoire in Provence gives us one view of reality, a topographic map another, a photograph another, and a painting by Cézanne, still another. Each is abstractive, impressionistic, partial, in terms of the totality it represents. And each, in one sense, is incomplete without the others.

How encouraging, then, that others – celebrated for achievements in other fields – should have been deeply interested in our planet. Our selections include two presidents of the United States – one a "Founding Father" – a nineteenth century English artist, author, critic and social reformer, one of America's greatest painters, and two of the greatest geniuses of all time.

4-1 Benjamin Franklin and the Gulf Stream – H. Stommel

Printer, author, philanthropist, inventor, statesman, diplomat, Founding Father, and scientist all describe the career of Benjamin Franklin (1706–1790). His interest in science led him as early as 1737 to write in the Pennsylvania Gazette *of earthquakes and he had equal curiosity about the nature of waves at sea, the cause of springs on mountains, fossil shells, variations in climate, the effect of oil on storm waves, and techniques of swamp drainage. While Postmaster General, Franklin recognized the existence of the Gulf Stream* and had a chart engraved and printed by the General Post Office. He made a series of surface temperature measurements while crossing the Atlantic and devised a method to attempt measurement of subsurface water temperatures. Henry Stommel (1920–1992) made many fundamental contributions to the field of physical oceanography, had a large role in shaping the field. In addition to his work as a scientist he was also a painter, "gentleman" farmer, and fiction writer. Our extract is taken from his book The Gulf Stream: a physical and dynamical description (1958).

In 1770 the Board of Customs at Boston complained to the Lords of the Treasury at London that the mail packets usually required two weeks longer to make the trip from England to New England than did the merchant ships. Benjamin Franklin was Postmaster General at the time and happened to discuss the matter with a Nantucket sea captain, Timothy Folger. The captain said he believed that charge to be true.

> "We are well acquainted with the stream because in our pursuit of whales, which keep to the sides of it but are not met within it, we run along the side and frequently cross it to change our side, and in crossing it have sometimes met and spoke with those packets who were in the middle of it and stemming it. We have informed them that they were stemming a current that was against them to the value of three miles an hour and advised them to cross it, but they were too wise to be councelled [sic] by simple American fisherman."

Franklin had Folger plot the course of the Gulf Stream for him and then had a chart engraved and printed by the General Post Office.
Franklin believed that the Gulf Stream was caused

> "by the accumulation of water on the eastern coast of America between the tropics by the trade winds. It is known that a large piece of water 10 miles broad and generally only 3 feet deep, has, by a strong wind, had its water driven to one side and sustained so as to become 6 feet deep while the windward side was laid dry. This may give some idea of the quantity heaped up by the American coast and the reason of its running down in a strong current through the islands into the Gulf of Mexico and from thence proceeding along the coasts and banks of Newfoundland where it turns off towards and runs down through the Western Islands."

By the time of Maury, in the middle of the nineteenth century, Franklin's estimates of the velocities of the Stream were regarded as excessive, but more recent studies tend to confirm them. Franklin did not give any details concerning the edge of the Stream. Starting in 1775, both Franklin and Charles Blagden independently, conceived the idea of using the thermometer as an instrument of navigation, and each made a series of surface temperature measurements while crossing the Atlantic. On Franklin's last voyage in 1785 he even attempted to measure subsurface temperatures to a depth of about 100 ft., first with a bottle and later with a cask fitted with a valve at each end.

4-2 Leonardo Da Vinci as a Geologist – Thomas Clements

The great art masterpieces The Last Supper, Mona Lisa, *and* Madonna of the Rocks *and the true nature of fossils, isostatic adjustment, and a modern concept of the immensity of geologic time might appear to be immiscible, but all are products of the genius of Leonardo Da Vinci (1452–1519). A scientist who rigorously applied the scientific method and placed faith only in his own research, Leonardo was centuries ahead of his time as a geologist. His contribution to geology has been less widely recognized than his accomplishments as a painter, canal builder, sculptor, inventor, and military engineer. In an article by Thomas Clements (1899–1996), Da Vinci receives due recognition for his geological endeavors. Clements, a pioneer of the field of engineering geology, completed his doctorate degree at California Institute of Technology and served on the faculty of the University of Southern California (USC) from 1929 to 1964. During his long tenure at USC, Clements chaired three different departments – Geology, Geography, and Petroleum Engineering. From his retirement from USC in 1964 into the late 1980s, Clements was an active and energetic engineering geology consultant, retaining in his later years his keen insight and observational skills, and an ability to climb hills with field geologists decades younger than him. Clements wrote this article in 1947 and it was published for the first time in* Language of the Earth *(1981).*

Leonardo has sometimes been thought of as dilettante or a dabbler as far as the sciences are concerned, because of his widely scattered interests. In the writer's opinion, this is not a just evaluation of his scientific work. It is true that he delved into many sciences, but Leonardo was an engineer; it was necessary for him to know certain fundamental facts in science, and when the knowledge was not already available, he attempted to work it out for himself. Furthermore, he had a tremendous amount of intellectual curiosity. While most of his research was of a practical nature, or begun with a practical application in view, this intellectual curiosity led him on to fields far from his original starting point. One idea suggested another, as indicated by the notes on varied subjects found on a single page of his manuscripts.

Although an original thinker, Leonardo did not disdain the learning of the past. That he read widely is suggested by lists of books mentioned in his manuscripts: on one page of the Codice Atlantico (Folio 210) there are the names of forty books, and it has been stated that he was familiar to a greater or lesser extent with the works of more than seventy authors of both the Classical and the Middle Ages. Nevertheless, he did not necessarily accept as final the writings of these men on any subject.

It is here that Leonardo is seen as the true scientist. Freely admitting that others had explored all of the various fields in which he himself was interested he insisted on testing their conclusions; he accepted no authority but the results of experimentation. He truly developed the inductive method, that approach to problems today commonly called the scientific method. He performed experiments; he drew conclusions; and he reasoned logically to arrive at general laws. . . .

It is particularly in the field of geology that Leonardo's great powers of observation, his controlled imagination, his sound reasoning, and his immense fund of common sense led him far beyond the beliefs of his time. He apparently early formed the habit of wandering in the fields and hills, seeing all things not only with the eye for beauty of the artist, but also with the inquiring eye of the scientist. His later work with rivers and canals gave him further opportunities to study geologic phenomena. His writings abound with references to geology, and his pictures, such as the *Madonna of the Rocks*, show a sound knowledge of the fundamental geology of landscapes.

Although originally accepting the common belief of the time that shells of marine organisms found far from the sea were the remains of animals left there by the deluge of Noah, his later observations and reasoning showed him the fallacy of such a belief. He argued that cockle shells occurring many miles from the sea could not have travelled that distance in the forty days and nights of the deluge. Nor could they have been washed there by the rising waters of the flood, since they occurred unmixed with other debris that a flood would have carried. That they were not an unnatural creation in place he knew from the presence of paired shells, and shells of varying stages of growth, as shown by their growth lines. He rightly reasoned that the distribution of land and sea had not always been as it was then, but that at one time the sea had covered those parts of the dry land where fossils now occur and "... above the plains of Italy, where now birds fly in flocks, fish were wont to wander in large shoals." Parsons called him the Father of Paleontology.

He made a study of wave action, and he acquired a wide knowledge of the work of running water. He realized that valleys were cut by the rivers occupying them, and varied as the streams varied. He observed that rocks transported by rivers gradually became rounded and were worn smaller and smaller. He knew that the material deposited in the sea by the rivers became cemented together to form the various types of sedimentary rocks, and that these later were uplifted again to form mountains.

His knowledge of the work of rivers led him naturally to a concept of geologic time that like so many other of his conclusions is remarkably modern. The teaching of the church in his day was that the earth was formed some five thousand years before the birth of Christ. Leonardo, however, had learned from observation how slowly river deposits form, and he declared that the deposits of the River Po had required two hundred thousand years to accumulate, and this, to him, by no means meant the whole of geologic time.

His study and experimentation in the field of mechanics led him to another strikingly modern geologic conclusion. He noted that rain and rivers were constantly gnawing at the mountains and carrying material to the sea, and yet the mountains were high, and the basin of the sea low. He says: "That part of the surface (of the earth) of whatever weight, is farthest from the center of gravity (of the earth) which is lightest. Therefore, the place from which weight is removed (by the rivers) is made lighter, and in consequence moves farther from the center of gravity (i.e., is uplifted)." This is a statement of isostatic adjustment, as defined by Dutton in 1889.

... Leonardo da Vinci's scientific inquiry led him over a wide range of interests. His early chroniclers were inclined to be apologetic for his scientific work: to them he was a painter and sculptor who let this foolish absorption with other matters interfere with the execution of his art commissions. The wonder is not that he did not finish so many of the paintings and sculpturings that he was commissioned to do, but rather that he completed as many as he did. All honor to Leonardo da Vinci as an artist; but more honor to him as an engineer and scientist!

4-3 Mineralogy, Geology, Meteorology – R. Magnus

Truly a man of all seasons, Johann Wolfgang von Goethe (1749–1832) had interests that embraced drama, poetry, painting, philosophy, politics, and various aspects of science. The creator of Faust *and the* Sorrows of Werther *was a member of nearly 30 scientific societies, an active participant in the Plutoist–Neptunist battle over the origin of rocks, foreshadowed portions of Darwin's theory of organic evolution, early recognized the concept of an "Ice Age," established a collection of rocks, minerals, and fossils of over 18,000 items, and it was in his honor that the familiar iron mineral, goethite, was named. The first geologic map of Germany was published in 1821 and the color system adopted was based on Goethe's proposals. Interestingly, modern geologic maps are prepared using essentially the same color code. Rudolf Magnus (1873–1927), famous for his founding work in neurology, wrote* Goethe als Naturforscher *in 1906. Our extract is from the 1949 translation by Heinz Norden* Goethe as a Scientist.

"I fear not the reproach that it must have been a spirit of contrariety that has led me from the contemplation and description of the human heart – the youngest, ficklest, most versatile, mobile and changeable aspect of creation – to the observation of nature's oldest, firmest, deepest and most steadfast scion."

Thus wrote Goethe in the year 1784 during his studies of granite. For more than fifty years we find the poet intensely preoccupied with the study of the earth's crust, its structure and origin. To trace down every last scientific detail in this field of research would transcend the framework of this book. All we can do is to give a general picture of his investigations and his views.

As we already know, Goethe's interest in mineralogy sprang from practical causes. There was the problem of reviving the long-dormant mining industry at Ilmenau, a problem which Goethe, as Chief Minister, began to tackle in 1777. He found that in Thuringia the way for geological study had already been paved. Interest in this field was stirring, especially because of the nearness of the Mining Academy at Freiberg and the work of the most noted geologist of the time, Abraham Gottlob Werner.

The mines at Ilmenau exploited seams. This posed the problem of identifying the geological strata carrying the seams. Goethe's attention was drawn to the great regularity in stratification that marked the earth's crust in Thuringia, evidence that to him bore out the teachings of Werner, an adherent of the so-called "Neptunian Theory," which attributed the formation of the earth's crust to the effects of water.

Even during this early period Goethe had occasion to display his practical bent in another direction. At his behest the Duke of Weimar in 1779 acquired the large Walch mineral collection. This was installed in the palace at Jena, under the direction of Lenz, later to become the first professor of mineralogy at Jena. The collection was organized on Werner's system and formed the core of the famous Museum of Mineralogy to come.

In order to secure a trained expert for the mining industry, Goethe persuaded the Duke to send J.C.W. Voigt – later likewise to become famous – to study at the Mining Academy. In 1780 Goethe drafted an elaborate set of instructions under which Voigt was to conduct a geological tour of inspection of the Duchy of Weimar.

Goethe's views on geology, rooted in the soil of Thuringia, greatly expanded on his numerous journeys. As early as his first and second Harz journey he had gathered geological observations and visited the mines at Goslar and Klausthal. His third journey to the Harz mountains, undertaken in the company of the artist Kraus, was wholly devoted to such studies. Goethe's geological diary covering these weeks has been preserved and gives evidence of the seriousness with which these inquiries were conducted. The Goethe House to this day contains the fine drawings by Kraus of the most noteworthy granite formations in the Harz outcroppings – for it was the study of granite that chiefly preoccupied the travelers.

His numerous sojourns in the watering resorts of Bohemia gave Goethe occasion to collect mineralogical data there too. In the course of time he grew intimately familiar with this section of Bohemia. He studied the nature of hot mineral springs, confirmed his "Neptunian" views, traced the distribution of the great coal deposits. In Switzerland he studied the form and distrubtion of glaciers and the effects of earlier glaciation. In Italy he climbed Aetna and Vesuvius and visited the Phlegraean Plain, thus gaining a firsthand view of volcanic activity. Even during the campaign in France he pursued his mineralogical researches. This, then, was the raw material on which Goethe based his studies in geology. Its significance lies in the fact that it was bound to strengthen the "Neptunian" views which Werner had instilled. Goethe himself pointed out that he might not have become such a strong adherent of this theory, had he conducted his initial studies in the volcanic Auvergne, for example.

He brought home from his journeys numerous specimens which he arranged systematically. Voigt had introduced him to mineralogical nomenclature. Thus there came into being the large collection, numbering more than 18,000 items, that still delights the visitor to the Goethe House by its wealth and the beauty of its rarities. The collection is largely devoted to specimens from Thuringia, the Harz mountains and Bohemia, but other sections of Germany, Italy and many foreign lands are likewise represented. Among other things Goethe kept a careful record of all fossils unearthed in Thuringia.

The collection assembled by one Joseph Müller, a lapidary of Karlsbad, came to have special importance for Goethe. In the course of his trade Müller had collected many mineral specimens from the vicinity of Karlsbad. His collection grew more and more and in 1806 he showed it to Goethe, who began to arrange the minerals in order. Proceeding from granite, Goethe worked out a continuous series of the various deposits and prepared a careful scientific catalogue. He then persuaded Müller to put his collecting activities on a broader basis and to market his specimens by Goethe's system. As early as 1806 Goethe advertised the collection in a literary periodical at Jena, and two years later he published the catalogue in Von Leonhard's *Pocketbook of Mineralogy.*

In this way a model collection of minerals was made available to scientists everywhere. Noeggerath, the Bonn mineralogist, described it as ideal for instructional purposes. Using this collection, Goethe published a careful mineralogical account of the region of Karlsbad. In 1821 he followed with a similar account of the region of Marienbad, based on a corresponding collection. He never lost his interest in the Karlsbad collection, and as late as 1832 he advertised a series of thermal-spring sinters marketed by Knoll, Müller's successor.

A collector and scientist himself, Goethe was in close touch with many experts in the field. On the subject of mineralogy he corresponded with his friends Merck and Von Knebel, with Von Trebra, Von Leonhard, August von Herder, Cramer, Count Sternberg, Grüner, and especially with Lenz. There was spirited barter in specimens among these men and even with others more remote. Goethe often figured in these exchanges, thus expanding his own collection and that at Jena. When Jena was unable to acquire the important Cramer collection for lack of funds, Goethe saw to it that it went to Heidelberg.

In 1796–98 Lenz established the Mineralogical Society at Jena, and it elected Goethe its first honorary member. Lenz was extremely active in enrolling new members for the Society and in expanding the Jena collection with their help. It grew to be one of the most important ones of its time, and savants from abroad came in large numbers to admire its treasures.

Mineralogy was at the time enlisting two important auxiliary sciences – analytical chemistry and crystallography. The initiative of the great Scandinavian chemist Berzelius had led to the chemical analysis of minerals. Goethe himself never carried out such analyses, though he followed the progress of the new science with great interest. He was kept abreast of development in the new chemistry first through Göttling and later through Döbereiner. The large collection of handbooks and textbooks of chemistry in his library gives eloquent testimony of his lively interest. His interest in crystallography, aroused especially by Soret, was less intense, though he did make occasional observations on the genesis, growth and size of crystals.

Goethe's researches in mineralogy and geology form no milestone in the history of these sciences, yet "he was up to the best of his contemporaries." In later years at least his observations were held in high esteem by scholars in the field. He collaborated in Von Leonhard's *Pocketbook of Mineralogy*, and Noeggerath wrote a most favorable review of his writings in the Jena press. On account of his researches in northern Bohemia he was elected an honorary member of the Bohemian scientific society founded by Count Sternberg. A scientific society in Edinburgh did likewise. . . .

Goethe devoted special attention to drift boulders and erratic blocks found in the Alps and the North German plains. He ascribed their origin to various causes. The granite blocks of the Rhone valley he thought had been transported to their resting places by glaciers. Many blocks in northern Germany, however, he regarded as remnants of an ancient mountain range that had somehow evaded weathering. As the most important example he cited the so-called Heiligendamm. In addition, he postulated that some of these boulders might have drifted across the sea from Scandinavia on ice floes and icebergs. This follows one of Voigt's hypotheses, and it should be noted that Preen actually observed Scandinavian rocks cast up on the Baltic coast by ice floes.

As a result of these studies, Goethe developed the idea of an Ice Age, and it appears that he was indeed the first to assume such an epoch. "It is my conjecture that all of Europe, at least, passed through an epoch of great cold." At the time, he thought, the ocean still extended over the continent to an altitude of a thousand feet, Lake Geneva was directly connected with the sea and the Alpine glaciers flowed down to Lake Geneva. These views reappear in his "Wilhelm Meister."

In his geological studies Goethe seems to have developed a whole series of concepts that did not achieve general recognition until much later. Apart from his views on the historical significance of fossils and his postulation of an Ice Age, there is especially his conviction that the forces that once shaped the earth and the mountains are the very same which we see at work every day, only in modified form. Hence his endeavor to find analogies to geological processes in ordinary observation – such as the formation of glacier ice. "As far as I am concerned, I am quite willing to believe that nature even today is capable of forming precious stones as yet unknown to us."

Indeed, Goethe extended this approach to the very rock that to him was the basis of everything – granite. "It is quite possible that granite may have been formed repeatedly." Since he thought that forces still at work were entirely adequate to explain the formation of the earth and that changes in the earth's surface in our own time take place only very slowly, he was bound to arrive at the conclusion, now generally accepted, that the periods of the earth's history were of extraordinary length. He put this conviction into the mouth of Thales in the second part of "Faust":

> Never did nature with her living might
> Depend on hours, on mere day and night.
> Each form in gentle temperance is wrought.
> Nor even the greatest change with violence fraught.

We see, therefore, that Goethe's geological studies led him to rather modern views on the formation of the earth.

The period that embraced Goethe's preoccupation with geology was dominated by the controversy between the "Neptunists" and the "Vulcanists" – between those who ascribed the main share in shaping the earth to water and those who favored volcanic forces. The controversy has long since been settled. Water and fire have each claimed their appropriate share in geological history. But at the time the battle was joined with the greatest vehemence. By and large Goethe avoided the exaggerations in both doctrines. The cast of his mind inclined more toward Werner's "Neptunian" view, and in the "Xeniae Tamed" he complained:

> Scarce noble Werner turns about,
> Poseidon's realm falls prey to loot.
> Hephaestus puts them all to rout –
> But not myself, to boot.
> With easy credit I grow pettish;
> Full many a credo have I spurned;
> Full well to hate them have I learned –
> New idol and new fetish.

But he reserved his grandest picture of this scientific controversy for the second part of his "Faust." In the "Classical Walpurgis Night" scene he has Thales and Anaxagoras roam the mountains and seas, hotly debating the issue – the former a "Neptunian," the latter a "Vulcanist." Here too he implied his partisanship on behalf of Thales, rejecting the doctrines of the headlong Vulcanists who would not shrink from having stones rain from the moon. Yet even in "Faust" the issue is not settled. Instead, the origin of the mountains is dramatically recited by Seismos (earthquake) along Vulcanist lines:

> In this, then, you must credit me –
> No other power chalked the score.
> So fair the world would never be,
> Had I not rocked it to the core.

But Thales clings to his Neptunian view, which he expresses in these magnificent lines:

> All sprang from out the briny deep!
> All does the water safe embrace!
> Ocean, grant us thy endless grace!
> If thou thy clouds us didst not send,
> If thou rich torrents didst not spend,
> The curving rivers didst not wend,
> To them thy bosom didst not lend,
> What were the hills, the plains, the world?
> 'Tis thou keeps life all fresh unfurled!

Goethe's own attitude toward the effect and significance of volcanic forces shifted in the course of his researches. He was never able to close his mind to their importance, yet he shrank from the excesses of the Vulcanist school. With Werner he sought to ascribe many alleged products of volcanic eruption to subterranean conflagrations which were supposed to have occurred especially in the great coal beds. He thought there was much evidence of this kind in the Bohemian deposits. The firing of the rocks was supposed to have been facilitated by the interspersed plant remains. Near Grünlass, for example, he found a bituminous shale that could be ignited by a flame. He studied the effect of combustion and annealing on a whole series of rocks – we have a list of thirty-eight different minerals which he had fired in a potter's kiln at Zwätzen in 1820 to study the results. He made repeated tests along these lines and imputed to them great importance in evaluating natural deposits.

4-4　Megalonyx, Mammoth, and Mother Earth – E.T. Martin

Thomas Jefferson (1743–1826), the third president of the United States, was also a reputable scientist, with interests that ranged from agronomy to botany and chemistry to paleontology. As a geologist, Thomas Jefferson's major contributions were in the field of vertebrate paleontology, but in his writings he speculated on the origin of the Earth, the formation of mountain ranges, the origin of the Natural Bridge of Virginia, the occurrence of shells far above sea level, and involved himself in the Plutonist–Neptunist controversy. While president of the American Philosophical Society of Philadelphia, the foremost scientific organization in America, he presented several learned geologic papers, one of which, "A Memoir on the Discovery of Certain Bones of a Quadruped of the Clawed Kind in the Western Part of Virginia," caused a mild furor in geologic circles. Later he commissioned General Clark, of Lewis and Clark fame, to excavate a mammoth in Kentucky. Our extract describes the Great-claw Megalonyx of Jefferson, mammoth bones in the White House, and details concerning the pioneer American vertebrate paleontologist. The present passage is taken from Edwin T. Martin's book, Thomas Jefferson: Scientist *(1952).*

In 1796 Jefferson came into possession of some fossilized bones of a large animal heretofore unknown to American scientists. The bones had been found in a saltpeter cave in Greenbriar County, Virginia, now West Virginia. Upon receipt of them Jefferson sent a brief description to David Rittenhouse, President of the American Philosophical Society. Rittenhouse was actually dead at the date of this letter, July 3, 1796, though the news had not yet reached Monticello. Jefferson ascribed the bones to "the family of the lion, tyger, panther etc. but as preeminent over the lion in size as the Mammoth is over the elephant." In fact, the size of the claw and general bulk of the animal led him to name this unknown creature "the Great-claw, or, Megalonyx." He expressed his ultimate intention to deposit the bones with the Society. . . .

Jefferson's discovery naturally created great interest among his scientific friends in Philadelphia. Benjamin Smith Barton wrote him on August 1 that his account of these bones would be "very acceptable" to the American Philosophical Society for publication in the forthcoming volume of the *Transactions*, if it were received in time.

Jefferson's excitement over this new discovery and his eagerness to have his account published in the *Transactions* glows through his letter to Dr. Benjamin Rush of January, 1797: "What are we to think of a creature whose claws were eight inches long, when those of the lion are not 1½ inches; whose thighbone was 6¼ diameter; when that of the lion is not 1½ inches? Were not the things within the jurisdiction of the rule and compass, and of ocular inspection, credit to them could not be obtained. . . . I wish the usual delays of the publication of the Society may admit the addition to our new volume, of this interesting article, which it would be best to have first announced under the sanction of their authority."

When Jefferson rode to Philadelphia to assume his duties as Vice-President of the United States, he carried with him his collection of bones and the article he had written about them. In Philadelphia, he discovered something that challenged his identification of this unknown animal, his *Megalonyx*, as a carnivore belonging to the lion or tiger family. This was "an account published in Spain of the skeleton of an enormous animal from Paraguay, of the clawed kind, but not of the lion class at all; indeed, it is classed with the sloth, ant-eater, etc., which are not of the carnivorous kinds; it was dug up 100 feet below the surface, near the river La Plata. The skeleton is now mounted at Madrid, is 12 feet long and 6 feet high."

The discovery Jefferson refers to had been made in South America in 1789. It was, in actuality, an extinct ground sloth of herbivorous habits whose habitat had been both North and South America. Its remains have been found in Pleistocene deposits.

Though his own identification was now in doubt, Jefferson went ahead with his original plans. His paper was read before the American Philosophical Society on March 10, 1797, and was published in the fourth volume of its *Transactions*, in 1799. The bones which he had – a radius, an ulna, the second, third, and fifth metacarpals, a second phalanx of the index finger, and a third phalanx of a thumb – he deposited with the Society. Jefferson added a postscript to his paper, commenting on the similarities between his animal and the giant sloth (*Megatherium*), and concluded that positive identification must await the discovery of missing parts of the skeleton. In the meantime he was unwilling to change his own identification.

This decision was consistent with Jefferson's own demand that scientific conclusions be reached with greatest caution after the most careful investigation of all facts. Perhaps he considered his postscript sufficient notice that the identification was tentative. Interestingly enough, an essentially accurate identification of Jefferson's animal as a species of ground sloth was made by Dr. Caspar Wistar, Philadelphia physician, later professor of anatomy at the University of Pennsylvania and a recognized authority in American vertebrate paleontology during this period, and was also published in the fourth volume of the *Transactions* of the American Philosophical Society. . . .

Jefferson's greatest popular fame in the field of paleontology came from his work with the so-called "mammoth," a huge prehistoric animal whose remains attracted international attention during Jefferson's day. Jefferson's interest in the mammoth appears to have begun when he was collecting material for his *Notes on Virginia*, that is, about 1780.

In his *Notes*, Jefferson points out that the mammoth must have been the largest of American quadrupeds. Remains "of unparalleled magnitude," he says, have been found rather recently on the Ohio River and in many parts of America further north. Indian tradition, Jefferson noted with satisfaction, has it that such a huge, carnivorous creature still roams the northern parts of America. Jefferson then argues that, contrary to certain European opinion, ascribing these remains to the hippopotamus or elephant, or to both, they cannot belong to either. . . .

The mammoth's great superiority over the elephant in size was insisted upon by Jefferson time and again during his ambassadorship to France. In arguing that the remains of this huge creature could not have belonged to the hippopotamus or to the elephant, he challenged the opinion of Daubenton, Buffon, and other leading naturalists of this period.

Further evidence of Jefferson's interest in this great creature and other prehistoric animals appeared during the early days of his presidency of the American Philosophical Society. At its first meeting over which he presided (May 19, 1797), "A Plan for Collecting Information Respecting the Antiquities of N. A." came under consideration. On April 6, 1798, a second report upon this plan was gone into in great detail and a committee consisting of Jefferson, George Turner, Caspar Wistar, Dr. Adam Seybert, C.W. Peale, James Wilkinson, and John Williams was appointed. By the end of the year this committee reported that a circular letter had been "extensively distributed" throughout the country soliciting help in accomplishing, among other things, the following: the procuring of "one or more entire skeletons of the Mammoth, so called, and of such other unknown animals as either have been, or hereafter may be discovered in America."

Fossils were among the treasures brought back by the Lewis and Clark Expedition of 1804–1806. At about this time Jefferson and Dr. Caspar Wistar also tried, but without success, to procure for the American Philosophical Society a large collection of bones dug out by a Dr. William Goforth at Big Bone Lick, famous source of prehistoric animal remains located in what is now the State of Kentucky. Jefferson expected Goforth's findings to include the skull of a mammoth – so desperately needed by the American Philosophical Society to complete its collection of the bones of this huge vertebrate. He was also hopeful (as he wrote to Dr. Samuel Brown, in July, 1806) that there would be bones of the Megalonyx. But Goforth sold his collection to an adventurer who took them to England. Therefore President Jefferson had to undertake a private paleontological venture. He arranged with the owner of the Big Bone Lick for Clark to go there in 1807 and do some excavating at Jefferson's expense. Clark "employed ten laborers several weeks" and unearthed some three hundred bones, which he shipped down the Mississippi to New Orleans, from whence they were forwarded to Jefferson in Washington.

On receiving Clark's letter with the news of the collection he had made, Jefferson immediately invited Dr. Caspar Wistar to Washington to see the bones as soon as they arrived. Jefferson confides to him his plans for their disposal. There is to be "a tusk and femur which General Clarke procured particularly at my request, for a special kind of cabinet I have at Monticello." But his primary purpose has been "to procure for the society as complete a supplement to what is already possessed as that Lick can furnish at this day." Hence Wistar is to select what bones the American Philosophical Society may desire, and the rest are to go to the National Institute of France.

The odd cargo arrived at the Presidential mansion in Washington early in 1803. "The bones are spread in a large room [the unfinished East Room]," Jefferson immediately informs Wistar, "where you can work at your leisure, undisturbed by any mortal, from morning till night, taking your breakfast and dinner with us. It is a precious collection, consisting of upwards of three hundred bones, few of them of the large kinds which are already possessed. There are four pieces of the head, one very clear, and distinctly presenting the whole face of the animal. The height of his forehead is most remarkable. In this figure, the indenture of the eye gives a prominence of six inches to the forehead. There are four jawbones tolerably entire, with several teeth in them, and some fragments; three tusks like elephants; one ditto totally different, the largest probably ever seen, being now from nine to ten feet long, though broken off at both ends; some ribs; an abundance of teeth studded, and also those of the striated or ribbed kind; a fore-leg complete; and then about two hundred small bones, chiefly of the foot. This is probably the most valuable part of the collection, for General Clarke, aware that we had specimens of the larger bones, has gathered up everything of the small kind. There is one horn of a colossal animal. ... Having sent my books to Monticello, I have nothing here to assist you but the *Encyclopédie Methodique*." All this during the desperate days of the Embargo!

4-5 Three Short, Happy Months – William A. Stanley

James McNeill Whistler's official biographer, Joseph Pennell, contended that the world has known but two supreme etchers, Rembrandt (1606–1669) and Whistler (1834–1903), and it is Whistler who is considered by some the greater. Few are aware that James Whistler learned the art of engraving and etching while employed by the United States Coast and Geodetic Survey. His first plate, made in 1854 of the coastline of Boston Bay, was used to demonstrate his aptitude for the art, and in the same year he completed United States Coast and Geodetic Survey Plate 414A of Anacapa Island, one of the Channel Islands of California. William A. Stanley of the Geodetic Survey in a revealing article, "Three Short, Happy Months" (1968), tells how Whistler became a member of the Survey, the circumstances under which he left the organization, and the influence these experiences had on the career of one of America's greatest artists.

How would the Coast and Geodetic Survey deal with a draftsman who was habitually late, frequently absent, given to graffiti, and inclined to doodle on his official charts?

You guessed it. But today, after more than a century, such a man remains one of that august agency's most highly regarded alumni.

He was James McNeill Whistler, famed for his magnificent portrait of his mother and for many other works of art. But to his superiors in the survey, where the first evidence of his artistic potential was expressed in its charts, he was trouble with a capital T.

A native of Lowell, Massachusetts, born July 10, 1834, he belonged to a family of soldiers, Scotch–Irish by descent. At the age of seventeen, after spending some time in St. Petersburg, where his father was carrying on an engineering assignment for the Czar, he entered West Point. He spent three years there, departing in his senior year due mainly to his disinterest in obeying rules, chief of which had been his lack of promptness.

His father's plan was to apprentice him to a friend who had connections with a locomotive works in Baltimore, Maryland, and where his stepbrother George was then employed. Whistler soon saw enough of the locomotive works, however, to know that he did not want to be an apprentice, and it was not long before he left Baltimore and headed for Washington, D.C.

In November 1854, after he tried to enter the Navy, his ability as a draftsman induced another friend of his father's to recommend him for a position in the engraving department of the United States Coast Survey. He was appointed on November 7, 1854, and at the age of 20 began a short but colorful employment with the Federal Government.

It has been reported that Whistler remarked, "I shall always remember the courtesy shown me by that fine Southern gentleman Jefferson Davis, through whom I got my appointment with the Coast Survey." Whistler continued, "It was after my little difference with the Professor of Chemistry at West Point that it was suggested – all in the most courteous and correct West Point manner – that perhaps I had better leave the Academy." With an introduction by Jefferson Davis he reported to Captain Benham, his new Coast Survey boss, who assigned him to the position of draftsman at the salary of $1.50 a day.

In the Coast Survey, then located on the northwest corner of New Jersey and C Street, SE., Whistler, the draftsman, was considered a playboy of sorts as he performed the duties of an assistant in the cartographic sections of the Survey; even in those permissive days when offices closed at 3:00 p.m. and it was quite the custom to send out the messenger for a bucket of ale. Whistler would engage his time by drawing caricatures and sketches in unconventional places somewhere in the building. It was said that he would bring an extra hat to the office and when a superior came looking for him he would find his hat on the peg, assuming then that James was in the building. However, it is said he was wearing the other hat on his way to a nearby tavern.

During his stay he was regarded as a person who could not adjust himself to any form of regular routine. He would say "It was not that I was late; the office opened too early." He worked only intermittently on his assignments and took great delight in occupying this time with odd sketches on any available fabric that presented a suitable surface – an envelope, or a copper-plate upon which he might have been assigned to engrave a chart.

Whistler sometimes drew uncomplimentary images of the Bureau officials on the bare white walls leading to the Superintendent's office. He would not overcome the temptation to stop frequently on his way up or down the stairs to correct, change, or add to his caricatures.

This was a time of uncertainty for him, when visions of his real career were beginning to form. In the brief period of less than four months Coast Survey employment, from November 1854 to February 1855, one fact was certain; he had no ambition to become a permanent government employee.

John Ross Key, fellow draftsman in the Coast Survey, and the grandson of Francis Scott Key, recalled in later years that he and Whistler roomed in a house of Thirteenth Street, near Pennsylvania Avenue and that Whistler usually dined in a restaurant closer by on Pennsylvania Avenue. He also lived for a while in a house at the north-east corner of Twelfth and E Streets, NW., a two-story brick building which is no longer standing. His rent was ten dollars a month at the rooming house, a considerable sum in the light of his small government pay.

He produced two works which have been referred to by historians as *Coast Survey No. 1* and *Coast Survey No. 2, Anacapa Island* (engraved by J. Whistler, J. Young, and C.A. Knight). The latter, which includes an etching by Whistler of the headland of the eastern extremity of Anacapa Island, California, has been reproduced and issued as *Coast and Geodetic Survey plate number 414A*, but there is no record of the former.

These two plates are not, however, Whistler's first efforts as a copperplate engraver in the Coast Survey. Upon his entrance to duty he received technical instructions in the art of etching and copper engraving, which are reflected in a series of practice sketches. First place should be accorded to those which are the initial efforts on copper of his fancy and invention, and the first genuine Whistler etchings. The sketches are in the shapes of little heads that intrude on the blank spaces of the copper above and around two neatly engraved views of portions of the coast of Boston Bay. At intervals, while doing the topographical view, he paused to etch on the upper part of the copperplate the vignette of a soldier's head, a suggestion of a portrait of himself as a Spanish hidalgo and an etching to a motherly figure which has attracted considerable interest in recent years. The original copper engraving is regarded as priceless. It was donated to the Freer Gallery of Art, where it is now displayed. . . .

Legend has it that on occasion he would also give play to his genius by inserting drawings of dignified and scientific characters that might not have been in the original plan of a seacoast; however, no one could prove it. It has also been reported, but this cannot be verified either, that late in 1854 he engraved a sketch of a portion of the Atlantic coast. This drawing in copper is said not only to include some graceful sea serpents and beautiful mermaids but also several large and smiling whales. His Coast Survey supervisor reportedly told him that if he ever again desecrated on the Survey's charts with animal life, he would be discharged. . . .

Due in large part to tardiness, James McNeill Whistler in mid-February 1855 terminated his employment with the United States Coast Survey and went off to Paris and London to achieve undying fame as one of the world's great artists. He is now known the world over, and by some authorities has been acclaimed the greatest of American artists. During his early training in the Coast and Geodetic Survey, he came in contact with some of the most talented engravers in the country.

4-6 Mountain-Worship – W.G. Collingwood

John Ruskin (1819–1900) – writer, critic, social reformer, and artist – was a giant of Victorian England. He had such profound influence on the public taste of the people of Britain that he came to be known as "The Great Victorian." Yet his early interests were geological and Ruskin was fascinated by minerals and crystallography, the architecture of the Alps, landscape, and the origin of agate, gneiss, breccias, and puddingstones. He studied with Buckland at Oxford, as a student conversed with Charles Darwin, and later in his career published a series of seven articles in the Geological Magazine. In 1869 he was elected Slade Professor of Art at the University of Oxford. His lectures reached an enormous public. He was later made an Honorary Fellow of Corpus Christi and still later endowed museums at Oxford and at Sheffield.

The early training, travels, and discoveries of this "almost-geologist" are described by William Gershom Collingwood (1854–1932). Collingwood was one of those rare individuals whose extraordinary intellectual talents were matched by an equally extraordinary personal goodness and commitment. His range of achievements was dazzling: he was a biographer, novelist, artist, poet, antiquarian, geologist, art historian, outdoorsman, translator and scholar of Icelandic history and literature. He published some thirty books, but it is chiefly as a Ruskin scholar that he is remembered today. Collingwood came under Ruskin's influence when he went up as a student to Oxford, where Ruskin – who was Slade Professor of Art – was then at the height of his powers. From then on, Collingwood acted as Ruskin's assistant and secretary, sacrificing his own career to support his mentor and caring for him during his mental breakdown and his last ailing and troubled years. Only after Ruskin's death, did Collingwood feel free to pursue his own career, accepting an appointment as a professor of fine art at University College, Reading. Our extract is taken from Collingwood's book The Life of John Ruskin *(1902), a revision of the first edition published in 1893.*

More interesting to him than school was the British Museum collection of minerals, where he worked occasionally with his Jamieson's Dictionary. By this time he had a fair student's collection of his own, and he increased it by picking up specimens at Matlock, or Clifton, or in the Alps, wherever he went, for he was not short of pocket-money. He took the greatest pains over his catalogues, and wrote elaborate accounts of the various minerals in a shorthand he invented out of Greek letters and crystal forms.

Grafted on this mineralogy, and stimulated by the Swiss tour, was a new interest in physical geology, which his father so far approved as to give him Saussure's 'Voyages dans les Alpes' for his birthday in 1834. In this book he found the complement of Turner's vignettes, something like a key to the 'reason why' of all the wonderful forms and marvellous mountain-architecture of the Alps. . . .

He soon wrote a short essay on the subject, and had the pleasure of seeing it in print, in Loudon's *Magazine of Natural History* for March, 1834, along with another bit of his writing, asking for information on the cause of the colour of the Rhine-water. . . .

In his second term he had the honour of being elected to the Christ Church Club, a very small and very exclusive society of the best men in the college: 'Simeon, Acland, and Mr. Denison proposed him; Lord Carew and Broadhurst supported.' And he had the opportunity of meeting men of mark, as the following letter recounts. He writes on April 22, 1937:

'My dearest Father,

'When I returned from hall yesterday – where a servitor read, or pretended to read, and Decanus growled at him, "Speak out!" – I found a note on my table from Dr. Buckland, requesting the pleasure of my company to dinner, at six, to meet two celebrated geologists, Lord Cole and Sir Philip Egerton. I immediately sent a note of thanks and acceptance, dressed, and was there a minute after the last stroke of Tom. Alone for five minutes in Dr. B.'s drawing-room, who soon afterwards came in with Lord Cole, introduced me, and said that as we were both geologists he did not hesitate to leave us together while he did what he certainly very much required – brushed up a little. Lord Cole and I were talking about some fossils newly arrived from India. He remarked in the course of conversation that his friend Dr. B.'s room was cleaner and in better order than he remembered ever to have seen it. There was not a chair fit to sit upon, all covered with dust, broken alabaster candlesticks, withered flower-leaves, frogs cut out of serpentine, broken models of fallen temples, torn papers, old manuscripts, stuffed reptiles, deal boxes, brown paper, wool, tow and cotton, and a considerable variety of other articles. In came Mrs. Buckland, then Sir Philip Egerton and his brother, whom I had seen at Dr. B.'s lecture, though he is not an undergraduate. I was talking to him till dinner-time. While we were sitting over our wine after dinner, in came Dr. Daubeny, one of the most celebrated geologists of the day – a curious little animal, looking through its spectacles with an air very *distinguée* – and Mr. Darwin, whom I had heard read a paper at the Geological Society. He and I got together, and talked all the evening.'

. . . Of less interest to the general reader, though too important a part of Mr. Ruskin's life and work to be passed over without mention, are his studies in Mineralogy. We have heard of his early interest in spars and ores; of his juvenile dictionary in forgotten hieroglyphics; and of his studies in the field and at the British Museum. He had made a splendid collection, and knew the various museums of Europe as familiarly as he knew the picture-galleries. In the 'Ethics of the Dust' he had chosen Crystallography as the subject in which to exemplify his method of education; and in 1867, after finishing the letters to Thomas Dixon, he took refuge, as before, among the stones, from the stress of more agitating problems.

4-7 Stanford University, 1891–1895 – Herbert C. Hoover

Herbert Clark Hoover (1874–1964) was the thirty-first president of the United States, serving from 1929 to 1933. Hoover was born in Iowa, the son of a blacksmith, and worked his way through Stanford University. Our extract, taken from The Memoirs of Herbert Hoover: years of adventure, 1874–1920 (1951), describes his undergraduate years at Stanford, where he worked as clerk-typist for the chairman of the geology department, and was taught by such leading geologists as John Branner, Waldemar Lindgren, and Joseph LeConte. By the time he was forty, Hoover had made a fortune as a mining engineer and subsequently became chairman of the Commission for Relief in Belgium during World War I and secretary of commerce under President Warren Harding. Hoover took up the presidency only seven months before the stock market crash of 1929. Retiring to California after his election loss to Franklin D. Roosevelt, Hoover devoted much of his time to writing, producing with his wife, Lou Henry Hoover – also a student of geology – a translation of George Agricola's De Re Metallica (1556). Late in his life, Hoover was recalled to Washington by President Harry Truman to head a bipartisan commission to study the organization of the executive branch of government.

The University opened formally on October 1, 1891. It was a great occasion. Senator and Mrs. Stanford were present. The speeches of Senator Stanford and Dr. David Starr Jordan, the first President, make dry reading today but they were mightily impressive to a youngster. Dr. John Branner, who was to preside over the Department of Geology and Mining, had not yet arrived, so with Professor Swain's guidance I undertook the preparatory subjects that would lead into that department later on. Upon Dr. Branner's arrival I came under the spell of a great scientist and a great teacher, whose friendship lasted over his lifetime.

My first need was to provide for myself a way of living. I had the $210 less Miss Fletcher's services, together with a backlog of some $600 which had grown from the treasured insurance of my father. The original sum had been safeguarded and modestly increased over the years by the devoted hands of Laurie Tatum. Professor Swain interested himself and secured me a job in the University office at $5 a week – which was enough of a supplement for the time. Soon after, Dr. Branner tendered me a job in his department because I could operate a typewriter. This increased my income to $30 a month. While dwelling on earnings I may mention that with two partners I had established a laundry agency and a newspaper route upon the campus, both of which, being sub-let, brought in constant but very small income.

The first summer vacation Dr. Branner obtained for me a job as an assistant on the Geological Survey of Arkansas where he had been State Geologist. The $60 a month and expenses for three months seemed like a fortune. During my sophomore and junior summer vacations I worked upon the United States Geological Survey in California and Nevada, where I saved all my salary. These various activities and the back-log carried me over the four years at the University. I came out with $40 in my pocket and no debts.

The work in Arkansas consisted of mapping the geologic outcrops on the north side of the Ozarks. I did my job on foot mostly alone, stopping at nights at the nearest cabin. The mountain people were hospitable but suspicious of all "government agents." Some were moonshiners and to them even a gawky boy might be a spy. There were no terms that could adequately explain my presence among them. To talk about the rocks only excited more suspicion. To say that I was making a survey was worse, for they wanted no check-up on their land-holdings. To say I was tracing the zinc- or coal-bearing formations made them fearful of some wicked corporate invasion. I finally gave up trying to explain. However, I never failed to find someone who would take the stranger in at nightfall and often would refuse any payment in the morning.

The living conditions of many of these people were just as horrible as they are today. Generations of sowbelly, sorghum molasses and cornmeal, of sleeping and living half a dozen in a room, had fatally lowered their vitality and ambitions. The remedy – then as today – is to regenerate racial vitality in the next generation through education and decent feeding of the children.

My work on the United States Geological Survey in the glorious High Sierra, the deserts of Nevada, and among the mining camps where vitality and character ran strong, was a far happier job. Dr. Waldemar Lindgren headed these survey parties and I was a cub assistant. When in the high mountains we camped out with teamsters, horses and pack mules, and, of equal importance, a good camp cook.

Most of the work was done on horseback. During those two summers I did my full lifetime mileage with that mode of transportation. In these long mountain rides over trails and through the brush, I arrived finally at the conclusion that a horse was one of the original mistakes of creation. I felt he was too high off the ground for convenience and safety on mountain trails. He would have been better if he had been given a dozen legs so that he had the smooth and sure pace of a centipede. Furthermore he should have had scales as protection against flies, and a larger water-tank like a camel. All these gadgets were known to creation prior to the geologic period when the horse was evolved. Why were they not used?

We had a foreign geologist visiting our party who had never seen a rattlesnake. On one hot day a rattlesnake alarm went off near the trail. The horse notified me by shying violently. I decided to take the snake's corpse to our visitor. I dismounted and carefully hit the snake on the head with a stick, then wrapped him in a bandana handkerchief and hung him on the pommel. Some minutes later while the horse and I were toiling along to camp, half asleep in the sun, the rattlesnake woke up and sounded another alarm. That was too much for the horse. After I got up and out of the brush I had to walk five miles to camp. It added to my prejudices against horses in general.

There was some uncertainty one summer as to whether I could get a Geological Survey job. Therefore when vacation came other students and I canvassed San Francisco for work at putting up or painting advertising signs along the roads. Our very modest rates secured a few hundred dollars of contracts, with which we bought a team and camp-outfit. We made for the Yosemite Valley, putting up eyesores, advertising coffee, tea and newspapers along the roads. We pitched our camp in the Valley intending to spend a few days looking the place over. Professor Joseph Le Conte was camped nearby, and I listened spellbound to his campfire talk on the geology of the Valley. A few days later I received a telegram advising me I could join the Survey party. There was not enough money left in the pockets of my sign-painting partners to pay my stage-fare to the railroad. I walked 80 miles in three days and arrived on time.

As the youngest member of the Geological party, I was made disbursing officer. It required a little time for me to realize this was not a distinction but a liability. I had to buy supplies and keep the accounts according to an elaborate book of regulations which provided wondrous safeguards for the public treasury. One morning high in the Sierras we discovered one of the pack-mules dead. I at once read the regulations covering such catastrophes and found that the disbursing officer and two witnesses must make a full statement of the circumstances and swear to it before a notary public. Otherwise the disbursing officer was personally responsible for the value of the animal. I was thus importantly concerned to the extent of $60. The teamsters and I held an autopsy on the mule. We discovered his neck was broken and that the caulk of one loose hind-shoe was caught in the neck-rope with which he was tied to a tree. We concluded that he had been scratching his head with his hind foot, had wedged his halter-rope in the caulk, had jerked back and broken his neck. When we reached civilization we made out an elaborate affidavit to that effect. About two months afterward I was duly advised from Washington that $60 had been deducted from my pay, since this story was too highly improbable. Apparently mules did not, according to the book, scratch their heads with their hind feet. Dr. Lindgren relieved my $60 of misery by taking over the liability, saying he would collect it from some d – bureaucrat when he got back to Washington in the winter.

For years I watched every mule I met for confirmation of my story. I can affirm that they do it. I even bought a statuette of a mule doing it. Some twelve years later I was privileged to engage Dr. Lindgren for an important job in economic geology in Australia. I met him at the steamer in Melbourne. His first words were, "Do you know that d – bureaucrat never would pay me that $60? And do you know I have since seen a hundred mules scratch their heads with their hind feet?" He would not take the $60 from me. . . .

Stanford is a coeducational institution, but I had little time to devote to coeds. However, a major event in my life came in my senior year. Miss Lou Henry entered Stanford and the geology laboratories, determined to pursue and teach that subject as a livelihood. As I was Dr. Branner's handy boy in the department, I felt it my duty to aid the young lady in her studies both in the laboratory and in the field. And this call to duty was stimulated by her whimsical mind, her blue eyes, and a broad grinnish smile that came from an Irish ancestor. I was not long in learning that she also was born in Iowa, the same year as myself, and that she was the daughter of a hunting-fishing country banker at Monterey who had no sons and therefore had raised his daughter in the out-of-door life of a boy. After I left college she still had three years to complete her college work. I saw her once or twice during this period. We carried on a correspondence.

All of these extra-curricular matters so crowded my life that I neglected to discharge those conditions on entrance "Credits" under which I had entered as a freshman. Had it not been for the active intervention of my friends, Dr. Branner and Professor J. Perrin Smith, who insisted among other things that I could write English, those implacable persons in the University office would have prevented my getting a diploma with my own class. Nevertheless it duly arrived. It has been my lot in life to be the recipient of honorary diplomas (often in exchange for Commencement addresses) but none ever had the sanctity or, in my opinion, the importance of this one.

Part 2

Interpreting the Earth

To experience the Earth is one thing: all of us – like it or not – experience it. But to interpret it is quite another thing and few take that second step. So this section of the present book (Part 2) and the next are devoted to those who do interpret the Earth; not only meteorologists, ecologists, conservationists, and Earth scientists of all kinds, but also writers, painters, musicians, miners, farmers, military leaders, architects, and planners. In Part 2, we consider the scientific interpretation of the Earth: in Part 3, the language of the Earth.

How do we make the transition from experience to interpretation, from encounter to explanation? How, in other words, does Earth – real, accessible to everyone, solid under our feet – become Earth science – intangible, inaccessible to most, cerebral in our minds? There is widespread supposition that scientific knowledge differs in some way from other areas of human experience, especially in its methods and conclusions. In two important respects scientific knowledge does differ in its emphasis: deliberate abstraction and pervasive quantification are essential features of science, though they are certainly not unknown in other human activities. Science is a statement of human experience, differing from others chiefly in its exclusive concern with the material world and in its degree of generality.

Science does indeed concern itself with what most of us assume to be a real world and an actual universe, but that does not provide it with some official imprimatur, certifying its authenticity by the authority of the cosmos at large. Science exists because there are people. It presents no final statement of truth, binding us and the rest of the universe to observe its dictates and its ordained boundaries. It is tentative, provisional, refinable, subject not only to minor correction and clarification, but also to the kind of partial rejection, wholesale restatement and sweeping reinterpretation which succeed such revolutions as those produced by the discovery of the heliocentric nature of the solar system, organic evolution, and sea floor spreading. Yet even these revolutions, which change the whole context and direction of scientific inquiry, are the results of the deeply personal insights and individual experiences of a Copernicus, a Darwin, a Hess, or a Vine. Knowledge grows neither by public polls nor by majority votes at scientific conferences, but by individual perception and by personal insight. It is public in its interest, verifiability, and utility; but it is private in its origin and source. It is not committees that lead us into new truth; it is individuals.

In the sections that follow we deal with four outstanding aspects of interpreting the Earth: the philosophy of geology – how reliable can a reconstruction of prehistoric events be when there was no one around to observe them?; the fossil record – our autobiography; geotectonics – how the Earth works; and, not least, controversies, for there is always more than one point of view.

All this represents the spectrum of our efforts to interpret the Earth, and the effectiveness of this interpretation, as well as the validity of conclusions based upon it, are of major social importance, having a potential impact on everything from energy supplies and mineral prospecting to protection from geologic hazards and natural disasters.

5. Philosophy

Geology, it is sometimes claimed, is unique among the sciences in being chiefly concerned with historical development, rather than present conditions, and by its lack of a distinctive methodology, such as those which are characteristic of other sciences. Unsatisfactory as both statements are as generalizations, they contain enough truth to merit our concern.

All sciences contain a historical element; it is the relative historical emphasis in which they differ. Meaningful discourse in biology or astronomy, for example, would be impossible without some historical awareness. G.G. Simpson has drawn a useful distinction between the immanent processes and products which are universal components of the Earth – and indeed of the whole material universe – and the particular state of the Earth or some part of it at any given moment. Such a state, though always contingent upon and the product of the interactions of immanent processes and products, is itself unique.

The goal of geology is so to describe and understand the present processes, constitution and composition of the Earth as to establish relationships between them, and thus to interpret earlier historical products and sequences. The study of existing states, processes and components involves the application of all of the sciences, including especially physics, chemistry and biology, even though the particular abstractions favored by geologists may not always be those chosen by these other scientists. The uniqueness of geology lies in its combination with and historical extrapolation to remote periods. It is not the events themselves of these periods that are thus studied, but rather their results and records. In this, geology infers causes from results, rather than predicts results from causes, as do the less-historical sciences.

Geology shares with other sciences the common goal of establishing generalizations (laws) of wide applicability which describe recurrent and invariable relationships between objects or states. But such laws, as Simpson has pointed out, are so massively abstractive in their basis, that it would be unwarranted to expect to derive them from the particular interaction of immanent variables which produce the unique sequences of events we call history. The very uniqueness of these events precludes their use as a basis for the establishment of the recurrent relationships involved in "laws."

The extracts that follow can do little more than introduce some aspects of these philosophical but fundamental aspects of geologic science. The interested reader can find no better introduction to the subject than Simpson's paper, and especially his discussion of predictive testing and predictability.

5-1 Concerning the System of the Earth, its Duration and Stability – James Hutton

James Hutton (1726–1797) was born in Edinburgh and qualified as a doctor after studying medicine in Edinburgh and Paris. He never practiced his profession, however, but instead devoted most of his life to practical agriculture and to study and writing concerning the history of the Earth. He is regarded as the father of modern geology. His System of the Earth, *from which the present extract is taken, was presented as an abstract to the Royal Society of Edinburgh in the spring of 1785. In it he argued cogently for two processes which account for the present composition and structure of the Earth: first, that the rocks of the Earth's surface are formed from loose materials which have been subsequently consolidated; and second, that sediments formed on the sea floor can subsequently be uplifted and transformed into land, at which time igneous rocks and veins may be intruded into the uplifted area. He argued for the widespread demonstration of terrestrial weathering and of the action of circulating water and heat in leading to liquefaction, and he also explored the questions of continental uplift, being convinced that volcanoes provided visible proof of the manifestation of the power required to produce such results. It was in this context that he was also led to argue for the need for an indefinite period of time and to make his often quoted statement, "with respect to human observation, this world has neither a beginning nor an end."*

The publication of Hutton's work, supplemented in 1788 by his Theory of the Earth, *and, after his death, by the publication of his friend John Playfair's book* Illustrations of the Huttonian Theory of the Earth *(1802), was of major importance in the establishment of geology as a science, not only because they provided the deathblow for the earlier claims that all rocks were the result of a single process (the Neptunist–Plutonist controversy), but also because Hutton, like Lyell after him, advocated the efficacy of existing processes and the immensity of geologic time.*

The purpose of this Dissertation is to form some estimate with regard to the time the globe of this Earth has existed, as a world maintaining plants and animals; to reason with regard to the changes which the earth has undergone; and to see how far an end or termination to this system of things may be perceived, from the consideration of that which has already come to pass.

As it is not in human record, but in natural history, that we are to look for the means of ascertaining what has already been, it is here proposed to examine the appearances of the earth, in order to be informed of operations which have been transacted in time past. It is thus that, from principles of natural philosophy, we may arrive at some knowledge of order and system in the economy of this globe, and may form a rational opinion with regard to the course of nature, or to events which are in time to happen.

The solid parts of the present land appear, in general, to have been composed of the productions of the sea, and of other materials similar to those now found upon the shores. Hence we find reason to conclude,

1st, That the land on which we rest is not simple and original, but that it is a composition, and had been formed by the operation of second causes.

2dly, That, before the present land was made, there had subsisted a world composed of sea and land, in which were tides and currents, with such operations at the bottom of the sea as now take place. And,

Lastly, That, while the present land was forming at the bottom of the ocean, the former land maintained plants and animals; at least, the sea was then inhabited by animals, in a similar manner as it is at present.

Hence we are led to conclude, that the greater part of our land, if not the whole, had been produced by operations natural to this globe; but that, in order to make this land a permanent body, resisting the operations of the waters, two things had been required; *1st*, The * consolidation of masses formed by collections of loose or incoherent materials; *2dly*, The elevation of those consolidated masses from the bottom of the sea, the place where they were collected, to the stations in which they now remain above the level of the ocean.

Here are two different changes, which may serve mutually to throw some light upon each other; for, as the same subject has been made to undergo both these changes, and as it is from the examination of this subject that we are to learn the nature of those events, the knowledge of the one may lead us to some understanding of the other.

Thus the subject is considered as naturally divided into two branches, to be separately examined: *First*, by what natural operation strata of loose materials had been formed into solid masses; *secondly*, By what power of nature the consolidated strata at the bottom of the sea had been transformed into land.

With regard to the *first* of these, the consolidation of strata, there are two ways in which this operation may be conceived to have been performed; first, by means of the solution of bodies in water, and the after concretion of these dissolved substances, when separated from their solvent; *secondly*, the fusion of bodies by means of heat, and the subsequent congelation of those consolidating substances.

With regard to the operation of water, it is *first* considered, how far the power of this solvent, acting in the natural situation of those strata, might be sufficient to produce the effect; and here it is found, that water alone, without any other agent, cannot be supposed capable of inducing solidity among the materials of strata in that situation. It is, *2dly*, considered, how far, supposing water capable of consolidating the strata in that situation, it might be concluded, from examining natural appearances, that this had been actually the case? Here again, having proceeded upon this principle, that water could only consolidate strata with such substances as it has the power to dissolve, and having found strata consolidated with every species of substance, it is concluded, that strata in general have not been consolidated by means of aqueous solution.

With regard to the other probable means, heat and fusion, these are found to be perfectly competent for producing the end in view, as every kind of substance may by heat be rendered soft, or brought into fusion, and as strata are actually found consolidated with every different species of substance.

A more particular discussion is then entered into: Here, consolidating substances are considered as being classes under two different heads, viz. Siliceous and sulphureous bodies, with a view to prove, that it could not be by means of aqueous solution that strata had been consolidated with those particular substances, but that their consolidation had been accomplished by means of heat and fusion.

Sal Gem, as a substance soluble in water, is next considered, in order to show that this body had been last in a melted state; and this example is confirmed by one of fossil alkali. The case of particular septaria of ironstone, as well as certain crystallized cavities in mineral bodies, are then given as examples of a similar fact; and as

Notes

* There are two senses in which the term *solidity* is used; one of these is in opposition to *fluidity*, the other to *vacuity*. When the change from a fluid state to that of solidity, in the first sense, is to be expressed, we shall employ the term *concretion*; consequently, the consolidation of a mass is only to be understood as in opposition to its vacuity, or porousness.

containing, in themselves, a demonstration, that all the various mineral substances had been concreted and crystallized immediately from a state of fusion.

Having thus proved the actual fusion of the substances with which strata had been consolidated, in having such fluid bodies introduced among their interstices, the case of strata, consolidated by means of the simple fusion of their proper materials, is next considered; and examples are taken from the most general strata of the globe, viz. siliceous and calcareous. Here also demonstration is given, that this consolidating operation had been performed by means of fusion.

Having come to this general conclusion, that heat and fusion, not aqueous solution, had preceded the consolidation of the loose materials collected at the bottom of the sea, those consolidated strata, in general, are next examined, in order to discover other appearances, by which the doctrine may be either confirmed or refuted. Here the changes of strata, from their natural state of continuity, by veins and fissures, are considered; and the clearest evidence is hence deduced, that the strata have been consolidated by means of fusion, and not by aqueous solution; for not only are strata in general found intersected with veins and cutters, an appearance inconsistent with their having been consolidated simply by previous solution; but, in proportion as strata are more or less consolidated, they are found with the proper corresponding appearances of veins and fissures.

With regard to the second branch, in considering by what power the consolidated strata had been transformed into land, or raised above the level of the sea, it is supposed that the same power of extreme heat, by which every different mineral substance had been brought into a melted state, might be capable of producing an expansive force, sufficient for elevating the land, from the bottom of the ocean, to the place it now occupies above the surface of the sea. Here we are again referred to nature, in examining how far the strata, formed by successive sediments or accumulations deposited at the bottom of the sea, are to be found in that regular state, which would necessarily take place in their original production; or if, on the other hand, they are actually changed in their natural situation, broken, twisted, and confounded, as might be expected, from the operation of subterranean heat, and violent expansion. But, as strata are actually found in every degree of fracture, flexure, and contortion, consistent with this supposition, and with no other, we are led to conclude, that our land had been raised above the surface of the sea, in order to become a habitable world; as well as that it had been consolidated by means of the same power of subterranean heat, in order to remain above the level of the sea, and to resist the violent efforts of the ocean.

This theory is next confirmed by the examination of mineral veins, those great fissures of the earth, which contain matter perfectly foreign to the strata they traverse; matter evidently derived from the mineral region, that is, from the place where the active power of fire, and the expansive force of heat, reside.

Such being considered as the operations of the mineral region, we are hence directed to look for the manifestation of this power and force, in the appearances of nature. It is here we find eruptions of ignited matter from the scattered volcanoes of the globe; and these we conclude to be the effects of such a power precisely as that about which we now inquire. Volcanoes are thus considered as the proper discharges of a superfluous or redundant power; not as things accidental in the course of nature, but as useful for the safety of mankind, and as forming a natural ingredient in the constitution of the globe.

The doctrine is then confirmed, by examining this earth, and by finding every-where, beside the many marks of ancient volcanoes, abundance of subterraneous or unerupted lava, in the basaltic rocks, the Swedish trap, the toadstone, ragstone, and whinstone of Britain and Ireland, of which particular examples are cited; and a description given of the three different shapes in which that unerupted lava is found.

The peculiar nature of this subterraneous lava is then examined; and a clear distinction is formed between this basaltic rock and the common volcanic lavas.

Lastly, The extension of this theory, respecting mineral strata, to all parts of the globe, is made, by finding a perfect similarity in the solid land through all the earth, although, in particular places, it is attended with peculiar productions, with which the present inquiry is not concerned.

A theory is thus formed, with regard to a mineral system. In this system, hard and solid bodies are to be formed from soft bodies, from loose or incoherent materials, collected together at the bottom of the sea; and the bottom of the ocean is to be made to change its place with relation to the centre of the earth, to be formed into land above the level of the sea, and to become a country fertile and inhabited.

That there is nothing visionary in this theory, appears from its having been rationally deduced from natural events, from things which have already happened; things which have left, in the particular constitutions of bodies, proper traces of the manner of their production; and things which may be examined with all the accuracy, or reasoned upon with all the light, that science can afford. As it is only by employing science in this manner that philosophy enlightens man with the knowledge of that wisdom or design which is to be found in nature, the system now proposed, from unquestionable principles, will claim the attention of scientific men, and may be admitted in our speculations with regard to the works of nature, notwithstanding many steps in the progress may remain unknown.

By thus proceeding upon investigated principles, we are led to conclude, that, if this part of the earth which we now inhabit had been produced, in the course of time, from the materials of a former earth, we should, in the examination of our land, find data from which to reason, with regard to the nature of that world, which had existed during the period of time in which the present earth was forming; and thus we might be brought to understand the nature of that earth which had preceded this; how far it had been similar to the present, in producing plants and nourishing animals. But this interesting point is perfectly ascertained, by finding abundance of every manner of vegetable production, as well as the several species of marine bodies, in the strata of our earth.

Having thus ascertained a regular system, in which the present land of the globe had been first formed at the bottom of the ocean, and then raised above the surface of the sea, a question naturally occurs with regard to time; what had been the space of time necessary for accomplishing this great work?

In order to form a judgment concerning this subject, our attention is directed to another progress in the system of the globe, namely, the destruction of the land which had preceded that on which we dwell. Now, for this purpose, we have the actual decay of the present land, a thing constantly transacting in our view, by which to form an estimate. This decay is the gradual ablution of our soil, by the floods of rain; and the attrition of the shores, by the agitation of the waves.

If we could measure the progress of the present land, towards it dissolution by attrition, and its submersion in the ocean, we might discover the actual duration of a former earth; an earth which had supported plants and animals, and had supplied the ocean with those materials which the construction of the present earth required; consequently, we should have the measure of a corresponding space of time, viz. that which had been required in the production of the present land. If, on the contrary, no period can be fixed for the duration or destruction of the present earth, from our observations of those natural operations, which, though unmeasurable, admit of no dubiety, we shall be warranted in drawing the following conclusions; *1st*, That it had required an indefinite space of time to have produced the land which now appears; *2dly*, That an equal space had been employed upon the construction of that former land from whence the materials of the present came; *lastly,* That there is presently laying at the bottom of the ocean the foundation of future land, which is to appear after an indefinite space of time.

But, as there is not in human observation proper means for measuring the waste of land upon the globe, it is hence inferred, that we cannot estimate the duration of what we see at present, nor calculate the period at which it had begun; so that, with respect to human observation, this world has neither a beginning nor an end.

An endeavour is then made to support the theory by an argument of a moral nature, drawn from the consideration of a final cause. Here a comparison is formed between the present theory, and those by which there is necessarily implied either evil or disorder in natural things; and an argument is formed, upon the supposed wisdom of nature, for the justness of a theory in which perfect order is to be perceived. For,

According to the theory, a soil, adapted to the growth of plants, is necessarily prepared, and carefully preserved; and, in the necessary waste of land which is inhabited, the foundation is laid for future continents, in order to support the system of this living world.

Thus either in supposing Nature wise and good, an argument is formed in confirmation of the theory, or, in supposing the theory to be just, an argument may be established for wisdom and benevolence to be perceived in nature. In this manner, there is opened to our view a subject interesting to man who thinks; a subject on which to reason with relation to the system of nature; and one which may afford the human mind both information and entertainment.

5-2 The Method of Multiple Working Hypotheses – T.C. Chamberlin

Thomas Chrowder Chamberlin (1843–1928) attended Beloit College to study theology, but later decided to follow a career in geology, subsequently attending the University of Michigan. In the course of a long career, he served as principal of a high school in Wisconsin, professor at the State Normal School at Whitewater, Wisconsin, and professor at Beloit, as well as Director of the Wisconsin Geological Survey. He later became successively Chief of the Glacial Division of the United States Geological Survey – taking part in the Peary Relief Expedition – and President of the University of Wisconsin, an office which he occupied for five years. Resigning in 1892 to return to active scientific work, he was appointed the first chairman of the Department of Geology at the University of Chicago, and taught there for the next twenty-six years. He was one of the founding members of the Geological Society of America and served as president of that body in 1905, and president of the American Association for the Advancement of Science in 1908. Chamberlin made substantial contributions to glacial geology, the study of diastrophism, the origin of the Earth, and the philosophy of scientific investigation. It is from his writings in connection with this last category that the present extract is taken, "The Method of Multiple Working Hypotheses" published in Science *in 1890. This famous paper – later altered slightly and published under the same name in the* Journal of Geology *(1897), a journal he founded in 1893 – is a useful summary of the method used to such effect by Chamberlin and his contemporaries, and continues to be used over a century later by many scientists.*

With this method the dangers of parental affection for a favourite theory can be circumvented.

There are two fundamental classes of study. The one consists in attempting to follow by close imitation the processes of previous thinkers, or to acquire by memorizing the results of their investigations. It is merely secondary, imitative, or acquisitive study. The other class is primary or creative study. In it the effort is to think independently, or at least individually, in the endeavor to discover new truth, or to make new combinations of truth, or at least to develop an individualized aggregation of truth. The endeavor is to think for one's self, whether the thinking lies wholly in the fields of previous thought or not. It is not necessary to this habit of study that the subject-material should be new; but the process of thought and its results must be individual and independent, not the mere following of previous lines of thought ending in predetermined results. The demonstration of a problem in Euclid precisely as laid down is an illustration of the former; the demonstration of the same proposition by a method of one's own or in a manner distinctively individual is an illustration of the latter; both lying entirely within the realm of the known and the old.

Creative study, however, finds its largest application in those subjects in which, while much is known, more remains to be known. Such are the fields which we, as naturalists, cultivate; and we are gathered for the purpose of developing improved methods lying largely in the creative phase of study, though not wholly so.

Intellectual methods have taken three phases in the history of progress thus far. What may be the evolutions of the future it may not be prudent to forecast. Naturally the methods we now urge seem the highest attainable. These three methods may be designated, first, the method of the ruling theory; second, the method of the working hypothesis; and, third, the method of multiple working hypotheses. . . .

As in the earlier days, so still, it is the habit of some to hastily conjure up an explanation for every new phenomenon that presents itself. Interpretation rushes to the forefront as the chief obligation pressing upon the putative wise man. Laudable as the effort at explanation is in itself, it is to be condemned when it runs before a serious inquiry into the phenomenon itself. A dominant disposition to find out what is, should precede and crowd aside the question, commendable at a later stage, "How came this so?" First full facts, then interpretations.

The habit of precipitate explanation leads rapidly on to the development of tentative theories. The explanation offered for a given phenomenon is naturally, under the impulse of self-consistency, offered for like phenomena as they present themselves, and there is soon developed a general theory explanatory of a large class of phenomena similar to the original one. This general theory may not be supported by any further considerations than those which were involved in the first hasty inspection. For a time it is likely to be held in a tentative way with a measure of candor. With this tentative spirit and measurable candor, the mind satisfies its moral sense, and deceives itself with the thought that it is proceeding cautiously and impartially toward the goal of ultimate truth. It fails to recognize that no amount of provisional holding of a theory, so long as the view is limited and the investigation partial, justifies an ultimate conviction. . . .

It is in this tentative stage that the affections enter with their blinding influence. Love was long since represented as blind, and what is true in the personal realm is measurably true in the intellectual realm. Important as the intellectual affections are as stimuli and as rewards, they are nevertheless dangerous factors, which menace the integrity of the intellectual processes. The moment one has offered an original explanation for a phenomenon which seems satisfactory, that moment affection for his intellectual child springs into existence; and as the explanation grows into a definite theory, his parental affections cluster about his intellectual offspring, and it grows more and more dear to him, so that, while he holds it seemingly tentative, it is still lovingly tentative, and not impartially tentative. . . .

Briefly summed up, the evolution is this: a premature explanation passes into a tentative theory, then into an adopted theory, and then into a ruling theory.

When the last stage has been reached, unless the theory happens, perchance, to be the true one, all hope of the best results is gone. To be sure, truth may be brought forth by an investigator dominated by a false ruling idea. His very errors may indeed stimulate investigation on the part of others. But the condition is an unfortunate one. Dust and chaff are mingled with the grain in what should be a winnowing process.

As previously implied, the method of the ruling theory occupied a chief place during the infancy of investigation. It is an expression of the natural infantile tendencies of the mind, though in this case applied to its higher activities, for in the earlier stages of development the feelings are relatively greater than in later stages.

Unfortunately it did not wholly pass away with the infancy of investigation, but has lingered along in individual instances to the present day, and finds illustration in universally learned men and pseudo-scientists of our time.

The defects of the method are obvious, and its errors great. If I were to name the central psychological fault, I should say that it was the admission of intellectual affection to the place that should be dominated by impartial intellectual rectitude. . . .

The first great endeavor was repressive. The advocates of reform insisted that theorizing should be restrained, and efforts directed to the simple determination of facts. The effort was to make scientific study factitious instead of causal. Because theorizing in narrow lines had led to manifest evils, theorizing was to be condemned. . . .

The inefficiency of this simply repressive reformation becoming apparent, improvement was sought in the method of the working hypothesis. This is affirmed to be *the* scientific method of the day, but to this I take exception. The working hypothesis differs from the ruling theory in that it is used as a means of determining facts, and has for its chief function the suggestion of lines of inquiry; the inquiry being made not for the sake of the hypothesis, but for the sake of facts. . . .

It will be observed that the distinction is not a sharp one, and that a working hypothesis may with the utmost ease degenerate into a ruling theory. Affection may as easily cling about an hypothesis as about a theory, and the demonstration of the one may become a ruling passion as much as of the other.

Conscientiously followed, the method of the working hypothesis is a marked improvement upon the method of the ruling theory; but it has its defects – defects which are perhaps best expressed by the ease with which the hypothesis becomes a controlling idea. To guard against this, the method of multiple working hypotheses is urged. It differs from the former method in the multiple character of its genetic conceptions and of its tentative interpretations. It is directed against the radical defect of the two other methods; namely, the partiality of intellectual parentage. The effort is to bring up into view every rational explanation of new phenomena, and to develop every tenable hypothesis respecting their cause and history. The investigator thus becomes the parent of a family of hypotheses: and, by his parental relation to all, he is forbidden to fasten his affections unduly upon any one. In the nature of the case, the danger that springs from affection is counteracted, and therein is a radical difference between this method and the two preceding. The investigator at the outset puts himself in cordial sympathy and in parental relations (of adoption, if not of authorship) with every hypothesis that is at all applicable to the case under investigation. Having thus neutralized the partialities of his emotional nature, he proceeds with a certain natural and enforced erectness of mental attitude to the investigation, knowing well that some of his intellectual children will die before maturity, yet feeling that several of them may survive the results of final investigation, since it is often the outcome of inquiry that several causes are found to be involved instead of a single one. In following a single hypothesis, the mind is presumably led to a single explanatory conception. But an adequate explanation often involves the co-ordination of several agencies, which enter into the combined result in varying proportions. The true explanation is therefore necessarily complex. Such complex explanations of phenomena are specially encouraged by the method of multiple hypotheses, and constitute one of its chief merits. We are so prone to attribute a phenomenon to a single cause, that, when we find an agency present, we are liable to rest satisfied therewith, and fail to recognize that it is but one factor, and perchance a minor factor, in the accomplishment of the total result. Take for illustration the mooted question of the origin of the Great Lake basins. We have this, that, and the other hypothesis urged by different students as the cause of these great excavations; and all of these are urged with force and with fact, urged justly to a certain degree. It is practically demonstrable that these basins were river-valleys antecedent to the glacial incursion, and that they owe their origin in part to the pre-existence of those valleys and to the blocking-up of their outlets. And so this view of their origin is urged with a certain truthfulness. So, again, it is demonstrable that they were occupied by great lobes of ice, which excavated them to a marked degree, and therefore the theory of glacial excavation finds support in fact. I think it is furthermore demonstrable that the earth's crust beneath these basins was flexed downward, and that they owe a part of their origin to crust deformation. But to my judgment neither the one nor the other, nor the third, constitutes an adequate explanation of the phenomena. All these must be taken together, and possibly they must be supplemented by other agencies. The problem, therefore, is the determination not only of the participation, but of the measure and the extent, of each of these agencies in the production of the complex result. This is not likely to be accomplished by one whose working hypothesis is pre-glacial erosion, or glacial erosion, or crust deformation, but by one whose staff of working hypotheses embraces all of these and any other agency which can be rationally conceived to have taken part in the phenomena.

5-3   Historical Science – George Gaylord Simpson

George Gaylord Simpson (1902–1984) was, in addition to being an outstanding field geologist (see page 49), also a philosopher of distinction. The present essay – "Historical Science" from The Fabric of Geology (1963) edited by C.A. Albritton – shows the refinement of the "simple" concept of causation which has taken place since the time of G.K. Gilbert. This essay is not easy reading, but it is of fundamental importance to anyone interested in the philosophical implications of historical science.

The simplest definition of history is that it is change through time. It is, however, at once clear that the definition fails to make distinctions which are necessary if history is to be studied in a meaningful way. A chemical reaction involves change through time, but obviously it is not historical in the same sense as the first performance by Lavoisier of a certain chemical experiment. The latter was a nonrecurrent event, dependent on or caused by antecedent events in the life of Lavoisier and the lives of his predecessors, and itself causal of later activities by Lavoisier and his successors. The chemical reaction involved has no such causal relationship and has undergone no change before or after Lavoisier's experiment. It always has occurred and always will recur under the appropriate historical circumstances, but as a reaction in itself it has no history.

A similar contrast between the historical and the nonhistorical exists in geology and other sciences. The processes of weathering and erosion are unchanging and non-historical. The Grand Canyon or any gully is unique at any one time but is constantly changing to other unique, nonrecurrent configurations as time passes. Such changing, individual geological phenomena are historical, whereas the properties and processes producing the changes are not.

The unchanging properties of matter and energy and the likewise unchanging processes and principles arising therefrom are *immanent* in the material universe. They are nonhistorical, even though they occur and act in the course of history. The actual state of the universe or of any part of it at a given time, its configuration, is not immanent and is constantly changing. It is *contingent* in Bernal's term, or *configurational*, as I prefer to say. History may be defined as configurational change through time, i.e., a sequence of real, individual but interrelated events. These distinctions betweens the immanent and the configurational and between the nonhistorical and the historical are essential to clear analysis and comprehension of history and of science. They will be maintained, amplified, and exemplified in what follows.

Definitions of science have been proposed and debated in innumerable articles and books. Brief definitions are inevitably inadequate, but I shall here state the one I prefer: Science is an exploration of the material universe that seeks natural, orderly relationships among observed phenomena and that is self-testing. Apart from the points that science is concerned only with the material or natural and that it rests on observation, the definition involves three scientific activities: the description of phenomena, the seeking of theoretical, explanatory relationships among them, and some means for the establishment of confidence regarding observations and theories. Among other things, later sections of this essay will consider these three aspects of historical science.

Historical science may thus be defined as the determination of configurational sequences, their explanations, and the testing of such sequences and explanations. (It is already obvious and will become more so that none of the three phases is simple or thus sufficiently described.)

Geology is probably the most diverse of all the sciences, and its status as in part a historical science is correspondingly complex. For one thing, it deals with the immanent properties and processes of the physical earth and its constituents. This aspect of geology is basically nonhistorical. It can be viewed simply as a branch of physics (including mechanics) and chemistry, applying those sciences to a single (but how complex!) object: the earth. Geology also deals with the present configuration of the earth and all its parts, from core to atmosphere. This aspect of geology might be considered nonhistorical insofar as it is purely descriptive, but then it also fails to fulfill the whole definition of a science. As soon as theoretical, explanatory relationships are brought in, so necessarily are changes and sequences of configurations, which are historical. The fully scientific study of geological configurations is thus historical science. This is the *only* aspect of geology that is peculiar to this science, that is simply geology and not also something else. (Of course I do not mean that it can be studied without reference to other aspects of geology and to other sciences, both historical and nonhistorical.) . . .

Uniformitarianism has long been considered a basic principle of historical science and a major contribution of geology to science and philosophy. In one form or another it does permeate geologic and historical thought to such a point as often to be taken for granted. Among those who have recently given conscious attention to it, great confusion has arise from conflicts and obscurities as to just what the concept is. To some, uniformitarianism (variously defined) is a law of history. Others, maintaining that it is not a law, have tended to deny its significance. Indeed, in any reasonable or usual formulation, it is not a law, but that does not deprive it of importance. It is commonly defined as the principle that the present is the key to the past. That definition is, however, so loose as to be virtually meaningless in application. A new, sharper, and clearer definition in modern terms is needed.

Uniformitarianism arose around the turn of the 18th to 19th centuries, and its original significance can be understood only in that context. It was a reaction against the then prevailing school of catastrophism, which had two main tenets: (1) the general belief that God has intervened in history, which therefore has included both natural and supernatural (miraculous) events; and (2) the particular proposition that earth history consists in the main of a sequence of major catastrophes, usually considered as of divine origin in accordance with the first tenet. Uniformitarianism, as then expressed, had various different aspects and did not always face these issues separately and clearly. On the whole, however, it embodied two propositions contradictory to catastrophism: (1) earth history (if not history in general) can be explained in terms of natural forces still observable as acting today; and (2) earth history has not been a series of universal or quasi-universal catastrophes but has in the main been a long, gradual development – what we would now call an evolution. (The term "evolution" was not then customarily used in this sense.) A classic example of the conflicting application of these principles is the catastrophist belief that valleys are clefts suddenly opened by a supernally ordered revolution as against the uniformitarian belief that they have been gradually formed by rivers that are still eroding the valley bottoms.

Both of the major points originally at issue are still being argued on the fringes of science or outside it. To most geologists, however, they no longer merit attention from anyone but a student of human history. It is a necessary condition and indeed part of the definition of science in the modern sense that only natural explanations of material phenomena are to be sought or can be considered scientifically tenable. It is interesting and significant that general acceptance of this principle (or limitation, if you like) came much later in the historical than in the nonhistorical sciences. In historical geology it was the most important outcome of the uniformitarian–catastrophist controversy. In historical biology it was the still later outcome of the Darwinian controversy and was hardly settled until our own day. (It is still far from settled among nonscientists.)

As to the second major point originally involved in uniformitarianism, there is no *a priori* or philosophical reason for ruling out a series of *natural* worldwide catastrophes as dominating earth history. However, this assumption is simply in such flat disagreement with everything we now know of geological history as to be completely incredible. The only issues still valid involve the way in which natural processes still observable have acted in the past and the sense in which the present is a key to the past. ...

Then what uniformity principle, if any, is valid and important? The distinction between immanence and configuration (or contingency) clearly points to one: the postulate that immanent characteristics of the material universe have not changed in the course of time. By this postulate all the immanent characteristics exist today and so can, in principle, be observed or, more precisely, inferred as generalizations and laws from observations. It is in this sense that the present is the key to the past.

Present immanent properties and relationships permit the interpretation and explanation of history precisely because they are *not* historical. They have remained unchanged, and it is the configurations that have changed. Past configurations were never quite the same as they are now and were often quite different. Within those different configurations, the immanent characteristics have worked at different scales and rates at different times, sometimes combining into complex processes different from those in action today. The uniformity of the immanent characteristics helps to explain the fact that history is not uniform. (It could even be said that uniformitarianism entails catastrophes, but the paradox would be misleading if taken out of context.) Only to the extent that past configurations resembled the present in essential features can past processes have worked in a similar way.

That immanent characteristics are unchanging may seem at first sight either a matter of definition or an obvious conclusion, but it is neither. Gravity would be immanent (an inherent characteristic of matter *now*) even if the law of gravity had changed, and it is impossible to prove that it has not changed. Uniformity, in this sense, is an unprovable postulate justified, or indeed required, on two grounds. First, nothing in our incomplete but extensive knowledge of history disagrees with it. Second, only with this postulate is a rational interpretation of history possible, and we are justified in seeking – as scientists we *must* seek – such a rational interpretation. It is on this basis that I have assumed on previous pages that the immanent is unchanging.

5-4   What is a Species? – Stephen Jay Gould

Stephen Jay Gould (1941–2002) was born in New York City and was both a versatile and prolific author as well as a first-rate Earth scientist. A graduate of Antioch and Columbia, he was Alexander Agassiz Professor of Zoology and Professor of Geology at Harvard. His scientific work ranged from studies of fossil molluscs to detailed analysis of evolutionary rates and styles. His most general writing – amounting to some twenty books and hundreds of essays on topics ranging from IQ to the nature of history and the implications of the scientific method – won him a huge following, and widespread acclaim. The extract we include here was written in 1992 in Discover Magazine.

I had visited every state but Idaho. A few months ago, I finally got my opportunity to complete the roster of 50 by driving east from Spokane, Washington, into western Idaho. As I crossed the state line, I made the same feeble attempt at humor that so many of us try in similar situations: "Gee, it doesn't look a bit different from easternmost Washington."

We make such comments because we feel the discomfort of discord between our mental needs and the world's reality. Much of nature (including terrestrial real estate) is continuous, but both our mental and political structures require divisions and categories. We need to break large and continuous items into manageable units.

Many people feel the same way about species as I do about Idaho – but this feeling is wrong. Many people suppose that species must be arbitrary divisions of an evolutionary continuum in the same way that state boundaries are conventional divisions of unbroken land. Moreover, this is not merely an abstract issue of scientific theory but a pressing concern of political reality. The Endangered Species Act, for example, sets policy (with substantial teeth) for the preservation of species. But if species are only arbitrary divisions in nature's continuity, then what are we trying to preserve and how shall we define it? I write this article to argue that such a reading of evolutionary theory is wrong and that species are almost always objective entities in nature.

Let us start with something uncontroversial: the bugs in your backyard. If you go out to make a complete collection of all the kinds of insects living in this small discrete space, you will collect easily definable "packages," not intergrading continua. You might find a kind of bee, three kinds of ants, a butterfly or two, several beetles, and a cicada. You have simply validated the commonsense notion known to all: in any small space during any given moment, the animals we see belong to separate and definable groups – and we call these groups species. In the eighteenth century this commonsense observation was translated, improperly as we now know, into the creationist taxonomy of Linnaeus. The great Swedish naturalist regarded species as God's created entities, and he gathered them into genera, genera into orders, and orders into classes, to form the taxonomic hierarchy that we all learned in high school (several more categories, families and phyla, for example, have been added since Linnaeus's time). The creationist version reached its apogee in the writings of America's greatest nineteenth-century naturalist (and last truly scientific creationist), Louis Agassiz. Agassiz argued that species are incarnations of separate ideas in God's mind, and that higher categories (genera, orders, and so forth) are therefore maps of the interrelationships among divine thoughts. Therefore, taxonomy is the most important of all sciences because it gives us direct insight into the structure of God's mind.

Darwin changed this reverie forever by proving that species are related by the physical connection of genealogical descent. But this immensely satisfying resolution for the great puzzle of nature's order engendered a subsidiary problem that Darwin never fully resolved: If all life is interconnected as a genealogical continuum, then what reality can species have? Are they not just arbitrary divisions of evolving lineages? And if so, how can the bugs in my backyard be ordered in separate units? In fact, the two greatest evolutionists of the nineteenth century, Lamarck and Darwin, both questioned the reality of species on the basis of their evolutionary convictions. Lamarck wrote, "In vain do naturalists consume their time in describing new species"; while Darwin lamented: "we shall have to treat species as . . . merely artificial combinations made for convenience. This may not be a cheering prospect; but we shall at last be freed from the vain search for the undiscovered and undiscoverable essence of the term *species*" (from the *Origin of Species*).

But when we examine the technical writings of both Lamarck and Darwin, our sense of paradox is heightened. Darwin produced four long volumes on the taxonomy of barnacles, using conventional species for his divisions. Lamarck spent seven years (1815–1822) publishing his generation's standard, multivolume compendium on the diversity of animal life – *Histoire naturelle des animaux sans vertebras*, or *Natural History of Invertebrate Animals* – all divided into species, many of which he named for the first time himself. How can these two great evolutionists have denied a concept in theory and then used it so centrally and extensively in practice? To ask the question more generally: If the species is still a useful and necessary concept, how can we define and justify it as evolutionists?

The solution to this question requires a preamble and two steps. For the preamble, let us acknowledge that the conceptual problem arises when we extend the "bugs in my backyard" example into time and space. A momentary slice of any continuum looks tolerably discrete; a slice of salami or a cross section of a tree trunk freezes a complexly changing structure into an apparently stable entity. Modern horses are discrete and separate from all other existing species, but how can we call the horse (*Equus caballus*) a real and definable entity if we can trace an unbroken genealogical series back through time to a dog-size creature with several toes on each foot? Where did this "dawn horse," or "eohippus," stop and the next stage begin; at what moment did the penultimate stage become *Equus caballus*? I now come to the two steps of an answer.

First, if each evolutionary line were like a long salami, then species would not be real and definable in time and space. But in almost all cases large-scale evolution is a story of branching, not of transformation in a single line – bushes, not ladders, in my usual formulation. A branch on a bush is an objective division. One species rarely turns into another by total transformation over its entire geographic range. Rather, a small population becomes geographically isolated from the rest of the species – and this fragment changes to become a new species while the bulk of the parental population does not alter. "Dawn horse" is a misnomer because rhinoceroses evolved from the same parental lineage. The lineage split at an objective branching point into two lines that became (after further events of splitting) the great modern groups of horses (eight species, including asses and zebras) and rhinos (a sadly depleted group of formerly successful species).

Failure to recognize that evolution is a bush and not a ladder leads to one of the most common vernacular misconceptions about human biology. People often challenge me: "If humans evolved from apes, why are apes still around?" To anyone who understands the principle of bushes, there simply is no problem: the human lineage emerged as a branch, while the rest of the trunk continued as apes (and branched several more times to yield modern chimps, gorillas, and so on). But if you think that evolution is a ladder or a salami, then an emergence of humans from apes should mean the elimination of apes by transformation.

Second, you might grasp the principle of bushes and branching but still say: Yes, the ultimate products of a branch become objectively separate, but early on, while the branch is forming, no clear division can be made, and the precursors of the two species that will emerge must blend indefinably. And if evolution is gradual and continuous, and if most of a species' duration is spent in this state of incipient formation, the species will not be objectively definable during most of their geologic lifetimes.

Fair enough as an argument, but the premise is wrong. New species do (and must) have this period of initial ambiguity. But species emerge relatively quickly, compared with their period of later stability, and then live for long periods – often millions of years – with minimal change. Now, suppose that on average (and this is probably a fair estimate), species spend one percent of their geologic lifetimes in this initial state of imperfect separation. Then, on average, about one species in a hundred will encounter problems in definition, while the other 99 will be discrete and objectively separate – cross sections of branches showing no confluence with others. Thus, the principle of bushes, and the speed of branching, resolve the supposed paradox: continuous evolution can and does yield a world in which the vast majority of species are separate from all others and clearly definable at any moment in time. Species are nature's objective packages.

5-5   Messages in Stone – Christine Turner

Christine Turner, who has served for over a quarter century with the United States Geological Survey, where she is a senior research scientist, argues that certain societal assumptions, derived in part from the frontier period of history, leave us ill-prepared for the complexity and interdependence of contemporary life. She asserts that a "geophilosphical perspective" is required to address this issue.

Turner, with a PhD in geology from the University of Colorado, specializes in integrated studies of sedimentary basins and ancient ecosystems, especially those of the late Jurassic. The present extract is from her essay Messages in Stone: field geology in the American West which appears in the book Earth Matters (2000). This book, edited by Robert Frodeman, has fifteen thoughtful essays that discuss connections between the Earth sciences and contemporary culture. In her essay, Turner gives a vivid picture of the geologist in the field.

As in many professions, it is more often in the "doing" that true learning takes place. As with all sciences, geology has its textbooks and procedures for how to interpret the rocks, and field equipment for taking certain types of measurements and for gathering data. A key characteristic of geology, however, is that we must interpret environments that have long since vanished. We must reconstruct the past from fragmentary evidence found within those strata. As an historical science, geology requires that we learn the skills of Sherlock Holmes, working with highly circumstantial evidence and hoping to convince the jury of our scientific peers that we have enough data to seek a verdict.

There is no way to return to the scene of the crime, however, because we are unraveling mysteries that occurred many millions of years ago. Nature has had ample time to perform a massive cover-up operation. Evidence has been modified by subsequent events or has been removed altogether. But as geologists we learn quickly to deal with incomplete data and with enormous uncertainty, and become so accustomed to extrapolating from fragmentary data that we forget how different our science is from chemistry and physics, where carefully designed and reproducible experiments are the final arbiter of theories.

Working intimately with "deep time," we have developed a repertoire based upon comparative approaches. We infer that "the present is the key to the past," and therefore assume that streams that deposited sand and mud a million years ago probably had the same hydraulic properties that characterize modern streams. But what we do not know for any particular ancient stream are all the particular variables that affect stream behavior, such as gradient, amount of vegetation, climate changes, and seasonality of precipitation, just to name a few. Deciphering rocks is an uncontrolled experiment: We cannot hold all of the variables constant and vary just one at a time. We can never know all of the factors that combined to produce the resulting strata – we only have the result and reasonable inferences gained from analogy with modern processes.

Perhaps the most important ability of the field geologist is the capacity to envision and even to "inhabit" ancient landscapes by way of deliberate day dreaming. A geologist climbs into an imaginary time machine, where every observation, every cryptic piece of evidence in the rocks is viewed not as part of a stonily silent cliff but as a landscape as vibrant as it was at the time the sediments were deposited. When a geologist finds a trace fossil, formed by a disruption of sediments when an organism burrowed into a river sand a hundred million years ago, he can imagine being that organism. He hears the waves crashing overhead, and imagines burrowing deeper. Strata left by an ancient stream migrating across a sandbar pulls the geologist back to a river 150 million years ago. He can envision gravel tumbling along the bar during flood stage, stinging his feet, his boots providing no protection against the deluge. Perhaps Sarah Andrews, a petroleum geologist and writer of mysteries with geologic themes, says it best: "Geologists have the ability to see into solid rock and imagine entire worlds."

5-6 Natural Science, Natural Resources, and the Nature of Nature – Marcia G. Bjørnerud

All science, all literature, all knowledge is based on assumptions – huge assumptions – concerning the nature of our world, and these assumptions – partly intuitive, partly rational – color every statement and conclusion. It is these assumptions and their implications that Marcia G. Bjørnerud examines in her essay "Natural Science, Natural Resources, and the Nature of Nature." At the extract's end, synthesized into one relatively short table, Bjørnerud presents the changing perceptions and paradigms surrounding the "nature of nature" and natural resources over the last few centuries. This essay was published in The Earth Around Us: maintaining a livable planet (2000), edited by Jill S. Schneiderman, which contains thirty-one essays by contemporary authors dealing with global environmental themes. Bjørnerud, who teaches at Lawrence University in Wisconsin, is a structural geologist whose field work has taken her to some of the more remote Arctic islands.

But the apocalypse has been postponed. The global famines predicted in *The Population Bomb* for the 1970s and 1980s did not occur. *The Limits to Growth* forecast depletion of copper and other strategic metals by the early 1990s, but reports of the dearth of these commodities appear to have been greatly exaggerated. In 1980, economist Julian Simon challenged ecologist Paul Ehrlich to a wager about how the prices of five metals would change in the coming decade. Ehrlich anticipated that increasing scarcity would drive prices higher. Simon predicted that prices would decrease, and he won the bet. What happened to finitude?

The Limits to Growth by the Massachusetts Institute of Technology-based wunderkinder in the "Club of Rome" was groundbreaking in its use of systems modeling to grapple with complexly linked entities. The "World Model" at the heart of *Limits*, with its profusion of circles and arrows, was by far the most ambitious attempt at the time to integrate natural, economic, and social variables into a single scheme. Like Burnet's *Sacred Theory of the Earth* nearly three centuries earlier, the World Model was an earnest, if naïve, effort to create a single unified vision of Earth. The limitations of *Limits*, some of which were recognized within a year of its publication, arose largely from its faith in the nineteenth-century view that Earth is measurable and mechanical. But the comforting concepts of measurement and mechanism were becoming unexpectedly problematic.

"How long is the coastline of Britain?" the provocative title of a 1967 paper by Benoit Mandelbrot of IBM grounded the new mathematics of fractals in the natural world. If your measuring stick is long, you will miss the firths and embayments, and you will conclude that the coastline is short. As you use shorter and shorter rulers, the coastline stretches. How long will copper last? How many people can Earth support? It depends. If you set the price higher, copper will last a few more decades. If most earthlings become vegetarians, the planet could support another billion. There is always more of what you are looking for if you look closer, if you pay more, if you learn to recycle and eat (only) your vegetables. As Joel E. Cohen, professor of populations at Rockefeller University and Columbia University, has recently argued, "Ecological limits appear not as ceilings but as trade-offs." The sensitive feedbacks of the global marketplace have made Earth effectively infinite again.

Measuring coastlines may no longer be a sure thing, but appraising the planet might be possible. In an audacious 1997 paper in *Nature*, Robert Costanza and 12 other ecologists and economists from around the world placed the value of the "world's ecosystem services and natural capital" at $33 trillion – about twice the global gross international product. If such an analysis seems crass, it at least has the virtue of using a system of measurement that is universally understood. Costanza and his coauthors concede that the undertaking is at some level absurd:

> The economies of the Earth would grind to a halt without the services of ecological life-support systems, so in one sense their total value to the economy is infinite. However, it can be instructive to estimate the "incremental" ... value of ecosystem services. ... We acknowledge that there are many conceptual and empirical problems inherent in producing such an estimate.

The international environmental summit meetings of the 1990s have been based on the premise that the only viable solutions to the global environmental problems will be those that harness the power of the global market. This is not to say that environmental decisions should be made strictly on economic grounds, but rather that the "hidden" environmental costs of, for example, fossil fuel combustion or nuclear power generation should be acknowledged in the market. Costanza and his coauthors assert that their analyses have two types of practical application: (1) helping nations and international bodies modify "systems of national accounting to better reflect the value of ecosystem services and natural capital" and (2) appraising governmental or commercial initiatives in which "ecosystem services lost must be weighted against the benefits of a specific project." Arguably, then, the most meaningful environmental measurements are those that can be converted to units of currency – as long as the contingent nature of these monetary "measurements" is recognized.

At the same time that Nature was beginning to resist measurement, the earth machine was also showing signs of unpredictable behavior. Like Laplace, many modern scientists have continued to embrace the idea that knowing Nature's "rules" – holding the user's manual – will eventually lead to omniscience. Biologist Edward O. Wilson, for example, has recently described his "conviction, far deeper than a mere working proposition, that the world is orderly and can be explained by a small number of natural laws." But even if the number of laws is small, the number of possible outcomes may be infinite, and order, if it exists, is probably ephemeral. These are the lessons emerging from studies of Earth's climate, for example, which is better characterized by the mathematics of chaos and nonlinear systems than by equilibrium thermodynamics. Cores from ice caps and deep-sea sediments record feverish oscillations in global temperature, catastrophic collapses of massive icesheets, and decade-scale reorganization of heat-transporting ocean currents. Biological evolution of species is no longer a stately march but a series of fits and starts. Contrary to Lyell's view, Nature apparently does not spend most of its time in "repose." Rather, "convulsions" large and small are the norm. What happened to equilibrium?

In some subdisciplines within the natural sciences, it is now common to invoke multiple, rather than unique, equilibrium states, separated by narrow thresholds. In these models, even a relatively small perturbation to an apparently stable system can trigger system-wide reorganization. Worse, the response may be sensitive to the history of the system, the scale of investigation, and the rate, as well as the magnitude, of the perturbation. If it was once placid and domesticated, Nature is now irritable and biting back.

Other scientists have begun to question the validity of ascribing even this qualified kind of equilibrium to natural systems. Anthropologist Paul Rabinow writes:

> Nature reflects the accumulation of countless accidents, not some hidden harmony. Things might have turned out quite differently. Ecosystems are ever changing, dissolving, transforming, recombining in new forms.

Jianguo Wu and Orie Loucks, both fervent conservationists, argue that sentimental attachment to the idea of the "balance of nature" has undermined the credibility of environmental policy:

> The classical equilibrium paradigm in ecology ... has failed not only because equilibrium conditions are rare in nature, but also because of our past inability to incorporate heterogeneity and scale multiplicity into our quantitative expressions for stability. The theories and models built around these equilibrium and stability principles have misrepresented the foundations of resource management, nature conservation, and environmental protection.

The earth machine has become too complex for the mechanics to understand. Is it time to acknowledge, with the Romantic poets, that the metaphor of mechanism fails to portray the full richness of Nature? The geosciences appear ready to return to Hutton's half-organic depiction of Earth as a "living machine." There are surprising similarities between Hutton's late eighteenth century vision and atmospheric scientist James Lovelock's late twentieth century Gaia hypothesis, which places Life and the center of the global climate system and geochemical cycles. Although Gaia has not yet been fully welcomed into the scientific fold, she has already found her way in the back gate in the less controversial guise of "geophysiology." If the manuals for the earth machine have begun to seem useless, perhaps it is because we should be consulting a medical reference instead. In any case, Earth seems to have recovered its strength and spirit. Unfortunately, the prognosis for the human race is less clear.

In many ways, scientific views of Earth have come full circle in 300 years. At the start, we feared the power and magnitude of a capricious and capacious Earth; then passed through a period of childlike self-aggrandizement in which we, like Antoine de Saint-Exupéry's Little Prince, were masters of a planet that was just the right size for us to explore and conquer. Rather suddenly, we found ourselves surrounded by the wreckage of our careless expeditions, on an Earth that seemed to be shrinking. Finally, we have paused long enough to see infinity once more, in grains of sand and nuclei of cells. We are again humbled in the presence of a venerable old planet that is at once benevolent and malevolent, comprehensible and complex, predictable and chaotic, robust and fragile.

CHANGING PERCEPTIONS OF THE NATURE OF NATURE AND NATURAL RESOURCES

	Pre-18th century	18th–19th centuries	20th century	21st century
Geological paradigms:	Catastrophism	Uniformitarianism	Plate tectonics	Gaia; Neo-catastrophism
Political paradigms:		Colonialism	Environmentalism	Globalism
SIZE AND SCOPE	**Infinite Earth** Raw materials are limitless; resource availability is restricted only by human labor and technology.	**Measurable Earth** Earth's form and face have been mapped; minerals have been classified; general patterns of resource occurrence have been characterized.	**Finite Earth** Resources are exhaustible in our lifetimes.	**Fractal Earth** Earth is infinite in its complexity at all scales. Previously unrecognized resources are to be found in the microbial world, the diversity of the rainforest, and the deep sea.
ORDER AND ORGANIZATION	**Chaotic Earth** Earth was shaped by cataclysm; geological phenomena are largely inexplicable except as Divine interventions.	**Mechanical Earth** Earth processes are subject to natural laws and are predictable, cyclical, uniformitarian, and gradual. Understanding these processes allows technological intervention in them, and more efficient exploitation of resources. Continents are fixed. Geology is land based. Geological change is gradual and progressionist.	Continents are mobile. Geology is ocean centered.	**Organic Earth** Earth is chaotic in the modern sense; characterized by interconnected systems, nonlinearity, and multiple stable states. Even if natural laws are known, outcomes may not be predictable. Geological change is punctuated and contingent.
BALANCE OF POWER BETWEEN PEOPLE AND PLANET	**Threatening Earth** Wilderness is the realm of beasts and savages; undeveloped land is wasteland.	**Subjugatable Earth** Earth can be cultivated, mined, and tamed on a global scale for human benefit.	**Fragile Earth** The earth machine can break down if abused by humans.	**Robust Earth** Earth will survive the environmental havoc wreaked by humans, but humans may not.

5-7 Does God Play Dice? – Ian Stewart

Ian Stewart, born in 1945, is Professor of Mathematics at the University of Warwick and at the Royal Institution. A Fellow and Faraday Medalist of the Royal Society, his mathematical work covers an extraordinary range of topics and applications. He is a prolific author, with more than sixty books, ranging from mathematics to science fiction, to his credit. The extract that follows is the prologue to the second edition (1997) of his book Does God Play Dice? The New Mathematics of Chaos.

You believe in the God who plays dice, and I in complete law and order.
                                    Albert Einstein, *Letter to Max Born*

There is a theory that history moves in cycles. But, like a spiral staircase, when the course of human events comes full circle it does so on a new level. The 'pendulum swing' of cultural changes does not simply repeat the same events over and over again. Whether or not the theory is true, it serves as a metaphor to focus our attention. The topic of this book represents one such spiral cycle: chaos gives way to order, which in turn gives rise to new forms of chaos. But on this swing of the pendulum, we seek not to destroy chaos, but to tame it.

In the distant past of our race, nature was considered a capricious creature, and the absence of pattern in the natural world was ascribed to the whims of the powerful and incomprehensible deities who ruled it. Chaos reigned and law was unimaginable.

Over a period of several thousand years, humankind slowly came to realize that nature has many regularities, which can be recorded, analysed, predicted, and exploited. By the 18th century science had been so successful in laying bare the laws of nature that many thought there was little left to discover. Immutable laws pre-scribed the motion of every particle in the universe, exactly and forever: the task of the scientist was to elucidate the implications of those laws for any particular phenomenon of interest. Chaos gave way to a clockwork world.

But the world moved on, and our vision of the universe moved with it. Today even our *clocks* are not made of clockwork – so why should our world be? With the advent of quantum mechanics, the clockwork world has become a cosmic lottery. Funda-mental events, such as the decay of a radioactive atom, are held to be determined by chance, not law. Despite the spectacular success of quantum mechanics, its probabil-istic features have not appealed to everyone. Albert Einstein's famous objection, in a letter to Max Born, is quoted at the head of this chapter. Einstein was talking of quantum mechanics, but his philosophy also captures the attitude of an entire age to classical mechanics, where quantum indeterminacy is inoperative. The metaphor of dice for chance applies across the board. Does determinacy leave room for chance?

Whether Einstein was right about quantum mechanics remains to be seen. But we do know that the world of classical mechanics is more mysterious than even Einstein imagined. The very distinction he was trying to emphasize, between the randomness of chance and the determinism of law, is called into question. Perhaps God can play dice, and create a universe of complete law and order, in the same breath.

The cycle has come full turn – but at a higher level. For we are beginning to discover that systems obeying immutable and precise laws do not always act in predictable and regular ways. Simple laws may not produce simple behavior. Deter-ministic laws can produce behavior that appears random. Order can breed its own kind of chaos. The question is not so much *whether* God plays dice, but *how* God plays dice.

This is a dramatic discovery, whose implications have yet to make their full impact on our scientific thinking. The notions of prediction, or of a repeatable experiment, take on new aspects when seen through the eyes of chaos. What we thought was simple becomes complicated, and disturbing new questions are raised regarding measurement, predictability, and verification or falsification of theories.

In compensation, what was thought to be complicated may become simple. Phenomena that appear structureless and random may in fact be obeying simple laws. Deterministic chaos has its own laws, and inspires new experimental techniques. There is no shortage of irregularities in nature, and some of them may prove to be physical manifestations of the mathematics of chaos. Turbulent flow of fluids, reversals of the Earth's magnetic field, irregularities of the heartbeat, the convection patterns of liquid helium, the tumbling of celestial bodies, gaps in the asteroid belt, the growth of insect populations, the dripping of a tap, the progress of a chemical reaction, the metabolism of cells, changes in the weather, the propagation of nerve impulses, oscillations of electronic circuits, the motion of a ship moored to a buoy, the bouncing of a billiard ball, the collisions of atoms in a gas, the underlying uncertainty of quantum mechanics – these are a few of the problems to which the mathematics of chaos has been applied.

It is an entire new world, a new kind of mathematics, a fundamental breakthrough in the understanding of irregularities in nature. We are witnessing its birth.

Its future has yet to unfold.

6.   The Fossil Record

Life: an unauthorized biography – the title of Richard Fortey's recent book on "the first 4,000,000,000 years of life on Earth" is a useful reminder of the broad scope of Earth science. For Earth science is concerned, not only with converging plates and erupting volcanoes – the physical forces and processes that govern the planet's development – but also with the profound changes in living things, which have shared the Earth's surface for over three quarters of its existence.

The first living things were probably inconspicuous, microscopic blue-green bacteria (cyanobacteria), which used photosynthesis to convert sunlight into useful energy. It was from such modest beginnings that the hordes of living things subsequently developed: whales and camels, frogs and eagles, tulips and dinosaurs, spiders and bananas. It is difficult to imagine our Earth without its forests of pine, meadows of wildflowers, schools of darting fish, herds of zebra, flights of geese, and all the teeming rest. But the young Earth was barren. We may trace the origins of all life back to these unlikely blue-green ancestors, whose own representatives still live on today, among the multitude of their descendants.

Nor is that all, for the development of life changed forever the character of the planet itself. The pre-organic Earth had an atmosphere devoid of oxygen. Its gaseous covering, derived from the outgassing of its myriad early volcanoes and fumaroles, probably consisted of carbon dioxide, ammonia, and steam and perhaps carbon monoxide, methane, and hydrogen cyanide: a profoundly toxic brew to most living things as we know them today. Only when the earliest blue-green bacteria used sunlight to break down the enveloping carbon dioxide into useful carbon and waste oxygen, did the atmosphere slowly change, as surplus oxygen became available as a support system for more advanced living things.

But it is not only the atmosphere which has been and is modified by living things. Sedimentation, stratification, rock deposition, streams, lakes, oceans, rock weathering and decomposition, soil formation – all these and more reflect life's quiet but global influence

So parent Earth – the source of life – has itself been profoundly modified by life, its offspring. The extracts that follow explore that interaction and the marvelous history of some of Earth's creatures.

6-1   Earth and Man – Frank H.T. Rhodes

This extract by Frank H.T. Rhodes – from the Wooster Alumni Magazine *(1972) – describes human history in evolutionary perspective.*

We share an Inhabited Planet. It is a remarkable thing that within our particular solar system, and the very limited area of which we have any direct knowledge, this planet alone is known to be inhabited. Of course, beyond our planetary system, many other such planets may well exist. One of the dramatic things as one reads the accounts of the astronauts returning to the Earth from the Moon is the incredible beauty of the planet Earth, in comparison with the darkness of space and the aridity and bleakness of the lunar landscape. As they approach nearer and nearer to our own planet, the bleakness and the grayness recede as they marvel at the colorful exuberance of the Earth. The thing that makes the difference is that ours is a living planet. The discovery that we share the Earth with other creatures is not a new one. Man has always known that he was part of the animal creation. He has always known that there were many different kinds of animals, and it requires no particular scientific skill to distinguish one from another. It has always been known that animals existed in great abundance. A swarm of locusts would convince anyone that this was the case. Yet, the question that constantly nagged our ancestors was the question, "Whence came this astonishing diversity of animals and plants?" There are over a million different species of animals, and over a third of a million different species of plants. How did they originate? Three kinds of answers were given by our great-grandparents.

The first suggestion was that living things arose by spontaneous generation and have existed since the creation of the planet. But the fossil record came to light, and somehow that had to be related to what was known of living creatures themselves. So, a second viewpoint developed to account for this record and for the origin of kinds of living things. It was supposed that there had been a series of great catastrophes within the history of the Earth, of which the flood of Noah was the most recent. New life came into existence spontaneously by divine command. But then a catastrophe overwhelmed the world, destroying it and producing massive extinction of living things, so that ultimately, there was a new creation, another episode of renewal. This, it was suggested, happened thirty or thirty-five times. Then, you remember, came those who in the 19th Century, and even earlier, offered another explanation, first of all hesitatingly, and later with more conviction. Lamarck, Alfred Russell Wallace, and Charles Darwin insisted that there was a continuity and an orderliness in the world of living things, as real as the orderliness in the world of non-living things. Darwin argued that reproduction produced variation, which was acted on by natural selection, so that those better fitted to the environment in which they lived gradually left more offspring than those that were less well-adapted. That was the secret behind the great diversity of living things, it was suggested. A hundred years of reflection has tended to confirm that particular hypothesis.

There are two implications of that evolutionary point of view. The first is that we are related to everything else in the universe, and our net effect as an agent of human action has scarcely been benevolent. There is an interdependence in which we are not only our brother's keeper, but, literally, the keeper of every other species on the planet. The second implication was even more profound, because the writings of Darwin threw the contemporary world into turmoil. You remember that the year after the publication of Darwin's book the British Associate for the Advancement of Science met at Oxford. You can still go to Oxford into the Natural History Museum to an upstiars room, now used to house collections, where a memorable debate took place. Four hundred people crowded inside a lecture theatre on a June evening in 1860 where the great debate of the century took place.

So it was that this debate was hostile from the start. The battle lines were drawn, and the nature of man became for fifty years the central theme of a furious debate between the emergent biological and geological sciences and established religion. You remember Tennyson having read "Chamber's Vestiges," crying out in anguish,

> Are God and Nature then at strife
> That Nature lends such evil dreams?
> So careful of the type she seems
> So careless of the single life.

Even a man such as George Bernard Shaw could remark, "If it could be proved that the whole universe has been produced by such natural selection, only fools and rascals could bear to live." Even Marxism, which was then dawning upon the world, had its own particular teleological view of history. In fact, Karl Marx offered to dedicate *Das Kapital* to Darwin, but Darwin politely declined the honor. It did seem, however, that the growth and understanding concerning the Inhabited Planet had overthrown all the security of man. The most remarkable thing about the publication of the *Origin of Species*, was not the reception which it received from the scientific public, but this worldwide confusion and outburst which it produced amongst men of all interests and persuasions. Philosophers, politicians, theologians, literary critics, historians, classical scholars, and the man in the street – all alike took it upon themselves to assess its worth. As so varied a group studied it, so their verdicts also varies – some accepted and respected Darwin's conclusions, others viewed them with suspicion, but most rejected them out of hand, and denounced both Darwinism and all its supposed implications. "As for the book, some treasured it, some burnt it, and some, undecided, like the Master of Trinity College, Cambridge, merely hid it!" Scientific theories, philosophies, political systems, ethic standards, revolutionary movements and social reforms, economic *laissez faire* – all these and more were established, modified, or justified upon Darwin's premises. Indeed, Darwinism soon became all things to all men.

In our own day, a century later, Darwin's theory, having found its place and left its mark in a host of different fields in inquiry, is now seen in its true perspective, and most of the clamor has subsided around it. The uninformed condemnation and hostility on the one hand, and the extravagances of many popular science writers on the other, have both been largely forgotten. There is, perhaps, only one area of human activity in which misunderstanding persists, for in the realm of religious belief, evolutionary theory still poses an "unresolved conflict," as Dr. David Lack called his recent book.

And yet, the truth is that the theory of evolution is neither anti-theistic nor theistic. So far as religion is concerned, it is strictly neutral, for it is a theory of the mechanism of descent with modification which seeks to explain *how* new species arise. It does indeed correct certain former ideas of the manner of creation, it does suggest that natural selection proceeds by natural laws, but like all other scientific theories, it provides no interpretation of the natural laws themselves, for it no more proves them to be the result of "pure" chance, than it proves them to be the servant of expression of purpose.

This conclusion should not surprise us, for it is true of science as a whole. And the neutrality arises, not because of some inadequacy in evolutionary theory, but because the student of evolution deliberately excludes from this explanation all reference to final cause. Unlike his Greek predecessor, he concerns himself not with the question "Why did life develop?" but with the question, "How did life develop?" To this latter question, we now begin to understand the answer. But the mind of man is such that, even as we understand the "How?" of life, we gaze towards the ultimate "Why?" To that question, evolutionary theory gives no answer.

This need not and does not imply that the question "Why?" is meaningless or irrelevant. It means only that we must look elsewhere for the answer, for the questions "How?" and "Why?" are not alternative and competitive, but rather complementary to one another. And this complementary relationship, which it has taken a century for the world to assimilate, brings us back to Darwin, for it is brilliantly summarized by the three short quotations from Whewell, Butler, and Bacon with which he prefaced the *Origin of Species*.

So, painfully, and slowly, Man has discovered himself as part of the inhabited planet. It is no accident that it was the young Charles Darwin, the geologist rather than the biologist, who provided us with the clue which was finally to solve the riddle of human relationships. As we look back in geological perspective today, we see a grand continuity, for biological evolution is but the last phase of a greater, threefold evolution, a continuous majestic process. First, there was a cosmic evolution, perhaps ten billion years ago in which the universe and all its elements came into existence. This was an evolution involving a low molecular order, but involving gigantic bodies of low structural complexity, such as stars, and galaxies. Then, perhaps five billion years ago, came the aggregation of cold planetary material around a star in this particular corner of our own local group galaxy. This was Earth's birthday. But there was a watershed around 3.5 billion years ago wherein that cosmic evolution which produced the Earth and stars provided a series of inorganically produced organic compounds, which came together for the first time to initiate a new kind of evolution: the organic evolution with natural selection about which we have been talking. We now understand at least some of the phases by which that change may have come. And so, for over three billion years, that particular evolutionary process complemented, but did not displace, the cosmic evolution that we have just described. But then a few thousand years ago, there emerged a third and new kind of evolution: a kind that Julian Huxley called psycho-social evolution, in which organic evolutionary change was partly displaced by tradition, and later by language and writing, so that each new generation does not have to learn by instinct or by firsthand experience, but where knowledge and precepts can be handed on from one generation to another. So the last generation stands on the shoulders of the accumulated wisdom of all its ancestors.

6-2   Flowering Earth – Donald Culross Peattie

Donald Culross Peattie (1898–1964) was born in Chicago and educated at the University of Chicago and Harvard. He began his career as a botanist at the United States Department of Agriculture, *but after two years became a freelance writer, contributing nature articles to the* Washington Evening Star *and the* Chicago Daily News. *He was a prolific author, producing* more than twenty non-fiction books, mostly about nature, as well as several works of fiction and over half-a-dozen history books for children. Our extract, taken from his book The Flowering Earth *(1939), is an elegant example of nature writing, linking the towering* Sequoia *forests of the High Sierra with their prehistoric forebears.*

What we love, when on a summer day we step into the coolness of a wood, is that its boughs close up behind us. We are escaped, into another room of life. The wood does not live as we live, restless and running, panting after flesh, and even in sleep tossing with fears. It is aloof from thoughts and instincts; it responds, but only to the sun and wind, the rock and the stream – never, though you shout yourself hoarse, to propaganda, temptation, reproach, or promises. You cannot mount a rock and preach to a tree how it shall attain the kingdom of heaven. It is already closer to it, up there, than you will grow to be. And you cannot make it see the light, since in the tree's sense you are blind. You have nothing to bring it, for all the forest is self-sufficient; if you burn it, cut, hack through it with a blade, it angrily repairs the swathe with thorns and weeds and fierce suckers. Later there are good green leaves again, toiling, adjusting, breathing – forgetting you.

For this green living is the world's primal industry; yet it makes no roar. Waving its banners, it marches across the earth and the ages, without dust around its columns. I do not hold that all of that life is pretty; it is not, in purpose, sprung for us, and moves under no compulsion to please. If ever you fought with thistles, or tried to pull up a cattail's matted rootstocks, you will know how plants cling to their own lives and defy you. The pond-scums gather in the cistern, frothing and buoyed with their own gases; the storm waves fling at your feet upon the beach the limp sea-lettuce wrenched from its submarine hold – reminder that there too, where the light is filtered and refracted, there is life still to intercept and net and by it proliferate. Inland from the shore I look and see the coastal ranges clothed in chaparral – dense shrubbery and scrubbery, close-fisted, intricately branched, suffocating the rash rambler in the noon heat with its pungency. Beyond, on the deserts, under a fierce sky, between the harsh lunar ranges of unweathered rock, life still, somehow, fights its way through the year, with thorn and succulent cell and indomitable root.

Between such embattled life and the Forests of Arden, with its ancient beeches and enchanter's nightshade, there is not great biological difference. Each lives by the cool and cleanly and most commendable virtue of being green. And though that is not biological language, it is the whole story in two words. So that we ought not speak of getting at the root of a matter, but of going back to the leaf of things. The orator who knows the way to the country's salvation and does not know that the breath of life he draws was blown into his nostrils by green leaves, had better spare his breath. And before anyone builds a new state upon the industrial proletariat, he will be wisely cautioned to discover that the source of all wealth is the peasantry of grass.

The reason for these assertions – which I do not make for metaphorical effect but maintain quite literally – is that the green leaf pigment, called chlorophyll, is the one link between the sun and life; it is the conduit of perpetual energy to our own frail organisms.

For inert and inorganic elements – water and carbon dioxide of the air, the same that we breathe out as a waste – chlorophyll can synthesize with the energy of sunlight. Every day, every hour of all the ages, as each continent and, equally important, each ocean rolls into sunlight, chlorophyll ceaselessly creates. Not figuratively, but literally, in the grand First Chapter Genesis style. One instant there are a gas and water, as lifeless as the core of earth or the chill of space; and the next they are become living tissue – mortal yet genitive, progenitive, resilient with all the dewy adaptability of flesh, ever changing in order to stabilize some unchanging ideal of form. Life, in short, synthesized, plant-synthesized, light-synthesized. Botanists say photosynthesized. So that the post-Biblical synthesis of life is already a fact. Only when man has done as much, may he call himself the equal of a weed.

Plant life sustains the living world; more precisely, chlorophyll does so, and where, in the vegetable kingdom, there is not chlorophyll or something closely like it, then that plant or cell is a parasite – no better, in vital economy, than a mere animal or man. Blood, bone and sinew, all flesh is grass. Grass to mutton, mutton to wool, wool to the coat on my back – it runs like one of those cumulative nursery rhymes, the wealth and diversity of our material life accumulating from the primal fact of chlorophyll's activity. The roof of my house, the snapping logs upon the hearth, the desk where I write, are my imports from the plant kingdom. But the whole of modern civilization is based upon a whirlwind spending of the plant wealth long ago and very slowly accumulated. For, fundamentally, and away back, coal and oil, gasoline and illuminating gas had green origins too. With the exception of a small amount of water [and atomic] power, a still smaller of wind and tidal mills, the vast machinery of our complex living is driven only by these stores of plant energy.

We, then, the animals, consume those stores in our restless living. Serenely the plants amass them. They turn light's active energy to food, which is potential energy stored for their own benefit. Only if the daisy is browsed by the cow, the maple leaf sucked of its juices by an insect, will that green leaf become of our kind. So we get the song of a bird at dawn, the speed in the hoofs of the fleeing deer, the noble thought in the philosopher's mind. So Plato's Republic was builded on leeks and cabbages.

Animal life lives always in the red; the favorable balance is written on the other side of life's page, and it is written in chlorophyll. All else obeys the thermodynamic law that energy forever runs down hill, is lost and degraded. In economic language, this is the law of diminishing returns, and it is obeyed by the cooling stars as by man and all the animals. They float down its Lethe stream. Only chlorophyll fights up against the current. It is the stuff in life that rebels at death, that has never surrendered to entropy, final icy stagnation. It is the mere cobweb on which we are all suspended over the abyss. . . .

So we made a prompt start that morning, with little ceremony about it but with some reverence preparing in us, for we went to visit the giants in the earth. Of all that has survived from the Mesozoic, which began two hundred million years ago and ended about 55,000,000 B.C., Sequoia is the king. It is so much a king that, deposed today from all but two corners of its empire, superseded, outmoded, exiled and all but exterminated, it still stands without rival. And from all over the world, those who can make the pilgrimage come sooner or later to its feet, and do it homage.

Of Sequoia there are two species left, though once they were as various and abundant as are today the pines, their lesser brothers. One is the coastal redwood of California, which is the tallest tree in the world, and the other is the Big Tree of the Sierra Nevada, which is the mightiest in bulk. These two surviving species were here before the last glacial period. But as a genus or clan of species Sequoia has its roots in a day of fabulous eld. This noble line knew the tyrant lizards; through its branches swept the pterodactyls on great batty wings. As they saw the coming of the first birds, crawling up out of lizard shapes, so the forebears of our Sequoia witnessed the evolution of the first mammals when they became scuttling rodents that perhaps, gnawing and sucking at dinosaur eggs, brought down that giant dynasty from its very base.

Sequoia as a tribe saw the rise of all the most clever and lovely types of modern insects – the butterflies and moths, the beetles and bees and ants. Yet since there were then none of the intricate inter-relationships that have developed between modern flower and modern bee, Sequoia sowed the wind. It had flowers of an antique sort, flowers by technical definition, at least; petals and scent they had none. But their pollen must have been golden upon that ancient sunlight, and the communicable spark of futurity was in it. For Sequoia towers still upon its mountain top, and I was going there. . . .

It is a long climb still through the foothills of the Sierra. But now I sit up, with a lifted face. Beyond, higher in the east, portent is gathering. It takes shape, cloud-colored, gleaming with a stern reality where the sun smites a rocky forehead. Then appears that eternally moving miracle – snow in the summer sky. Sierra Nevada. . . .

The forests march upon the car; the ruddy soaring trunks of the sugar pines close around in escort. One hundred and two hundred feet overhead, their foliage is not even visible, screened by the lower canopy spread by western yellow pines which are giants in themselves. Groves of white fir, smelling like Christmas morning, troop between the yellow pines. Aisles of incense cedar with gracious down-sweeping boughs and flat sprays of gleaming foliage invite the eye down colonnaded avenues, fragrance drifting from their censers that appear to smoke with the long afternoon light. It grows darker with every mile, darker and deeper in moss and lichen, dim with the dimness of a vanished era. We have got back into earliest spring, at this altitude, and the blossoming dogwood troops along, illuminating the dusky places with a white laughter. . . .

[N]ow, as the land of sunny levels has fallen remotely out of sight, there is a prescience in the cold air, of grandeur. We have climbed into the shadows; the drifts of snow are thicker between great roots, and richer grows the livid green mantle of staghorn lichen that clothes all Sierra wood in green old age. The boles of the sugar pines, which are kings, give place before the coming of an emperor. The sea sound of the forest deepens a tone in pitch. The road is twisting to find some way between columns so vast they block the view. They are not in the scale of living things, but geologic in structure, fluted and buttressed like colossal stone work, weathered to the color of old sandstone. They are not the pillars that hold up the mountains. They are Sequoia. The car has stopped, and I am standing in the presence.

Centuries of fallen needles make silence of my step, and the command upon the air, very soft, eternal, is to be still. I am at the knees of gods. I believe because I see, and to believe in these unimaginable titans strengthens the heart. Five thousand years of living, twelve million pounds of growth out of a tiny seed. Three hundred vertical feet of growth, up which the water travels every day dead against gravity from deep in the great root system. Every ounce, every inch was built upward from the earth by the thin invisible stream of protoplasm that has been handed down by the touch of pollen from generation to generation, for a hundred million years. Ancestral Sequoias grew here before the Sierra was uplifted. Today they look down upon the plains of men. No one has ever known a Sequoia to die a natural death. Neither insects nor fungi can corrupt them. Lightning may smite them at the crown and break it; no fire gets to the heart of them. They simply have no old age, and the only down trees are felled trees.

In their uplifted hands they permit the little modern birds, the passerine song birds, vireos and warblers, tanagers and thrushes, to nest and call. I heard, very high above me in the luminous glooms, voices of such as these. I saw, between the huge roots that kept a winter drift, the snowplant thrust through the earth its crimson fist.

A doe – so long had I stood still – stepped from behind the enormous bole and, after a long dark liquid look, ventured with inquiring muzzle to touch my outheld hand. Bright passing things, these nestle for an hour in the sanctuary of the strong and dark, the vast and incalculably old.

That day I stood upon a height in time that let me glimpse the Mesozoic. It followed the Coal Age, the age of the fern forests, and it was itself the age of Gymnosperms. Sequoias are Gymnosperms. So are the pines, the larches, spruces, fir, yew, cypress, cedar – all that we call conifers, though there are other Gymnosperms that do not bear cones.

The Gymnosperms are, literally translating, "the naked-seeded" plants. For their seed is not completely enclosed in any fruit or husk, as it is in the higher modern plants that truly fruit and flower. Neither is the Gymnosperm egg or ovule completely enclosed in an ovary, as in the true flowers. To make an analogy, you could say that the Gymnosperms are plants without wombs, while the Angiosperms, the true flowering plants with genuine fruits, are endowed with that engendering sanctuary.

But though the seeds of the Gymnosperms are naked, they are seeds, and the seed is mightier than the spore. For the seed contains an embryo. Spores are very many and very small; they blow lightly about the world and find a lodging easily. But the seed is weighted with a great thing. Within even the tiniest lies the germ of a fetal plantlet, its fat cotyledons or first baby leaves still crumpled in darkness, its primary rootlet ready to thrust and suckle at the breast of earth.

This vital secret was inherited from the seed-ferns, back in misty days when the ferns were paramount. The conifers bore it forward; the true flowering plants were to carry it on and spread it in blossoming glory. Of that there was no sign in the Mesozoic forests. They must have been dark with an evergreen darkness, upright with a stern colonnaded strength. For they developed the power of building wood out of earth, not the punky wood of the tree ferns, but timber as we know it.

And we know no timber like the conifers'. No other trees are cut on such a scale. Where they grow, wooden cities swiftly rise, railroads are bent to them, mushroom fortunes arise from them, great fleets built to export them. Scandinavia is one vast lumber camp, supplying western Europe; Port Oxford cedar of Oregon crosses the ocean in a perpetual stream of logs, supplying Japan and China; Kauri pines of New Zealand feed the wood hunger of barren Australia. The world's books and newspapers are printed on coniferous pulp; it is driving silk and cotton to the wall, as sources of cellulose and textile fibre. For beautiful grains, for capacity to take stains, the evergreen woods are incomparable. The living conifers are to us what the dead coal forests are.

But they can be replenished. They can be grown and cut as crops, and they yield a profit on poor sandy and rocky soil, or in swampy lands where no other crop could be hopefully tilled. Thrifty, fertile, tough, industrial, they are of all tress the most practical. Ancient in lineage beyond all others, they rise tall and straight in the pride of their aristocracy. Sea-voiced, solemn, penciled against the sky, their groves are poetic as no leafier places. Conifers stand in the sacred temple yards of Japan, where, with venerating care, their old limbs are supported by pillars. They line the solemn approaches to the tombs of the Chinese emperors at Jehol. Solomon sought them in the peaks of Lebanon for his temple. But in all the world there are none like those in our western states.

And it was in the Black Hills of Wyoming that a fragment of the Mesozoic lay hidden till the days when the West came to be called new country. Miners on their way to Deadwood, cowboys riding herd, found strange stone shapes, and broke off fragments. What lay in those calloused brown fingers, turned over curiously, ignorantly, was once sprung in the Gothic glooms of the Mesozoic forests. These were cycads, a kind of Gymnosperm which must have formed the undergrowth of those prehistoric coniferous woods, hundreds and hundreds of species of them. A few linger today, scattered thinly over the tropics of the world. Some call them sago-palms; they have an antique look, stiff, sparse and heavy; crossed in pairs upon a coffin, they impart a funebrial dignity. Cretin of stature, for the most part, growing sometimes only six feet in a thousand years, they are beloved in the Japanese dwarf horticulture, cherished in family pride there, since a cycad of even moderate size may represent a long domestic continuity.

What pride, then, and what a ring of age was there in the first set of fossil cycads from the Black Hills rim to reach the men of science at the National Museum in 1893! Professor Lester Ward hastened to the field, and what he found there, besides the bones of a great dinosaur and the petrified logs of old conifers, were not imprints but complete petrifactions. Atom by atom the living tissue had been replaced by stone. Here were hundreds of fruits, all the leaves a gloating paleobotanist could desire, perfect trunks, every detail of wood structure preserved, and dozens of species, some dwarf, some colossal.

Ward took back with him what he could. Other students hurried to the find; Yale and the Universities of Iowa and Wyoming have great collections from Deadwood, and the government museums too. Tourists carted away entire specimens, and what remained might have been utterly scattered and destroyed, had not Professor G.R. Wieland saved the last rich tract in the Black Hills. Close to the mountain where Borglum carved his heroic profiles, the scientist filed on the area under the homestead laws, and then presented his claim to his country. It has since been made Cycad National Monument.

These cycads, when the world was young and they were flourishing, must have brought into the dark monotony of the evergreen forests the first bright splashes of color. For the seeds of cycads are gorgeous scarlet or yellow or orange, borne on the edge of the leaf or commonly in great cones. They are sweet and starchy to the taste, and perhaps Archaeopteryx, that first feathered bird in all time, crunched them in the teeth that he still kept, a reminder of his lizard ancestry. So, it may be, the earliest animals came to aid in the dissemination of plants, as squirrels do today, and birds. Somehow, at least, the cycads over-ran the world. Their reign had grandeur, but its limits narrowed. There is evidence that some of the Mesozoic cycads flowered only at the end of their immensely long lives – a thousand years, perhaps. Then, after one huge cone of fruit was set, the plant died to the very root. A hero's death, but a plan ill fit to breed a race of heroes. In the cupped hand of the future lay other seeds, with a fairer promise.

6-3 Habits and Habitats – Robert Claiborne

Robert Watson Claiborne (1919–1990) was born in Buckinghamshire, England, and educated at MIT, Antioch College, and New York University. After working as a lathe operator, union official, folk singer and music teacher, he pursued a successful career as an editor and author. He wrote or edited over a dozen books, ranging from works on the English language to medical genetics. In the following extract from Climate, Man, and History *(1970) Claiborne gives a lively, though simplified, account of the lifestyle of some early hominids.*

We pick up man's fossil record at just about the time that the Pliocene was giving way to the Glacial or Pleistocene epoch. What the fossils reveal is a creature – several creatures, in fact – unequivocally adapted to ground life. The long drying out of the Pliocene had done its work well. When we examine the skulls of these animals, we find that the foramen magnum – the hole through which the nerves of the spinal cord enter the skull – lies almost at the bottom, rather than toward the rear as in tree-dwelling apes, meaning that the creatures' heads were balanced on the tops of their spines, like our own, rather than hunched forward. Leg, foot, and pelvis bones tell the same story: the animals stood and moved erect, or nearly so, though their gait was apparently a trot or waddle rather than the striding walk of our own species. Their feet were no longer capable of gripping a tree-branch, as their ancestors' feet presumably had been; doubtless they could and did climb trees for food or safety, even as men do today, but they did so no more efficiently than we do.

I have said "animals" – but were these creatures merely animals? Here we must drag in that old cliché of academic dialogue: Define your terms. What is a mere animal? Or, putting it in a more meaningful way, what is a man?

We have already noted that the Pliocene ground apes almost certainly used tools. Chimpanzees do – and not simply as missiles. They have been seen to use sticks for digging and rocks for cracking nuts; they poke thin twigs into termite nests and then withdraw them covered with delicious termites; on occasion they even carry the tools around with them. . . . These observations force us to define man as an animal that makes tools *to a pattern*. It does not simply modify objects, but does so in a uniform, customary, one might almost say habitual way. . . .

Judge by this criterion, the apelike creatures that inhabited southern and eastern Africa in the early Pleistocene stood on the very threshold of humanity, for they fashioned tools out of stones and, in some cases, fashioned them to a pattern. The tools are terribly primitive: pebbles ranging from ping-pong to billiard ball size, with a few chips knocked off one end to make a crude cutting or chopping edge. Some of them, indeed, were apparently chipped by nature – e.g., being rolled about in the bed of a stream – and are identifiable as tools only through being dug up on "living floors" miles from where similar pebbles are found today. Others are so similar to the "nature-made" group as to recall the classic remark of a French prehistorian: "Man made one, God made ten thousand. God help the man who has to distinguish the one from the ten thousand!" Yet by carefully examining the appearance and angle of the chipping, we can be reasonably sure that man, not God, made them. . . .

And here we get into one of those hassles over terminology which reminds us again that science is not quite as rational as some of its practitioners would have us believe. When the first skull of one of these man-apes was discovered in South Africa, nobody, not even its discoverer, was expecting to find a manlike creature in that region. Accordingly it was christened merely *Australopithecus* – the "southern ape." Subsequently, other specimens were discovered with tools which made it clear that they were more than apes, and were christened *Paranthropus* ("very like man"), *Plesianthropus* ("nearly man") and *Zinjanthropus* ("man of Zinj" – an old word for East Africa). Obviously the splitters have been at work here. Indeed, they have pushed one group of fossils into still another genus, or own – *Homo*, meaning "man" without trimmings. This step Elwyn Simons, who might be called the lumper's lumper, calls "a conjecture derived from a hypothesis based on an assumption." Simons and the other lumpers have tried to bring some order out of this terminological chaos by reducing all the genera to one. But they were stuck with the original generic name coined in 1924, so that the earliest creature that was definitely not an ape is still known as the southern ape – Australopithecus.

Much of what we know about Australopithecus we owe to the famous diggings at Olduvai Gorge in Tanzania, East Africa. The gorge, a sort of miniature Grand Canyon, was cut out by a now-vanished river which exposed fossil deposits covering, with only one important gap, something like a million and a half years of prehistory. Moreover, by a singular bit of geological luck, the gorge lies close to two now-extinct volcanoes, Ngorongoro and Lemagrut, whose periodic eruptions during the Pleistocene laid down numerous layers of ash by which, using the potassium–argon technique, the Olduvai deposits can be dated. Due largely to the skillful and devoted labors of Louis Leakey and his wife Mary, Olduvai has yielded up a singularly detailed and dated record of the ground apes' emergence into humanity. Their findings have also helped to make chronological sense out of man–ape fossils from South Africa's Transvaal, which, though quite as abundant as those in Olduvai, did not have the luck to be laid down in a volcanic area.

Perhaps the most interesting fact about the Australopithecuses of Olduvai and the Transvaal is that they came in at least two (Leakey would say three or four) species, the Australopithecus africanus, was a slender four footer, no bigger than a modern pygmy. Its cousin, Australopithecus robustus (Paranthropus, Zinjanthropus) was, as its name suggests, bigger – about five feet – and burlier. From all the evidence, the two species seem to have been more or less contemporary (at Olduvai, much more than less) during the earlier part of their careers on earth. The simultaneous existence of two closely related man–ape species is puzzling, because it is exceptional. There is today only one species of gorilla, one species of chimpanzee, of orangutan, and, for that matter, of man. The reason seems to lie in a general principle of ecology: Where two closely related species occupy the same territory (and at Olduvai we are pretty certain they did), they must be adapted to different ways of life. If both occupy the same "ecological niche," either one species will displace the other in fairly short order or, much more likely, they will not evolve into two species in the first place.

One clue to the explanation of the two Australopithecine species comes from the bones themselves. I have on my living room table a plaster cast of a robustus skull which I have doctored with furniture stains so that it makes a fairly plausible-looking fossil. It does not, needless to say, look much like a human skull; perhaps the most striking difference emerges when I turn it over and examine the upper teeth (the lower jaw was missing in the specimen from which the cast was made). The biting teeth (incisors) are not very remarkable; they are, if anything, smaller than our own. But the molars are enormous – massive, and so broad that they look as if they had been set in the jaw sideways.

The top and back of the skull are also missing, but from descriptions of it I know that robustus had a bony ridge along the top of his cranium, similar to, but smaller than, that on the skull of a modern gorilla. Mechanically, the function of such a ridge is to give a strong anchorage to the massive thick muscles which move the lower jaw. Thus the ridge with its vanished muscular attachments, tells the same story as the massive molars: robustus lived on a bulky diet which required lots of chewing. It seems very likely, then, that like the gorilla he was a vegetarian, or close to it, living largely or entirely on the traditional primate diet of fruits and shoots during the rainy season and on roots and seeds during the dryer parts of the year. This conjecture is strengthened by microscopic studies of robustus teeth; these reveal scratches apparently made by sand grains, which presumably were carried into his mouth with the roots.

Australopithecus africanus, with neither skull ridge nor massive molars, seems to have adapted to a considerably more catholic diet. From bones found on his living floors he ate lizards, tortoises, various kinds of small mammals, and the young of some medium-sized species such as antelopes – along with, we can be sure, such vegetable foods as he could get.

This hypothesis (and I should stress that it is still only a hypothesis) explains a number of other apparent facts about the two species. From what I have said about their contrasting life patterns, it will be obvious that while both could survive in a fairly moist savannah (which the Olduvai area seems to have been some two million years ago), only africanus could have managed to make a living on a dry savanna or steppe, since the vegetable foods to which robustus' anatomy – and perhaps other things – adapted him would have been too sparse for survival.

The South African deposits seem to confirm this, notably in the sand found in the various layers. Where we find africanus fossils, we find sand whose grains are rounded – evidently by being blown along for considerable distances, perhaps from the Kalahari some hundreds of miles to the west. And wind-blown sand, obviously, means a dry climate. The robustus fossils, on the other hand, are found with sand having sharp, angular grains – evidence that the climate was wetter. Still further confirmation comes from another curious fact. Robustus did not evolve significantly between his first appearance in the fossil record, some two million years ago, and his final bow, perhaps as much as a million and a half years later. During the same period, however, africanus changed into quite a different creature – larger, brainier, and in other respects more human – to the point where he had become a full-fledged member of the genus *Homo*.

It seems at least plausible that the difference stemmed from diet. I have already noted the important mental differences between herbivores and predators. Vegetable food, provided it is tolerably plentiful, offers little mental challenge, since all you have to do is find it and eat it. Animals, however, must not only be found but also caught – which, unless you are as superbly equipped by nature as the cats and dogs, means living by your wits. Africanus, it seems, must have been smarter than robustus. It is quite possible, indeed, that of the two species only he was a toolmaker. At any rate, no fabricated tools have ever been found with robustus bones – unless africanus bones were also present. Africanus fossils, on the other hand, often occur without robustus remains – but with tools.

Thus it appears that not all the ground apes learned the lesson of the savanna equally well. Robustus, sticking to a vegetable diet and adapting himself to it anatomically, remained tied to a particular climate, one that was neither too dry nor too wet. Africanus, on the other hand, evolved in the direction of less anatomical specialization, not more. From an evolutionary standpoint, he cultivated his mind, and reaped his reward by ultimately becoming the one animal that can survive amid rain forest, desert, or polar ice. Robustus remained a manlike ape, ultimately driven to the evolutionary wall (and quite possibly eaten) by his more adaptable cousin. Africanus became man.

6-4 Diplodocus, The Dinosaur – James A. Michener

James Albert Michener (1907–1997) had an extraordinary career. After a youthful stint as a member of a traveling roadshow and as a sports columnist, he graduated from Swarthmore and what is now the University of Northern Colorado, became a teacher, college professor, naval officer and naval historian in World War II, popular writer, congressional candidate, government advisor, and television producer. He did not begin his fictional writing until his late thirties but he subsequently produced not only twenty plus novels and five volumes of short stories – one, Tales of the South Pacific (1947), won the Pulitzer Prize – but also an equal number of works of non-fiction, on topics ranging from social studies to Japanese art, the Kent State confrontation, American values, the electoral system, history, sports, and travel. One of the most prolific and popular authors of the twentieth century, his lengthy epic novels often describe the origin and geological history of the land in which the plots and family sagas are set.

The extracts we present here and on page 152 are from Creatures of the Kingdom (1993), a collection of animals presented in his earlier works. The extract below – originally written for Centennial (1974) – is a first-hand account of "Diplodocus, The Dinosaur."

Toward dusk on a spring evening one hundred and thirty-six million years ago a small furry animal less than four inches long peered cautiously from low reeds growing along the edge of a tropical lagoon that covered much of what was to be the present state of Colorado. It was looking across the surface of the water as if waiting for some creature to emerge from the depths, but nothing stirred. From among the fern trees to the left there was movement, and for one brief instant the little animal looked in that direction.

Shoving its way beneath the drooping branches and making considerable noise as it awkwardly approached the lagoon for a drink of water, came a medium-sized dinosaur, lumbering on two legs and twisting its short neck from side to side as it watched for larger animals that might attack.

It was about three feet tall at the shoulders and not more than six feet in length. Obviously a land animal, it edged up to the water carefully, constantly jerking its short neck in probing motions. In paying so much attention to the possible dangers on land, it overlooked the real danger that waited in the water, for as it reached the lagoon and began bending down in order to drink, what looked like a fallen log lying partly submerged in the water lunged toward him. It was a crocodile, armored in heavy skin and possessing powerful jaws lined with deadly teeth. But it had moved too soon. Its well-calculated grab at the reptile's right foreleg missed by a fraction, for the dinosaur managed to withdraw so speedily that the great snapping jaw closed not on the bony leg, as intended, but only on the soft flesh covering it.

There was a ripping sound as the crocodile tore off a strip of flesh, and a sharp guttural click as the wounded dinosaur responded to the pain. Then peace returned. The dinosaur could be heard for some moments retreating. The disappointed crocodile swallowed the meager meal it had caught, then resumed its log camouflage.

As the furry little animal returned to its earlier preoccupation of staring at the surface of the lagoon, it became aware, with a sense of panic, of wings in the darkening sky. At the very last moment it threw itself behind the trunk of a gingko tree, flattened itself out and held its breath as a large flying reptile swooped down, its sharp-toothed mouth open, and just missed its target. Still flat against the moist earth, the little animal watched in terror as the huge reptile banked low over the lagoon and then came straight at the crouching animal, but, abruptly, it had to swerve away because of the gingko roots. Dipping one wing, the reptile turned gracefully in the air, then swooped down on another small creature hiding near the crocodile, unprotected by any tree. Deftly it opened its beak and caught its prey, which shrieked as it was carried aloft. For some moments the little animal hiding behind the gingko tree watched the flight of its enemy as the reptile dipped and swerved through the sky like a falling feather, finally vanishing with its catch.

The little watcher could breathe again. Unlike the great reptiles, which were cold-blooded and raised their babies from hatching eggs, it was warm-blooded and came from the mother's womb. It was a pantothere, one of the earliest mammals and progenitor of later types like the opossum, and it had scant protection in the swamp. Watching cautiously lest the flying hunter return, it ventured forth to renew its inspection of the lagoon, and after a pause, spotted what it had been looking for.

About ninety feet out into the water a small knob had appeared on the surface. It was only slightly larger than the watching animal itself, about six inches in diameter. It seemed to be floating on the surface, unattached to anything, but actually it was the unusual nose of an animal that had its nostrils on top of its head. The beast was resting on the bottom of the lagoon and breathing in this unique manner.

Now, as the watching animal expected, the floating knob began slowly to emerge from the waters. It was attached to a head, not extraordinarily large but belonging obviously to an animal markedly bigger than either the first dinosaur or the crocodile. It was not a handsome head, nor graceful either, but what happened next displayed each of those attributes.

The head continued to rise from the lagoon, higher and higher in one long beautiful arch, until it stood twenty-five feet above the water, suspended at the end of a long and graceful neck. It was like a ball extended endlessly upward on a frail length of wire, and when it was fully aloft, with no body visible to support it, the head turned this way and that in a delicate motion, surveying the world that lay below.

The head and neck remained in this position for some minutes, sweeping in lovely arcs of exploration. Apparently what the small eyes on either side of the projecting nose had seen reassured the beast, for now a new kind of motion ensued.

From the surface of the lake an enormous mound of dark flesh began slowly to appear, muddy waters falling from it as it rose. The body of the great reptile looked as if it were about twelve feet tall, but how far into the water it extended could not be discerned; it surely went very deep. Now, as the furry animal on shore watched, the massive beast began to move, slowly and rhythmically. Where the neck joined the great dark bulk of the body, little waves broke and slid along the flanks of the beast. Water dripped from the upper part of the animal as it moved ponderously through the swamp.

The reptile appeared to be swimming, its neck probing in sweeping arcs, but actually it was walking on the bottom, its huge legs hidden by water. And then, as it drew closer to shore and entered shallower water, an enormous tail emerged. Longer than the neck and disposed in more delicate lines, it extended forty-four feet, swaying slightly on the surface of the lagoon. From head to tip of tail, the reptile measured eighty-seven feet.

Up to now it had looked like a snake, floundering through the lagoon, but the truth was revealed when the massive legs became visible. They were enormous, four pillars of great solidity but attached to the torso by such crude joints that although the creature was amphibious, it could not easily support herself on dry land, where water did not buoy it up.

With slow, lumbering strides the reptile moved toward a clear river that emptied into the swamp, and now its total form was visible. Its head reared thirty-five feet; its shoulders were thirteen feet high; its tail dragged aft some fifty feet; it weighed nearly thirty tons.

It was a diplodocus, not the largest of the dinosaurs and certainly not the most fearsome. This particular specimen was a female, seventy years old and in the prime of life. She lived exclusively on vegetation, which she now sought among the swamp waters. Moving her head purposefully from one plant to the next, she cropped off such food as she could find. This was not an easy task, for she had an extremely small mouth studded with minute peglike front teeth and no back ones for chewing. It seems incomprehensible that with such ineffectual teeth she could crop enough food to nourish her enormous body, but she obviously did.

After finishing with such plants as were at hand, she moved into the channel. The mammal, still crouching among the roots of the gingko tree, watched with satisfaction as she moved past. It has been afraid that she might plant one of her massive feet on its nest as another dinosaur had done, obliterating both the nest and its young. Indeed, diplodocus had left underwater footprints so wide and so deep that fish used them as nests. One massive footprint might be many times as wide as the fish was long.

And so diplodocus moved away from the lagoon and the apprehensive watcher. As she went she was a veritable poem in motion. Placing each foot carefully and without haste, and assuring herself that at least two were planted solidly on the bottom at all times, she moved like some animated mountain, keeping the main bulk of her body always at the same level, while her graceful neck swayed gently and her extremely long tail remained floating on the surface.

The various motions of her great body were always harmonious; even the plodding of the four gigantic feet had a pleasant rhythm. With the graceful neck and the tapering tail, this large reptile epitomized the beauty of the animal kingdom as it then existed.

She was looking for a stone. For some time she had instinctively known that she lacked a major stone, and this distressed her. Keeping her head low, she scanned the bottom of the stream but found no suitable stones.

This forced her to move upstream, the delicate motion of her body conforming to the shifting bottom as it rose slightly before her. Now she came upon a wide selection of stones, but prudence warned her that they were too jagged for her purpose, and she ignored them. Once she stopped and turned a stone over with her blunt nose. She scorned it – too many cutting edges.

Preoccupied with her futile search, she failed to notice the approach of a rather large land-based dinosaur that walked on two legs. He did not come close to approaching her in size, but he was quicker of motion, and had a large head, a gaping mouth and a ferocious complement of jagged teeth. He was a meat-eater, always on the watch for the giant water-based dinosaurs who ventured too close to land. He was not large enough to tackle a huge animal like diplodocus if she was in her own element, but he had found that usually when the large reptiles came into the stream, there was something physically wrong with them, and twice he had been able to hack one down.

He approached diplodocus from the side, stepping gingerly on his two powerful hind feet, keeping his two small front feet ready like hands to grasp her should she prove to be in a weakened condition. He was careful to keep clear of her tail, for this was her only weapon.

She remained unaware of her would-be attacker, and continued to probe the river bottom for an acceptable stone. The carnivorous dinosaur interpreted her lowered head as a sign of weakness. He lunged at the spot where her vulnerable neck joined the torso, only to find that she was in no way incapacitated, for at the moment of attack she twisted adroitly and presented him with the broad and heavy side of her body. This made him stumble back. As he did so, diplodocus stepped forward, and slowly swung her tail in a mighty arc, hitting him with such force that he was thrown off his feet and sent crashing into the brush.

One of his small front feet was broken by the blow and he uttered a series of awk-awks, deep in his throat, as he shuffled off. Diplodocus gave him no more attention and resumed her search for the stone.

Finally she found what she wanted. It weighed about three pounds, was flattish on the ends and both smooth and rounded. She nudged it twice with her snout, satisfied herself that it suited her purposes, then lifted it in her mouth, raised her head to its full majestic height and swallowed the stone. It slid easily down her long throat into her gullet and from there into her grinding gizzard, where it joined six smaller stones that rubbed together gently and incessantly as she moved. This was how she chewed her food, the seven stones serving as substitutes for the molars she lacked.

With awkward yet oddly graceful motions she adjusted herself to the new stone, and could feel it find its place among the others. She felt better all over and hunched her shoulders, then twisted her hips and flexed her long tail.

6-5 A Window on the Oligocene – Berton Roueché

The Florissant fossil beds of Colorado are unique for the abundance of delicate plant and insect fossils they contain. That the area should be preserved for study and enjoyment for future generations as a National Park or National Monument seems obvious. Yet to do so proved a long and arduous task. Initial attempts were made as early as 1921, and it was only in 1971 that partial success finally was attained. The political procedures and maneuverings, and the fragile beauty of Florissant fossils, are all part of the story in Berton Roueché's contribution, "A Window on the Oligocene." Berton Roueché (1911–1994) was born in Kansas City, Missouri, graduated from the University of Missouri, and was a journalist and author, who was associated for over fifty years with the New Yorker. He wrote some twenty books, including five novels, as well as travel and nature works. Our extract is taken from a 1971 article by Roueché that appeared in the New Yorker as part of his widely read feature series, The Annals of Medicine.

There are few more beautiful natural parks than the mountain valley that encloses the village of Florissant (pop. 102), in Teller County, Colorado, about a hundred miles southwest of Denver, on the western slope of the Front Range of the Rockies. This valley has been variously known as South Park and Castello's Ranch, but it now takes its name from the village. It is ten miles long and a mile or two wide, and it follows a gently winding course, with many sudden, serpentine arms. The valley floor is a rolling meadow. It flows, deep in grass, around an occasional thrust of rust-red granite outcrop, and it is framed on the east and west by low hills crowned with open groves of ponderosa pine. Florissant is high country – almost nine thousand feet – but above its highest hills, encircling the infinite distance, are the peaks of higher mountains. There are antelope in the Florissant valley, and deer and elk and bear and badgers and coyotes and porcupines, and in July the meadows are a garden of wild flowers – bluebells, yellow clover, purple locoweed, pink trumpet phlox, alpine daisies, brown-eyed Susans, flaming Indian paintbrush. The beauty of Florissant, however, is not only its natural glory. It has another, a rarer and a richer one. Beneath the flowering Florissant meadows lies a treasury of paleontological history. Florissant is one of the richest repositories of plant and insect fossils in the world. . . .

The Florissant fossil beds were formed by a series of volcanic eruptions that began some thirty-eight million years ago. This places the beds, by geologic time, in the Oligocene epoch of the Cenozoic era. Florissant was then, as now, a valley, but it was a highland rather than a mountain valley, and it had a mild, humid, probably subtropical climate. A wandering stream flowed south through the valley, and the banks of the stream and the slopes above it were covered with willow and oak and pine and beech and giant sequoia. The first, progenitive volcanic eruption changed that distant scene. Its convulsions deposited a levee of volcanic mud at the mouth of the valley, blocking the outward flow of the stream, and the stream thus dammed became a pond, and the pond became a lake. Fish flourished there, and insects swarmed, and trees and shrubs grew dense along the shore. The volcano presently came to life again, and for perhaps three million years there were frequent wild eruptions. With every eruption, a cloud of fine volcanic ash blew into Florissant. The ash fell thickly on the lake, and as it settled it carried with it to the bottom a multitude of leaves and insects and tiny fish and stored them there in a thin, impressionable casing. Millennia passed, and the layered deposits of silt and ashy shale accumulated. . . .

The first Florissant fossils to be properly studied were those collected by Scudder in 1877. That visit, though brief, was prophetically productive. In the mere five days that Scudder spent in the beds, he and three companions (two fellow-scientists and a local rancher Adam Hill) gathered a total of around five thousand specimens. The quality of the material also was prophetically high. "The shales of Florissant," Lesquereux enthusiastically noted, "have preserved the most delicate organisms – feathers, insects, small flies, petals, even anthers and stamens of flowers.". . .

The paleontological importance of Florissant that Scudder first proclaimed has been ardently confirmed by many other paleontologists. At least a dozen major scientific expeditions have worked the Florissant beds and assembled large and comprehensive collections. . . . The respect of the owners of Florissant for science has undoubtedly been a factor in the preservation of its fossil beds for continued scientific study. It has not, however, been an important one. Florissant has been largely preserved by the accident of nature that made it cattle country.

It has long been felt that the fossil riches of Florissant are too fragile to be so lightly guarded. The desirability of some defense against the human compulsion to rearrange nature which would be stronger than a long winter and a short summer was briefly considered as early as 1921. In that year, a group of interested scientists and others suggested that Florissant be brought into the protective embrace of the National Park Service. The suggestion received a politely sympathetic hearing, but no action was taken, and the interest of even its sponsors faded away and finally vanished under the impact of the Depression and the Second World War. It was revived, in 1953, by Edmund B. Rogers, then (he has since retired) the superintendent of Yellowstone National Park, with a testimonial based on first-hand exploration and entitled "Florissant Fossil Shale Beds, Colorado." Rogers' recommendation excited sufficient influential interest to set in motion the standard salvational procedure. This involved a pilot study, a preliminary report, and a formal proposal, and it proceeded at the standard pace. A formal proposal to preserve six thousand of the twelve thousand acres at Florissant as a Florissant Fossil Beds National Monument was approved by the director of the National Park Service in 1962. The following year, an enabling bill was introduced in the House of Representatives. It died, ignored. A similar bill was introduced in 1965, and it was similarly received. Two years later, in 1967, the House was offered a third Florissant bill. An amendment to it reduced the size of the proposed Monument from six thousand to one thousand acres. This truncated preserve attracted a wave of penny-pinching support, and the bill was approved with little opposition and sent along to the Senate. There it once more died. Another year went by and another Florissant bill was conceived. The new proposal differed from its predecessor in that it restored the cut in acreage and originated in the Senate. It was introduced on February 4, 1969. A few days later, an identical bill was introduced in the House. The winter passed, and also the spring. Then, on June 20th, the Senate approved its Florissant bill. The House bill remained in committee. Summer came on. . . .

Two weeks went by. I watched the New York papers, but there was no mention of the Florissant affair. Then, around the middle of August, I got a letter from a friend in Denver. The letter enclosed a clipping from the Washington bureau of the *Rocky Mountain News*. It read, "The House Monday approved on a voice vote a bill creating the 6,000-acre Florissant Fossil Beds National Monument 35 miles west of Colorado Springs. The bill now goes to the Senate, where approval of technical amendments is expected. The Senate earlier passed a measure identical in substance to the House version. Following anticipated Senate passage, the measure will go to President Nixon for his signature. . . ." The story had no dateline, but the date of issue was printed at the top of the clipping. It was August 5th. That had left a week for the unleashing of the bulldozer. Too bad, but not irreparable. On August 20th, President Nixon signed the bill bringing Florissant under the protection of the National Park Service. A few days later, the bulldozer had vanished and the culvert liner had been hauled away.

6-6   A Fish Caught in Time – Samantha Weinberg

Samantha Weinberg is a writer, journalist, and former feature editor of Harpers and Queen, *who, though born in London in 1966 and educated at Cambridge, is of South African extraction. That provides a link to her dramatic account of the discovery of the coelacanth in her book* A Fish Caught in Time *(2000), "part biography . . . part natural history, part adventure story."*

> Consider now the Coelacanth,
> Our only living fossil,
> Persistent as the amaranth,
> And status quo apostle.
> It jeers at fish unfossilized
> As intellectual snobs elite;
> Old Coelacanth, so unrevised
> It doesn't know it's obsolete.
> —Ogden Nash

I saw my first coelacanth on February 5, 1992. It was a sultry day in the Comoros, a remote and beautiful archipelago on the western fringe of the Indian Ocean. I was researching a book about a French mercenary, Bob Denard, who had fallen in love with the tiny outpost, and staged successive coups over more than a decade, in a strong-arm attempt to make the islands his own. Only a few weeks into a six-month stay, I too was well on the way to becoming entrapped by the mysterious charm of the Comoros.

After several hours exploring the warren of narrow lanes of the medieval old town of Moroni, the capital, I was hot and bathed in sweat. I took refuge inside the grandly named Centre Nationale des Réchèrches Scientifiques – the country's only museum – hoping to find a cool space to catch my breath. But it was just as hot – only the bank has air-conditioning – and as I was about to leave, a strange exhibit caught my eye: a glass tank containing a large, stuffed, on the face of it ugly fish. It was unlike any fish I had seen before – its body was covered in scaly armor and its fins were attached by fat limb-like protuberances. It had large, yellowy-green eyes, and a surprisingly gentle expression on its prehistoric-looking face.

The sign beside the exhibit identified the creature as a coelacanth, the marine equivalent of a dinosaur – only older – and a present-day inhabitant of Comoran waters. But there was little more information. I studied the coelacanth for a long time, rolling the strange syllables of its name around my tongue. It conjured up distant memories of nature classes at school, and picture books of bizarre, long-lost creatures. How, I wondered, had it survived, virtually unchanged, for all that time, lost at the bottom of the vast, cold ocean, watching silently as other creatures evolved and became extinct? *Homo sapiens* first walked on earth only a hundred thousand years ago; the fish before me, suspended in murky formalin, pre-dated modern man by 399.9 million years.

The coelacanth stayed in my mind during the rest of my time on the Comoro Islands. There was a shabby hotel named after it; I visited a taxidermist on Anjouan – the second-largest island – who had one on display. In my book about Denard, I used the coelacanth to illustrate the remoteness of the islands, and when I got home, I wrote a story using it as a metaphor for innocence, its discovery a metaphor for colonization. Since then, it has stayed in my thoughts, insinuating itself into conversation, trespassing through my dreams.

At the end of 1997, I decided to do something about it. I started looking into the story of the coelacanth's discovery – digging up documents and papers from the bowels of the Natural History Museum in London – and the more I discovered, the more I was hooked. When my idea for a book about this most unlikely of subjects was commissioned, I was overjoyed, and immediately set out for my beloved islands.

I spent several months in the Comoros, over two visits, much of the time with the night fishermen from the south of Grande Comore, the largest of the four islands, and a known habitat of coelacanths. One night, I was taken fishing by Hassani Ahamada, a veteran of many years of lonely nights on the ink-black ocean.

We met on a black, lava rock beach near the village of Itsoundzou, in time to launch our small wooden pirogues in the last, pink light of day. Two to each boat, we pushed them off the rocks, then paddled quickly out to sea. My fellow fishermen were a quiet group: small and wiry, they wore tatty T-shirts and holey shoes, and battered palm frond hats on their heads, even though the sun was below their sightline. A short way from the shore, Hassani stopped briefly to let out a line, and within a minute he brought it in again, with a small silvery fish on the hook, to use as bait.

He then squatted on the front of two narrow benches that cross the narrow boat, knees under his armpits, beckoning to me to do the same. While I struggled to keep upright, Hassani rowed out to sea with strong, deft strokes, wielding his single short paddle first to one side, then to the other. His movements were swift, but controlled – even with its pair of spidery outriggers, the small pirogue is surprisingly unstable, and a sudden movement to one side could easily topple it. . . .

About five hundred meters from shore, Hassani carefully laid his paddles across the boat and prepared the line. He tied two flat black stones, collected from the beach, about eighteen inches above the baited hook, then let out the single line until it touched the sea bed, many hundreds of meters below. When the stone sinkers hit the bottom, he raised the line and jerked free the rocks. He then jiggled the bait along the ocean bed with a quick seesawing action of his arms, feeling its movement through his fingers.

I watched quietly, drinking in the silence of the warm night. We were both waiting, Hassani and I, for a quick tug on his line. We stayed out on the ocean for half a night, beneath the star-filled southern skies, adjusting our position from time to time, occasionally hooking a small fish. Although I knew we had less that a sliver of a chance of catching a coelacanth, I was still half hoping, half dreading that we would haul the magnificent, man-sized fish up from the ocean depths. But a coelacanth never so much as flicked a fin in our direction, and I think I was pleased.

And when a barely perceptible lightening of the sky indicated the coming of dawn, we paddled swiftly back to shore, pulled our boat over the smooth, rounded black lava rocks, and carried our catch to the village to wait for the market to open at daybreak.

I could have gone out with the night fishermen again and again: the rhythm of their lives seeps into one's soul. They have been fishing in the same way for hundreds, perhaps thousands of years – like the coelacanth, the ebbs and swells of the ocean are as familiar as heartbeats. I hope, wish, pray that both man and fish will be allowed to continue their quiet lives for a hundred, a thousand, perhaps a million more years.

6-7 Ape-like Ancestors – Richard E. Leakey

Richard Leakey, a paleoanthropologist born in Kenya in 1944 – the son of Louis and Mary Leakey, themselves both distinguished paleoanthropologists – has made major contributions to the study of human origins. Leakey dropped out of high school to start a safari business, but gradually resumed the interest in early hominids developed from his parents in his youth. He has led a series of successful field parties to Lake Turkana, Kenya, which provided evidence that Australopithecus was a contemporary of early members of our own genus Homo. He is a museum curator and author and has also had an active political career in Kenya. Our extract, taken from the book The Making of Mankind *(1981), describes a remarkable find by his mother, Mary Leakey (1913–1996), at Laetoli in Tanzania, about 25 miles south of Olduvai Gorge.*

By an astonishing combination of circumstances, a very ordinary event which happened some three-and-three-quarter million years ago led to what is probably the most dramatic archaeological discovery of this century. Three hominids left a trail of footprints that have been clearly preserved, presenting us with an amazing picture of a few moments in the lives of some of our ancestors.

The place where it happened is now called Laetoli, a wooded area near a volcanic mountain, Sadiman, some 40 kilometres (25 miles) south of the present-day Olduvai

Gorge in Tanzania. The dry season was probably nearing its end and the sight of gathering rain clouds promised welcome relief from months of drought. For a week or two the volcano had rumbled restlessly, occasionally belching out clouds of grey ash that settled over the surrounding countryside. Nothing violent or startling, just a steady background of subterranean stirrings such as is still experienced today from the Oldoinyo Lengai, 70 kilometres (45 miles) southeast of Olduvai. Like the ash from Lengai, the ash that Sadiman produced had a chemical composition that made it set like cement when dampened slightly and then dried by the sun.

As luck would have it, the end of that dry season was signalled by a scattering of brief showers. Large raindrops splashed onto the newly fallen ash, leaving tiny craters as on a miniature moonscape. The clouds passed; the promised downpour was yet to come. The carpet of ash was now in perfect condition for taking clear impressions: less rain and the ash would have blown about in the breeze, more rain and any impressions would have been washed away.

Following the rain, various animals left their tracks in the damp volcanic ash as they went their ways. Spring hares, guinea fowl, elephants, pigs, rhinoceroses, buffaloes, hyaenas, antelopes, a sabre-tooth tiger and dozens of baboons all made their marks. And so did three hominids. A large individual, probably a male, walked slowly towards the north. Following behind, then or a little later, was a smaller individual who for some reason placed his or her feet in the prints of the first individual. A youngster skipped along by their side, turning at one point to look to its left. The sun soon baked the prints into rock-hard impressions. More ash, rain and windblown sand covered and preserved the prints until they were discovered by a lucky chance in 1976.

'They are the most remarkable find I have made in my entire career,' says my mother, who is directing the excavations. 'When we first came across the hominid prints I must admit I was sceptical, but then it became clear that they could be nothing else. They are the earliest prints of man's ancestors, and they show us that hominids three-and-three-quarter million years ago walked upright with a free-striding gait, just as we do today.'

6-8   The Relic Men – Loren Eiseley

At one time or another, almost every geologist has been presented with a peculiarly formed rock mass which is interpreted by the finder to be almost anything a well-rounded imagination permits. Loren Corey Eiseley (1907–1977), with whimsy and a touch of pathos, relates how he reacted to a specimen of unusual proportion and contours. Eiseley was educated at the University of Nebraska and the University of Pennsylvania and taught anthropology at the University of Kansas, Oberlin and the University of Pennsylvania, where he served for thirty years in a variety of academic and administrative roles. He was both a distinguished anthropologist and a gifted writer, with particular skill in writing for the general reader. He published many popular essays, poems, and fourteen books. Our extract "The Relic Men" is from Eiseley's compilation of essays The Night Country *(1971).*

In the proper books, you understand, there is no such thing as a petrified woman, and I insist that when I first came to that place I would have said the same. It all happened because bone hunters are listeners. They have to be.

We had had terrible luck that season. We had made queries in a score of towns and tramped as many canyons. The institution for which we worked had received a total of one Oligocene turtle and a bag of rhinoceros bones. A rag picker could have done better. The luck had to change. Somewhere there had to be fossils.

I was cogitating on the problem under a coating of lather in a barbershop with an 1890 chair when I became aware of a voice. You can hear a lot of odd conversation in barbershops, particularly in the back country, and particularly if your trade makes you a listener, as mine does. But what caught my ear at first was something about stone. Stone and bone are pretty close in my language and I wasn't missing any bets. There was always a chance that there might be a bone in it somewhere for me.

The voice went off into a grumbling rural complaint in the back corner of the shop, and then it rose higher.

"It's petrified! It's petrified!" the voice contended excitedly.

I managed to push an ear up through the lather.

"I'm a-tellin' ya," the man boomed, "a petrified woman, right out in that canyon. But he won't show it, not to nobody. 'Tain't fair, I tell ya."

"Mister," I said, speaking warily between the barber's razor and his thumb, "I'm reckoned a kind of specialist in these matters. Where is this woman, and how do you know she's petrified?"

I knew perfectly well she wasn't, of course. Flesh doesn't petrify like wood or bone, but there are plenty of people who think it does. In the course of my life I've been offered objects that ranged from petrified butterflies to a gentleman's top hat.

Just the same, I was still interested in this woman. You can never tell what will turn up in the back country. Once I had a mammoth vertebra handed to me with the explanation that it was a petrified griddle cake. Mentally, now, I was trying to shape that woman's figure into the likeness of a mastodon's femur. This is a hard thing to do when you are young and far from the cities. Nevertheless, I managed it. I held that shining bony vision in my head and asked directions of my friend in the barbershop.

Yes, he told me, the woman was petrified all right. Old man Buzby wasn't a feller to say it if it 'tweren't so. And it weren't no part of a woman. It was a *whole* woman. Buzby had said that, too. But Buzby was a queer one. An old bachelor, you know. And when the boys had wanted to see it, 'count of it bein' a sort of marvel around these parts, the old man had clammed up on where it was. A-keepin' it all to hisself, he was. But seein' as I was interested in these things and a stranger, he might talk to me and no harm done. It was the trail to the right and out and up to the overhang of the hills. A little tarpapered shack there.

I asked Mack to go up there with me. He was silent company but one of the best bone hunters we had. Whether it was a rodent the size of a bee or an elephant the size of a house, he'd find it and he'd get it out, even if it meant that we carried a five-hundred-pound plaster cast on foot over a mountain range.

In a day we reached the place. When I got out of the car I knew the wind had been blowing there since time began. There was a rusty pump in the yard and rusty wire and rusty machines nestled in the lee of a wind-carved butte. Everything was leaching and blowing away by degrees, even the tarpaper on the roof.

Out of the door came Buzby. He was not blowing away, I thought at first. His farm might be, but he wasn't. There was an air of faded dignity about him.

Now in that country there is a sort of etiquette. You don't drive out to a man's place, a bachelor's, and you a stranger, and come up to his door and say: "I heard in town you got a petrified woman here, and brother, I sure would like to see it." You've got to use tact, same as anywhere else.

You get out slowly while the starved hounds look you over and get their barking done. You fumble for your pipe and explain casually you're doin' a little lookin' around in the hills. About that time they get a glimpse of the equipment you're carrying and most of them jump to the conclusion that you're scouting for oil. You can see the hope flame up in their eyes and sink down again as you explain you're just hunting bones. Some of them don't believe you after that. It's a hard thing to murder a poor man's dream.

But Buzby wasn't the type. I don't think he even thought of the oil. He was small and neat and wore – I swear it–pince-nez glasses. I could see at a glance that he was a city man dropped, like a seed, by the wind. He had been there a long time, certainly. He knew the corn talk and the heat talk, but he would never learn how to come forward in that secure, heavy-shouldered country way, to lean on a car door and talk to strangers while the horizon stayed in his eyes.

He invited us, instead, to see his collection of arrowheads. It looked like a good start. We dusted ourselves and followed him in. It was a two-room shack, and about as comfortable as a monk's cell. It was neat, though, so neat you knew that the man lived, rather than slept here. It lacked the hound-asleep-in-the-bunk confusion of the usual back-country bachelor's quarters.

He was precise about his Indian relics as he was precise about everything, but I sensed after a while a touch of pathos – the pathos of a man clinging to order in a world where the wind changed the landscape before morning, and not even a dog could help you contain the loneliness of your days.

"Someone told me in town you might have a wonderful fossil up here," I finally ventured, poking in his box of arrowheads, and watching the shy, tense face behind the glasses.

"That would be Ned Burner," he said. "He talks too much."

"I'd like to see it," I said, carefully avoiding the word *woman*. "It might be something of great value to science."

He flushed angrily. In the pause I could hear the wind beating at the tarpaper.

"I don't want any of 'em hereabouts to see it," he cried passionately. "They'll laugh and they'll break it and it'll be gone like – like everything." He stopped, uncertainly aware of his own violence, his dark eyes widening with pain.

"We are scientists, Mr. Buzby," I urged gently. "We're not here to break anything. We don't have to tell Ned Burner what we see."

He seemed a little mollified at this, then a doubt struck him.

"But you'd want to take her away, put her in a museum."

I noticed the pronoun, but ignored it. "Mr. Buzby," I said, "we would very much like to see your discovery. It may be we can tell you more about it that you'd like to know. It might be that a museum would help you save it from vandals. I'll leave it to you. If you say no, we won't touch it, and we won't talk about it in the town, either. That's fair enough, isn't it?"

I could see him hesitating. It was plain that he wanted to show us, but the prospect was half-frightening. Oddly enough, I had the feeling his fright revolved around his discovery, more than fear of the townspeople. As he talked on, I began to see what he wanted. He intended to show it to us in the hope we would confirm his belief that it was a petrified woman. The whole thing seemed to have taken on a tremendous importance in his mind. At that point, I couldn't fathom his reasons.

Anyhow, he had something. At the back of the house we found the skull of a big, long-horned, extinct bison hung up under the eaves. It was a nice find, and we coveted it.

"It needs a dose of alvar for preservation," I said. "The museum would be the place for a fine specimen like this. It will just go slowly to pieces here."

Buzby was not unattentive. "Maybe, Doctor, maybe. But I have to think. Why don't you camp here tonight? In the morning –"

"Yes?" I said, trying to keep the eagerness out of my voice. "You think we might–?"

"No! Well, yes, all right. But the conditions? They're like you said?"

"Certainly," I answered. "It's very kind of you."

He hardly heard me. That glaze of pain passed over his face once more. He turned and went into the house without speaking. We did not see him again until morning.

The wind goes down into those canyons also. It starts on the flats and rises through them with weird noises, flaking and blasting at every loose stone or leaning pinnacle. It scrapes the sand away from pipy concretions till they stand out like strange distorted sculptures. It leaves great stones teetering on wineglass stems.

I began to suspect what we would find, the moment I came there. Buzby hurried on ahead now, eager and panting. Once he had given his consent and started, he seemed in almost a frenzy of haste.

Well, it was the usual thing. Up, Down, Up, Over boulders and splintered deadfalls of timber. Higher and higher into the back country. Toward the last he outran us, and I couldn't hear what he was saying. The wind whipped it away.

But there he stood, finally, at a niche under the canyon wall. He had his hat off and, for a moment, was oblivious to us. He might almost have been praying. Anyhow I stood back and waited for Mack to catch up. "This must be it," I said to him. "Watch yourself." Then we stepped forward.

It was a concretion, of course – an oddly shaped lump of mineral matter – just as I had figured after seeing the wind at work in those miles of canyons. It wasn't a bad job, at that. There were some bumps in the right places, and a few marks that might be the face, if your imagination was strong. Mine wasn't just then. I had spent a day building a petrified woman into a mastodon femur, and now that was no good either, so I just stood and looked.

But after the first glance it was Buzby I watched. The unskilled eye can build marvels of form where the educated see nothing. I thought of that bison skull under his eaves, and how badly we needed it.

He didn't wait for me to speak. He blurted with a terrible intensity that embarrassed me, "She – she's beautiful, isn't she?"

"It's remarkable," I said. "Quite remarkable." And then I just stood there not knowing what to do.

He seized on my words with such painful hope that Mack backed off and started looking for fossils in places where he knew perfectly well there weren't any.

I didn't catch it all; I couldn't possibly. The words came out in a long, aching torrent, the torrent dammed up for years in the heart of a man not meant for this place, nor for the wind at night by the windows, nor the empty bed, nor the neighbors twenty miles away. You're tough at first. He must have been to stick there. And then suddenly you're old. You're old and you're beaten, and there must be something to talk to and to love. And if you haven't got it you'll make it in your head, or out of a stone in a canyon wall.

He had found her, and he had a myth of how she came there, and now he came up and talked to her in the long afternoon heat while the dust devils danced in his failing corn. It was progressive. I saw the symptoms. In another year, she would be talking to him.

"It's true, isn't it, Doctor?" he asked me, looking up with that rapt face, after kneeling beside the niche. "You can see it's her. You can see it plain as day." For the life of me I couldn't see anything except a red scar writhing on the brain of a living man who must have loved somebody once, beyond words and reason.

"Now Mr. Buzby," I started to say then, and Mack came up and looked at me. This, in general, is when you launch into a careful explanation of how concretions are made so that the layman will not make the same mistake again. Mack just stood there looking at me in that stolid way of his. I couldn't go on with it. I couldn't even say it.

But I saw where this was going to end. I saw it suddenly and too late. I opened my mouth while Mr. Buzby held his hands and tried to regain his composure. I opened my mouth and I lied in a way to damn me forever in the halls of science.

I lied, looking across at Mack, and I could feel myself getting redder every moment. It was a stupendous, a colossal lie.

"Mr. Buzby," I said, "that – um – er – figure is astonishing. It is a remarkable case of preservation. We must have it for the museum."

The light in his face was beautiful. He believed me now. He believed himself. He came up to the niche again, and touched her lovingly.

"It's okay," I whispered to Mack. "We won't have to pack the thing out. He'll never give her up."

That's where I was a fool. He came up to me, his eyes troubled and unsure, but very patient.

"I think you're right, Doctor," he said. "It's selfish of me. She'll be safer with you. If she stays here somebody will smash her. I'm not well." He sat down on a rock and wiped his forehead. "I'm sure I'm not well. I'm sure she'll be safer with you. Only I don't want her in a glass case where people can stare at her. If you can promise that, I –"

"I can promise that," I said, meeting Mack's eyes across Buzby's shoulder.

"And if I come there I can see her!"

I knew I would never meet him again in his life.

"Yes," I said, "you can see her there." I waited, and then I said, "We'll get the picks and plaster ready. Now that bison skull at your house . . ."

It was two days later, in the truck, that Mack spoke to me. "Doc."

"Yeah."

"You know what the Old Man is going to say about shipping that concretion. It's heavy. Must be three hundred pounds with the plaster."

"Yes, I know."

Mack was pulling up slow along the abutment of a bridge. It was the canyon of the big Piney, a hundred miles away. He got out and went to the rear of the truck. I didn't say anything, but I followed him back.

"Doc, give me a hand with this, will you?"

I took one end, and we heaved together. It's a long drop in the big Piney. I didn't look, but I heard it break on the stones.

"I wish I hadn't done that," I said.

"It was only a concretion," Mack answered. "The old geezer won't know."

"I don't like it," I said. "Another week in that wind and I'd have believed in her myself. Get me the hell out of here – maybe I do, anyhow. I tell you I don't like it. I don't like it at all."

"It's a hundred more to Valentine," Mack said,

He put the map in the car pocket and slid over and gave me the wheel.

7. Geotectonics

Tectonics involves the study of the form, arrangement and structure of the rocks of the Earth's crust. Its scale is comprehensive, ranging from concern with micro-faulting and minute folding to the study of the major architectural features of the Earth. Geotectonics involves these larger features: the grand arrangement of mountain chains, ocean trenches, volcanoes and earthquakes, and the forces that produce them. The interpretation of such great features depends upon a range of studies of lesser scale, including such things as the pattern of remnant magnetism in oceanic rocks, the distribution and nature of volcanoes, earthquake activity and intensity, patterns of heat flow, and many others.

It is only within the last thirty years that any coherent explanation of how the Earth works has been developed. Until then, the synthesis, such as it was, involved the repeated sequence of marine transgression and regression, the development and downsinking of elongated troughs (geosynclines) which, under compression, became uplifted into mountain ranges. Though mapped in methodical detail, and catalogued with exquisite refinement, these structures lacked both any clear relationship and any credible explanation, still less any global pattern.

In fact, those who had offered such explanations had been regarded with suspicion and even hostility, especially if they had the misfortune to belong to a profession outside the geological guild. Our extracts include two brief accounts of these contro-versies, and a review of other major issues of geotectonics.

Our present geotectonic model – plate tectonics – ranks as one of the great syntheses of science, so audacious in its concept that Harry Hess – one of its creators – described his first account of it as "geopoetry." Nothing in Earth science is untouched by it. Nothing will ever be quite the same again. Its predictive power is startling. It is so beautiful, so simple, that, like all great science, we should have guessed it long ago. But we did not, not only because it required the discovery of new data, but also because it involved new and unorthodox ways of looking at old data. And we are generally very orthodox in our scientific thinking.

We now conclude that the face of our planet, its continents and oceans, its mountain ranges and rift valleys are the surface expressions of a fundamental global architecture, each part of a larger unity, brought into existence by an ongoing process maintained and modified by the engine of Earth's interior. Not the least beautiful expression of this harmony is the view of our Home Planet from space, a view that carries for our generation both privilege and burden. It has been said that the greatest benefit from the whole space program may be this view of our Home Planet, brown and white, green and blue, veiled here and there by shifting cloud patterns of delicate beauty; solid but fragile, ancient but also constantly new, familiar yet seen for the first time.

7-1 From the Boundless Deep & the Birth of the Rockies – James A. Michener

The love of the larger dimension is well reflected in many books written by James Michener (1907–1997), whom we first met on page 137. The following, also from the anthology Creatures of the Kingdom *(1993), exemplifies*

Michener's skill in geological descriptions. We include here a description of the ocean taken from Hawaii *(1959) and an account of the interior of the Earth taken from* Centennial *(1974).*

Millions upon millions of years ago, when the continents were already formed and the principal features of the earth had been fixed, there was, then as now, one aspect of the world that dwarfed all others. It was a mighty ocean, lying to the east of the largest continent, a restless, ever-changing, gigantic body of water that would later be described as pacific.

Over its brooding surface immense winds swept back and forth, whipping the waters into towering waves that crashed down upon the world's seacoasts, tearing away rocks and eroding the land. In its dark bosom, strange life was beginning to form, minute at first, then gradually of a structure now lost even to memory. Upon its farthest reaches birds with enormous wings came to rest, and then flew on.

Agitated by a moon stronger then than now, immense tides ripped across this tremendous ocean, keeping it in a state of torment. Since no great amounts of sand had yet been created, the waters where they reached shore were universally dark, black as night.

Scores of millions of years before man rose from the shores of the ocean to perceive its grandeur and to venture forth upon its turbulent waves, this eternal sea existed, larger than any other of the earth's features, more enormous than the sister oceans combined, wild, terrifying in its immensity and imperative in its universal role.

How utterly vast it was! How its surges modified the very balance of the earth! How completely lonely it was, hidden in the darkness of the night or burning in the dazzling power of a sun younger than ours.

At recurring intervals the ocean grew cold. Ice piled up along its extremities and pulled vast amounts of water from the sea, so that the wandering shoreline of the continents sometimes jutted miles farther out than before. Then, for a hundred thousand years, the ceaselessly turbulent ocean would tear at the exposed shelf of the continents, grinding rocks into sand and incubating new life.

Later the fantastic accumulations of ice would melt, setting cold waters free to join the heaving ocean, and the coasts of the continents would lie submerged. Now the restless energy of the sea deposited upon the ocean bed layers of silt and skeletons and salt. For a million years the ocean would build soil, and then the ice would return; the waters would draw away; and the land would lie exposed. Winds from the north and the south would howl across the empty seas and lash stupendous waves upon the shattering shore. Thus the ocean continued its alternate building and tearing down.

Master of life, guardian of the shorelines, regulator of temperatures and sculptor of mountains – the great ocean was all these. . . .

When the earth was already ancient, of an age incomprehensible to man, an event of basic importance occurred in the area of the North American continent that would later be known as Colorado. To appreciate its significance, one must understand the structure of the earth, and to do this, one must start at the vital center.

Since the earth is not a perfect sphere, the radius from center to surface varies. At the poles it is 3,950 miles and at the equator 3,963. At the time we are talking about, Colorado lay about the same distance from the equator as it does now, and its radius was 3,956 miles.

At the center then, as today, was a ball of solid material, very heavy and incredibly hot, made up mostly of iron; this extended for about 770 miles. Around it was a cover about 1,375 miles thick, which was not solid but which could not be called liquid either, for at that pressure and that temperature, nothing could be liquid. It permitted movement, but it did not easily flow. It transmitted heat, but it did not bubble. It is best described as having characteristics with which we are not familiar, perhaps like a warm plastic.

Around this core was fitted a mantle of dense rock 1,784 miles thick, whose properties are difficult to describe, though much is known of them. Strictly speaking, this rock was in liquid form, but the pressures exerted upon it were such as to keep it more rigid than a bar of iron. The mantle was a belt that absorbed both pressure and heat from any direction and was consequently under considerable stress. From time to time the pressure became so great that some of the mantle material forced its way toward the surface of the earth, undergoing marked change in the process. The resultant body of molten liquid, called magma, would solidify to produce the igneous rock, granite, but if it was still in liquid form as it approached the surface, it would become lava. It was in the mantle that many of the movements originated that would determine what was to happen next to the visible structure of the earth; deep beneath the surface, it accumulated stress and generated enormous heat as it prepared for its next dramatic excursion toward the surface, producing the magma that would appear as either granite or lava.

At the top of the mantle, only twenty-seven miles from the surface, rested the earth's crust, where life would develop. What was it like? It can be described as the hard scum that forms at the top of a pot of boiling porridge. From the fire at the center of the pot, heat radiates not only upward, but in all directions. The porridge bubbles freely at first when it is thin, and its motion seems to be always upward, but as it thickens, one can see that for every slow bubble that rises at the center of the pan, part of the porridge is drawn downward at the edges; it is this slow reciprocal rise and fall that constitutes cooking. In time, when enough of this convection has taken place, the porridge exposed to air begins to thicken perceptibly, and the moment the internal heat stops or diminishes, it hardens into a crust.

This analogy has two weaknesses. The flame that keeps the geologic pot bubbling does not come primarily from the hot center of the earth, but rather from the radioactive structure of the rocks themselves. And as the liquid magma cools, different types of rocks solidify: heavy dark ones rich in iron settle toward the bottom; lighter ones like quartz move to the top.

The crust was divided into two distinct layers. The lower and heavier, twelve miles thick, was composed of a dark, dense rock known by its made-up name of 'sima,' indicating the predominance of silicon and magnesium. The upper and lighter layer, fifteen miles thick, was composed of lighter rock known by the invented word 'sial,' indicating silicon and aluminum. The subsequent two miles of Colorado's rock and sediment would eventually come to rest on this sialic layer.

Three billion six hundred million years ago the crust had formed, and the cooling earth lay exposed to the developing atmosphere. The surface as it then existed was not hospitable. Temperatures were too high to sustain life, and oxygen was only beginning to accumulate. What land had tentatively coagulated was insecure, and over it winds of unceasing fury were starting to blow. Vast floods began to sweep emerging areas and kept them swamplike, rising and falling in the agonies of a birth that had not yet materialized. There were no fish, no birds, no animals, and had there been, there would have been nothing for them to eat, for grass and trees and worms did not exist.

Even under these inhospitable conditions, there were elements like algae from which recognizable life would later develop, but the course of their future development had not yet been determined.

The earth, therefore, stood at a moment of decision: would it continue as a mass with a fragile covering incapable of sustaining either structures or life, or would some tremendous transformation take place that would alter its basic surface appearance and enlarge its capacity?

Sometime around three billion six hundred million years ago, the answer came. Deep within the crust, or perhaps in the upper part of the mantle, a body of magma began to accumulate. Its concentration of heat was so great that previously solid rock partially melted. The lighter materials were melted first and moved upward through the heavier material that was left behind, coming to rest at higher elevations and in enormous quantities.

Slowly but with irresistible power it broke through the earth's crust and burst into daylight. In some cases, the sticky, almost congealed magma may have exploded upward as a volcano whose ash would cover thousands of square miles or, if the magma was of a slightly different composition, it would pour through fissures as lava, spreading evenly over all existent features to a depth of a thousand feet.

As the magma spread, the central purer parts solidified into pure granite. Most of it, however, was trapped within the crust, and slowly cooled and solidified into rock deep below the surface. What amount of time was required for this gigantic event to complete itself? It almost certainly did not occur as a vast one-time cataclysm although it might have, engulfing all previous surface features in one stupendous wrenching that shook the world. More likely, convective movements in the mantle continued over millions of years. The rising internal heat accumulated eon after eon, and the resultant upward thrust still continues imperceptibly.

The earth was at work, as it is always at work, and it moved slowly. A thousand times in the future this irresistible combination of heat and movement would change the aspect of the earth's surface.

This great event of three billion six hundred million years ago was different from many similar events for one salient reason: it intruded massive granite bodies which, when the mountains covering it were eroded away, would stand as the permanent basement rock. In later times it would be penetrated, wrenched, compressed, eroded and savagely distorted by cataclysmic forces of various kinds. But through three billion six hundred million years ago, down to this very day, it would endure. Upon it would be built the subsequent mountains; across it would wander the rivers; high above its rugged surface animals would later roam; and upon its solid foundations homesteads and cities would rest.

A relatively short distance below the surface of the earth lies this infinitely aged platform, this permanent base for action. How do we know of its existence? From time to time, in subsequent events, blocks of this basement rock were pushed upward, where they could be inspected, and tested and analyzed, and even dated. At other memorable spots throughout Colorado this ancient rock was broken by faults in the earth's crust, and large blocks of it uplifted to form the cores of present-day mountain ranges.

This rock is beautiful to see – a hard, granitic pink or gray-blue substance as clean and shining as if it had been created yesterday. You find it unexpectedly along canyon walls, or at the peaks of mountains or occasionally at the edge of some upland meadow, standing inconspicuously beside alpine flowers. It is a part of life, an almost living thing, with its own stubborn character formed deep in the earth, once compressed by enormous forces and heated to hundreds of degrees. It is a poem of existence, this rock, not a lyric but a slow-moving epic whose beat has been set by eons on the world's experience.

Often the basement rock appears not as granite but as unmelted gneiss, and then it is even more dramatic, for in its contorted structure you can see proof of the crushing forces it has undergone. It has been fractured, twisted, folded over to the breaking point and reassembled into new arrangements. It tells the story of the internal tumult that has always accompanied the genesis of new land forms, and it reminds us of the wrenching and tearing that will be required when new forms rise into being, as they will.

7-2   When Pigs Ruled the Earth – Anna Grayson

Anna Grayson, a freelance journalist, vividly remembers seeing her first fossil while on a childhood holiday in Dorset. Her brother, in a fit of temper, threw a rock over a hedge. It split open, revealing a perfect fossil leaf. This began a lifelong passion for geology, which she studied at St. Andrews University. She became a popular radio and TV reporter, specializing in geology and educational programs. The mother of two young sons, she recently received the R.H. Worth Prize and the Glaxo-Wellcome Science Writers Award for her work in scientific broadcasting. Our extract "When Pigs Ruled the Earth" is taken from an essay in her book Equinox: the Earth (2000), based on a series of programs on BBC's Channel 4. In this extract she discusses the link between the distribution of 250-million-year-old fossil, mammal-like reptiles and the reconstruction of a vast supercontinent, Pangea.

Two hundred and fifty million years ago, fifty million years before the start of the Jurassic period – that great age in Earth history when dinosaurs ruled the world – another set of weird and wonderful reptiles held sway. This was the world of the Permian period, when landscapes were a mosaic of lush green forests and orange desert, and when all the continents had joined together to make one supercontinent, the vast land mass of Pangaea.

Some of the Permian reptiles were the ancestors of the dinosaurs, and showed glimpses of the giants to come in their bodies and behaviours. Another large group of reptiles was in a way more interesting, for they were the group of animals that gave rise to the mammals, and so eventually to us. These reptiles are termed the mammal-like reptiles and showed tantalizing glimpses of mammalian features inside an essentially reptilian framework. Within this group was a highly successful creature called *Lystrosaurus*, which looked and behaved rather like a modern pig. *Lystrosaurus* was not our direct ancestor, more of a great-great-aunt on a side chain, whose progeny eventually died out with no issue. But, despite it not being a direct forebear, its success illustrates, and to a large extent explains, two of the great phenomena of Earth's history – continental drift and survival after a mass extinction.

Lystrosaurus and its relatives, a whole group of mammal-like reptiles called therapsids, are known from fossils all over the world, and most particularly from the Karoo desert of South Africa where their fossilized remains are found in large numbers, and in many varieties. The British vertebrate palaeontologist Mike Benton of Bristol University has studied them as the first group of animals that lived in complex ecosystems parallel to those we see in the world today: 'There were small ones the size of a mouse, through dog-sized ones, right up to some animals the size of a rhinoceros.' Over in the USA, at Washington State University, Peter Ward is interested in their relationship to us and the lessons we can learn from them; as he says, 'Mammal-like reptiles are our distant ancestors – the precursors to mammals.'

The finding and interpretation of such fossils is the nearest thing science has to a time machine, enabling palaeontologists to revisit and reconstruct an ancient world. But it is more than that – the messages read from fossils of the past can tell stories of how the Earth works, and how animals and plants have responded to natural climate changes and global disasters. So, scientists can read messages and signs, locked within stone, which may tell us something about our future on this planet. . . .

Anyone who has ever looked at a map of the world will have noticed that Africa and the Americas seem to fit into one another like pieces of a jigsaw puzzle. The first scientist to write about this was a Frenchman, Antonio Snider, who published maps showing a united Africa, Europe and the Americas in a book entitled *La Création et ses mystères dévoilés* in 1858. No one took the slightest notice – the idea was considered absurd.

It was not until 1915 that a (literally) ground-breaking book was published suggesting that there may indeed once have been one large continent that split up and drifted apart. The author of that paper was Alfred Wegener, a meteorologist and astronomer rather than a geologist – and his being an 'outsider' may well have contributed to the extreme hostility encountered by his theory. Yet now, Wegener is considered probably the most significant contributor to the Earth sciences of the twentieth century.

It was not just the shape of the continents that drew Wegener to the conclusion that they had once all been joined together in one large supercontinent. There was fossil evidence too in the form of the seed-fern *Glossopteris* which was found on all continents. Rocks matched up on different continents, in particular a very distinctive rock-type called a till which is deposited by ice sheets. Finding tills that fitted like the pictures printed on jigsaw pieces was very compelling evidence: not only did this add evidence to the fit, but it also suggested that when the continents had been joined, they had been in a different position, near the South Pole. Clearly, it seemed to Wegener, the continents had split up and drifted northwards. Wegener went so far as to name this supercontinent Pangaea (the name we still use today), and to work out that it had split up in two stages – South America from Africa during the age of the dinosaurs; Australia from Antarctica and Europe from North America some time later.

The opposition and hostility to which Wegener was subjected must have been very hard for him. Only a handful of geologists took him seriously, and came up with other explanations for the distribution of fossils, the favourite being land-bridges that had long since disappeared. The fact that there was no evidence for land-bridges, nor any explanation for their disappearance, did not subdue the hostility; neither did the fact that land-bridges failed to explain the distribution of glacial tills in the tropics and equatorial regions. The geologists had closed ranks against him – a stance that remains a permanent scar on the profession.

The big problem for Wegener was his failure to find a mechanism that could move continents. The best he could come up with was some kind of magnetic 'flight from the poles' or a slow nudging formed by oceanic tides. Wegener died during a meteorological expedition to Greenland in 1930 and never knew that his theory would be proved right by a series of courageous and more open-minded geologists.

The first of these was Arthur Holmes, a remarkably astute British geologist working in Edinburgh, who the year before Wegener's death suggested a workable mechanism. Holmes, realizing that the Earth had been cooling down since its creation, suggested that convection currents – carrying heat from the Earth's interior towards the surface, and then (having cooled) sinking back down again – could provide enough force to move mountains.

Then in 1937 a South African geologist, Alexander Logie du Toit, published a work dedicated to Wegener, and entitled *Our Wandering Continents*. He cited many more fossil plants, vertebrates and insects that were common to all the continents Wegener had cited as parts of Pangaea. In particular, du Toit noted occurrences of *Lystrosaurus* all over the globe – in 1934 they were recorded as far away as China. How could a squat, pig-like herbivore have crossed oceans and colonized corners of the globe as far apart as Antarctica and Russia?

The only rational answer was for the continents to have once nestled together so that *Lystrosaurus* could walk from its presumed birthplace of South Africa immediately across to all its other current resting-places.

Of all the fossil evidence, *Lystrosaurus*'s conquering of the continents was the most compelling. Undoubtedly other scientists had to take this evidence seriously. Nonetheless, according to the eminent South African palaeontologist Colin McCrae, writing in his book *Life Etched in Stone*, du Toit's support of Wegener earned him 'much censure and hostile comment.' McCrae describes his diligent and thorough field-working style: 'He would send his donkey wagon on ahead on the few passable roads and then cover the area on foot or by bicycle . . . his observations and maps were of an outstanding standard and many of his maps cannot be improved today despite all the modern geological aids available.' Now, of course, du Toit is considered a hero of South African geology, and he was eventually made a Fellow of the Royal Society of London, which was a great honour for someone living and working outside Britain.

Du Toit died in 1948, living long enough to see Arthur Holmes's work on convection currents gain respectability and a body of support after Holmes's publication in 1944 of the famous textbook, *Principles of Physical Geology*. In this book Holmes eloquently argues that the idea of continental drift should be taken seriously. In the final paragraph of this book, however, Holmes inserts a caveat to his proposal of a convection-current mechanism: '. . . many generations of work may be necessary before the hypothesis can be adequately tested.' How wonderful, then, it must have been for Holmes to find that proof came within his own lifetime – in the 1960s. And how tragic that Wegener and du Toit did not live to see their work vindicated.

In 1963 a young research student, Fred Vine, together with his supervisor, Drummond Matthews, discovered a phenomenon called sea-floor spreading. The idea had been put in Vine's mind a year earlier by an American scientist and ex-naval officer, Harry Hess. During voyages at sea, Hess had put some thought to Holmes's ideas of convection. He wondered whether the mid-ocean ridges – which were just revealing themselves to those who were mapping the ocean floors – were places where Holmes's convection currents rose. This indeed turned out to be the case.

All ocean-floor rocks are the same – black basalt, a volcanic rock that comes up from the Earth's interior as hot molten lava. As it cools and solidifies, magnetic minerals form within it. These 'mini-magnets' retain a record of Earth's magnetic field at the time of cooling. For the Earth's magnetic field is not static – it wanders about, and every so often completely flips over, so that the North Pole becomes a magnetic South Pole, and the South Pole a magnetic North Pole.

Reading magnetic traces recorded by instruments on board ship crossing over the mid-ocean ridges, Fred Vine and Drummond Matthews found a remarkable symmetry. Either side of the ridges were identical 'stripes' of magnetic polarity – matching alternate bands of north–south then south–north polarity. Also the age of the basalts became progressively older as they moved away from the ridges: at the edges of continents the basalts were of the order of 50 million years old, but by the ridges they were real youngsters – erupted yesterday in geological terms.

So what Vine and Matthews had in front of them was the basaltic equivalent of a tape recording – a magnetically coded recording of the Earth's magnetic history. As there were two of these rocky recordings, sitting symmetrically either side of the mid-ocean ridges, getting older further away, Vine and Matthews concluded that the mid-ocean ridges must be the source of these basalts.

What was happening – and is still happening – was that basalt was erupting along the mid-ocean ridges, and then falls to the sides, pushing older basalt to one side. So, to think of an ocean as a whole, it gradually gets wider as more basalt is formed at its middle. As the ocean widens, so the continents on either side of it get pushed further apart – and there you have it, a mechanism for continental drift. Vine and Matthews had proved Wegener's theory to be right all along. The great continent of Pangaea had been a reality. *Lystrosaurus* did indeed walk its way round to colonize the supercontinent.

Continents are not always moving away from each other, of course. The whole point is that they had once moved closer and joined up to form the one massive continent – Pangaea, where *Lystrosaurus* and its friends lived 250 million years ago. Oceans cannot continue to get wider forever. At the margins of some oceans, such as the Pacific today, the black basalt ocean floor takes a nose-dive back into the Earth – because it has cooled and become heavier. So the ocean floor gets swallowed up and, as it does so, the ocean gets progressively narrower. All round the 'Ring of Fire' surrounding the Pacific, the ocean floor is returning to the bowels of the Earth – the Pacific is shrinking, and in around another 100 million years, the west coast of the Americas will collide with Kamchatka, Japan, the Philippines and Australia to form another supercontinent.

7-3   The Living Planet – David Attenborough

David Frederick Attenborough is best known to the general public as a broadcaster, explorer and television host on PBS travel and natural history programs. Born in London in 1926 and educated at Cambridge, where he graduated in geology and biology, he had a long career in the BBC before becoming an independent television producer. He has undertaken numerous scientific and ethnographic filming expeditions to Africa and Asia and is the author of more than twenty books on natural history. Our extract is taken from The Living Planet: a portrait of the earth *(1984).*

And now another extraordinary characteristic of the Kali Gandaki becomes apparent. It seems to be flowing the wrong way. Rivers, after all, normally rise in the mountains, flow down their slopes, gathering water from their tributary streams as they go, and then continue down to the plains. The Kali Gandaki does the reverse. It rises on the edge of the great plains of Tibet and heads straight for the mountains. It worms and wriggles its way downwards through the giant interlocking buttresses as the mountains on either side of it grow higher and higher. Only after it has found its way right through them does it reach a relatively flat plain and unite with the Ganges to flow down to the sea. When you stand close to its source, high on the wall of its valley, tracing its course with your eye as it writhes away like a silver snake into the distant mountains, you cannot believe that the river could have cut its way through the mountains by itself. How, then, did it ever come to follow such a course?

Clues to the answer lie at your feet, scattered among the rubble. The rock here is a crumbling, easily-split sandstone and in it lie thousand upon thousand of coiled shells. Most are only a few inches across. Some are as big as cartwheels. They are ammonites. No ammonite is alive today, but a hundred million years ago, they flourished in vast numbers. From their anatomy and the chemical constituents of the rocks in which their fossilised remains are found, we can be quite certain that they lived in the sea. Yet here, in the centre of Asia, they are not only 800 kilometres from the sea but some 4 vertical kilometres above its level.

How this came about was, until only a few decades ago, the subject of great controversy among geologists and geographers. Now, at last the broad outlines of the explanation have been deduced beyond dispute. Once, between the great continental mass of India to the south and Asia to the north, there lay a wide sea. In its waters lived the ammonites. Rivers flowing from the two continents brought down layer upon layer of sediments. As the ammonites died, so their shells fell to the bottom of the sea and were covered by fresh deposits of mud and sand. But the sea was becoming narrower and narrower for, year after year and century after century, India was moving closer to Asia. As it neared, the sediments on the sea floor began to ruck and crumple so that the sea became increasingly shallow. But still the continent of India advanced. The sediments, now compacted into sandstones, limestones and mudstones, rose to form hills. Their elevation was infinitesimally slow. Nonetheless, some of the rivers that had been flowing south from Asia were unable to maintain their course over the slopes that were rising in front of them. Their waters were diverted eastwards and avoided the infant Himalayas by running around their eastern end, eventually joining the Brahmaputra. But the Kali Gandaki had enough strength to cut through the soft rocks as fast as they rose so that they formed the great cliffs of crumpled strata that can now be seen on either side of its valley.

The process continued for millions of years. Tibet, which before the collision of the continents had been a well-watered plain along the southern edge of Asia, was not only pushed upwards but gradually deprived of its rainfall by the young mountains and so changed into the high cold desert that it is today; the upper reaches of the Kali Gandaki lost much of the rain that had given the river its initial erosive power and shrank inside its vast valley; and on the site of the ancient sea there now stood the highest and newest mountains in the world containing, within their fabric, the remains of ammonites. Nor has this process stopped. India is still moving north at the rate of about 5 centimetres a year, and each year the rocky summits of the Himalayas are a millimetre higher.

This transformation of sea into land began some 65 million years ago. Though this seems inconceivably distant to us, a species that has only been in existence for less than half a million years, in terms of the history of life as a whole it was a comparatively recent event. It was, after all, some 600 million years ago that simple animals began to swim in the ancient seas; and over 200 million years since amphibians and reptiles invaded the land. Birds developed feathers and wings and took to the air a few million years afterwards and mammals evolved fur and warm blood around the same time. Sixty-five million years ago, the reptiles fell into their still-mysterious decline and mammals assumed the dominance of the land which they still hold today. So 50 million years ago, as the island continent of India approached Asia, all the major groups of animals and plants that we know today, and indeed almost all the large families within those groups, were already in existence. Each of the continents had its own multitudinous complement of inhabitants, though India, having been isolated as an immense island since just after the decline of the reptiles, was undoubtedly much poorer in advanced groups of animals than Asia. When the two eventually met and the new mountains began to rise some 40 million years ago, the animals and plants from the two old continents began to spread into the new uncolonised extension to their territory. . . .

We have to recognise that the old vision of a world in which human beings played a relatively minor part is done and finished. The notion that an ever-bountiful nature, lying beyond man's habitations and influence, will always supply his wants, no matter how much he takes from it or how he maltreats it, is false. We can no longer rely on providence to maintain the delicate interconnected communities of animals and plants on which we depend. Our success in controlling our environment, that we first achieved 10,000 years ago in the Middle East, has now reached its culmination. We now, whether we want to or not, materially influence every part of the globe.

The natural world is not static, nor has it ever been. Forests have turned into grassland, savannahs have become desert, estuaries have silted up and become marshes, ice caps have advanced and retreated. Rapid though these changes have been, seen in the perspective of geological history, animals and plants have been able to respond to them and so maintain a continuity of fertility almost everywhere. But man is now imposing such swift changes that organisms seldom have time to adapt to them. And the scale of our changes is now gigantic. We are so skilled in our engineering, so inventive with chemicals, that we can, in a few months, transform not merely a stretch of a stream or a corner of a wood, but a whole river system, an entire forest.

If we are to manage the world sensibly and effectively we have to decide what our management objectives are. Three international organisations, the International Union for the Conservation of Nature, the United Nations Environmental Programme, and the World Wildlife Fund, have cooperated to do so. They have stated three basic principles that should guide us.

First, we must not exploit natural stocks of animals and plants so intensively that they are unable to renew themselves, and ultimately disappear. This seems such obvious sense that it is hardly worth stating. Yet the anchoveta shoals were fished out in Peru, the herring has been driven away from its old breeding grounds in European waters, and many kinds of whales are still being hunted and are still in real danger of extermination.

Second, we must not so grossly change the face of the earth that we interfere with the basic processes that sustain life – the oxygen content of the atmosphere, the fertility of the seas – and that could happen if we continue destroying the earth's green cover of forests and if we continue using the oceans as a dumping ground for our poisons.

And third, we must do our utmost to maintain the diversity of the earth's animals and plants. It is not just that we still know so little about them or the practical value they might have for us in the future – though that, too, is so. It is, surely, that we have no moral right to exterminate for ever the creatures with which we share this earth.

As far as we can tell, our planet is the only place in all the black immensities of the universe where life exists. We are alone in space. And the continued existence of life now rests in our hands.

7-4   The Road to Jaramillo – William Glen

William Glen, born in 1932, a graduate of the University of Nevada at Reno, began his career as a geophysicist, but developed into a distinguished historian of contemporary science. At the time of the publication of the book from which our extract is taken – The Road to Jaramillo (1982) – Glen was a research associate in the Office for History of Science and Technology at Berkeley. The book describes not only the overall development of plate tectonics, but also the work of some of the individuals whose work established the theory, among more than 100 of whom Glen recorded over 500 hours of interviews.

"You owe me a martini," read the telegram. It had reached Allan Cox in a remote Eskimo village on the Yukon, and he was "pretty sure it meant that Dick Doell had done the heating experiments and there was no indication of self-reversal." He was right: the rock samples had not somehow reversed their own polarity. He knew, too, what that meant: the Rock Magnetics group, back in Menlo Park, had confirmed that the earth's magnetic field reversed itself 900,000 years ago.

"Allan was a little doubtful about the Jaramillo event at the time," Doell recalled years later. "He thought we were going too far out, to be basing this event just upon a few rocks. 'But hell,' we argued with him, 'we had no more rocks than that when we hypothesized the whole polarity-reversal time scale in the beginning!' "

For more than four years, Cox, Doell, and Brent Dalrymple had been collecting rocks from around the world and carefully determining the age and magnetic polarity of each rock sample. From these data they had been assembling a polarity-reversal time scale – a calendar stretching back millions of years from the present, in which were identified intervals when the earth's magnetic field was of normal polarity, as it is today, alternated with others of reverse polarity. Throughout those four years, the geological community paid little heed to their efforts, or to the work of their competitors in Australia, Ian McDougall and Don Tarling. Both the importance and the validity of the research were questioned (after all, could the earth actually have reversed its own magnetic field, on several occasions?), and it was clear in any case that the time scale was incomplete.

Each time a new reversal, or "event," was discovered, the Americans or the Australians had published a new, more refined, version of the scale – to modest applause. But now a crucial datum, from rocks collected near Jaramillo Creek, in the Jemez Mountains of New Mexico, had fallen into place. That newly found short interval of normal polarity of the earth's field, recorded in the rocks almost a million years ago, was the basis for the eleventh version of the time scale – the version that found immediate corroboration in data from the seafloor and became the key to a revolution.

The discovery of the event was first revealed privately, at a meeting of the Geological Society of America in November 1965. Fred Vine, who would make the most of the discovery, vividly recalled that day: "The crucial thing at that meeting was that I met Brent Dalrymple for the first time. He told me in private discussion between sessions, 'We think we've sharpened up the polarity-reversal scale a bit, but in particular we've defined a new event – the Jaramillo event.' I realized immediately that with the new time scale, the Juan de Fuca Ridge [in the Pacific, southwest of Vancouver Island] could be interpreted in terms of a constant spreading rate. And that was fantastic, because we realized that the record was more clearly written than we had anticipated. Now we had evidence of constant seafloor spreading."

In February, Vine went to the Lamont–Doherty Geological Observatory of Columbia University to visit Neil Opdyke. "Neil was poring over papers on a light table. He must have been drawing up magnetic profiles. . . . In the same room, on the wall, Walter Pitman had pinned up all the *Eltanin* profiles from the South Pacific; he was working on interpretation of the magnetic anomalies. While talking to me about research in general, Neil said, 'We've discovered a new event, and we're just about to publish it. Look, we've got all the details of the time scale of Cox, Doell, and Dalrymple; and we've got a new event at 0.9 million years.' I said 'Neil, I hate to tell this, but Cox, Doell, and Dalrymple have already defined it and named it the Jaramillo event.' Neil just about fell of his chair – he said they'd given it another name."

The disappointment could not have troubled Opdyke long, for the discovery his group had just made went beyond confirming Jaramillo. They had been studying profiles showing variations in the magnetism of seafloor rock – diagrams of alternating polarity stretching across hundreds of kilometers of seafloor. One of the profiles had just been drawn up, from data they had recorded the previous fall aboard the vessel *Eltanin* in crossing the East Pacific Rise on leg 19. Once they had interpreted that profile, the *Eltanin* 19 – by using time scale number eleven as their key – there was no mistaking its importance. Walter Pitman confided later that after working all night running out magnified profiles, he "pinned up all the profiles of *Eltanin* 19, 20, and 21 on Neil Opdyke's door and went home for a bit of rest." When he came back, Opdyke (like Vine, an advocate of continental drift) had seen them and "was just beside himself! He knew that we'd proved seafloor spreading! It was the first time you could see the total similarity between the profiles – the correlation, anomaly by anomaly." (Pitman recalled also that Vine, whose theory of magnetization of the spreading seafloor had recently been published, had not felt comfortable with his own theory until he saw *Eltanin* 19.)

Simultaneously, the Jaramillo/*Eltanin* revelations triggered the development of a still greater idea, one that has transformed the earth sciences during the past two decades just as profoundly as the propositions of Copernicus, Darwin, and Einstein overturned their worlds and discipline. For if seafloors could spread, continents could drift. This powerful new theory, which we know as plate tectonics, explained continental drift and seafloor spreading and went beyond both. It opened up new visions of how the planet works, with implications beyond our ken. During the tumultuous half-decade beginning with that confirmation of seafloor spreading in 1966, the centuries-old belief that continents and oceans were fixed in position was displaced, almost overnight, by the contradictory hypothesis that continent-size slabs or plates of rigid crustal shell move continuously about, opening and closing ocean basins, thrusting up mountains or forming deep-sea trenches where they collide, and forming and reforming the major features of the earth's surface.

This grand unifying theory has final interconnected a great storehouse of isolated geological facts that had been accumulating for centuries. Plate tectonics hold that the lithosphere – the rigid outer shell of the earth, as much as 150 kilometers thick – is divided into about 25 huge plates of diverse expanse and shape. These plates ride or float on a plastic zone beneath them in the weak, upper part of the earth's mantle called the asthenosphere. The plates move about independently, and those that make up the seafloor are propelled outward from the axes of very long, mid-ocean submarine ridges: by the almost constant injection of lava from below the ridge, new crust is added along the two opposing plate edges as they pull steadily apart along the line of spreading; and far to each side, on the two plate edges thousands of kilometers away from the mid-ocean ridge, the plate boundaries dive into deep-sea trenches to be consumed in the earth's mantle. Moving from spreading axis to trench at the rate of only several centimeters per year, the entire seafloor is recycled every hundred million years or so.

Some of the other plates slide past each other laterally, along great transform faults. California's San Andreas Fault is such a fault, the land to its west moving steadily north with respect to the larger North American landmass. The continents, for so long thought immutable, are in fact only passengers riding the plates, and often comprise parts of several plates, as is the case in the lands about the Mediterranean. The major mountain belts, earthquake zones, and chains of volcanoes – including the great "Ring of Fire" about the Pacific – all tend to occur at plate boundaries.

Once these ideas were grasped, progress was rapid in learning how plates grow and are destroyed, and when and where they moved in the geologic past. What it is that drives the plates, however, is not well understood. Heat-diffusing convection currents moving at several centimeters per year in the plastic upper mantle, and acting somehow in concert with lesser forces, seem at present the best answer to a still difficult question. That problem notwithstanding, plate tectonics has been shown to be a theory of enormous breadth and predictive power, offering convincing answers to a host of centuries-old questions about how the earth works – and to still other questions no one had thought to ask. Almost two decades after its advent, the theory continues to be confirmed daily in the practice of earth science.

7-5 Mao's Almanac: 3,000 Years of Killer Earthquakes – J. Tuzo Wilson

J. Tuzo Wilson (1908–1993) was born in Ottawa, educated at the Universities of Toronto, Cambridge and Princeton, and was one of the founders of plate tectonic theory. It was Wilson who developed the concepts of hot spots and transform faults. After three years with the Geological Survey of Canada, he served in the Canadian army throughout World War II, attaining the rank of colonel. He spent the whole of his academic career at the University of Toronto, becoming principal of one of its constituent colleges – Erindale. Our extract is taken from "Mao's Almanac," an article he wrote for the Saturday Review in 1972.

Earthquakes have always been one of China's great scourges. In the last three millenniums, there has been a major quake every six years on the average, including what is probably the most destructive temblor ever recorded anywhere. It shook

Kansu Province so severely in A.D. 1556 that falling buildings and walls killed more than 820,000 persons recorded by name. Nobody knows how many were killed whose names were not recorded. Small wonder then that the regime of Mao Tse-tung has instructed the Institute of Geophysics and the Institute of Geology of China's Academy of Sciences to concentrate on the predicting of earthquakes. . . .

Of greatest interest are the Chinese discoveries about the distribution of earthquakes and the cycles of seismic activity. These are based upon records that go back more than 3,000 years and are considered reliable over the past 2,752 years. From them Chinese seismologists have put together a four-stage model of the process that first accumulates and then releases the energy that causes earthquakes. The only other study that is at all comparable covers the seismic history of the eastern Mediterranean over the past 2,000 years, and it does not reveal the clear cycles of activity shown in the Chinese records.

If the rate at which earthquakes release energy could be controlled, the damage they inflict might be considerably reduced. Before control can be seriously thought of, however, the nature of quakes must be better understood. This understanding requires international cooperation. . . .

My wife, Isabel, and I entered China via Rangoon, flying along the great Irrawaddy River to Mandalay, then over the Burma "hump" and across the plains of China. Monsoon clouds hid most of the landscape until we neared Shanghai at dusk, but I was surprised by the extent of water – rivers, lakes, canals, and flooded paddy fields – that reflected the bright moonlight off the vast delta. I was even more surprised by the twinkles of tiny electric lights that marked the countryside: none of the brilliance of Western cities, but proof of extensive electrification.

Shanghai's airport resembled that of a city in the central plains of North America except for one thing: It was virtually deserted. The absence of planes, automobiles, and bustle focused our attention on a long double line of cheerful young Chinese dressed in blue track suits, enthusiastically clapping. They formed a welcoming corridor for our traveling companions, a group of young Africans and Arabs bound for the great Afro–Asian Table Tennis Friendship Invitational Tournament soon to begin in Peking. Loudspeakers boomed martial music and exhilarating choruses.

The sound was more Russian than Chinese. It combined with the architecture of the airport's imposing new terminal and the four Ilyushin planes on the tarmac to suggest to us that we might have flown over the wrong mountains and landed in the Soviet Union. Frequently during the following weeks we had reason to recall the close connections that had existed between China and the Soviet Union up to 1960. Our hosts often reminded us, however, that those connections had been completely broken.

Isabel and I were met by half a dozen charming and attentive scientists, officials, and interpreters. They expressed sympathy for our fatiguing, long journey and led us into the terminal, where we sat in easy chairs and were offered Chinese tea and cigarettes. I had scarcely been asked for the baggage checks and my passport when the bags appeared. Isabel's passport, our health certificates, and all questions of foreign currency were waived. We were foreign friends, guests of the academy. It was the simplest and most gracious entry I recall making into any country.

We spent the night at a wonderful old hotel in Shanghai. It was identified only by a street number, but I have surmised, since leaving China, that it must originally have been Victor Sassoon's Cathay Hotel. It overlooked the Whampo River near a clock tower that chimed "The East Is Red."

The next morning we boarded a crack steam train, one of the more appealing aspects of a bygone age. The engine had four scarlet drive wheels, three red flags between its eyebrows, and gold Chinese characters on its tender. As we rolled past the fields and villages of the Yangtze plain on our way to Peking, we could see people

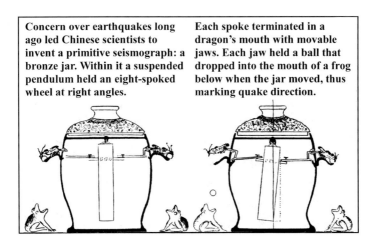

Concern over earthquakes long ago led Chinese scientists to invent a primitive seismograph: a bronze jar. Within it a suspended pendulum held an eight-spoked wheel at right angles. Each spoke terminated in a dragon's mouth with movable jaws. Each jaw held a ball that dropped into the mouth of a frog below when the jar moved, thus marking quake direction.

working everywhere. A young interpreter who pointed out improvements on every side apologized rather unnecessarily for his lack of fluency in English. He mentioned his greater familiarity with Russian, from which he had recently switched because he could foresee no future for Russian interpreters in China.

We were met at the Peking station by the acting director of the Institute of Geophysics, the deputy director of the Institute of Geology, a senior scientist named Mrs. Woo, several officials from the foreign relations section of the Academy of Sciences, two interpreters, and a Canadian Embassy counselor whose black-felt hat towered above the blue Mao caps. We were greeted cordially and then driven to our hotel, where we were told to "rest yourselves a little and then we shall return to discuss the program."

I had hoped to be able to return to the places I had visited in 1958, in order to assess the changes wrought by the Cultural Revolution. It had swept through the academic life of China like a typhoon, shutting off all scientific publications, essentially closing universities for four years, and sending the professors out to plant rice, work in factories, or otherwise gather firsthand knowledge of workers and peasants so as to be able to apply academic knowledge to the problems of the countryside while undergoing re-education in the thought of Chairman Mao.

My 1958 visit had taken me to Peking, Siam, Lanchow, and Canton. However, I was told that Lanchow was out of the question for 1971, apparently because cities as far inland as it is are unprepared to receive visitors with traditional Chinese gracious-ness. The substitute itinerary we agreed upon included all the other cities I had seen in 1958 plus Yenan (famous as the point where Mao's monumental "long march" in the early 1930s ended), Nanking, Shanghai, and Hangchow. Relatively, this would expose me to the Chinese equivalent of a chunk of North America bounded by the cities of New York, Atlanta, New Orleans, Chicago, and Winnipeg. . . .

[T]he historical study of large earthquakes . . . has long been the specialty of Dr. Lee Shan-pang, a former California Institute of Technology student of Professor Beno Gutenberg and author of a three-volume classic on the subject. . . . The earliest records he had found were from 1198 B.C. These he did not consider reliable. But he plotted the location and magnitude of 530 major earthquakes that had occurred in China since 780 B.C. By his estimation, all of these had reached a magnitude of six or greater on the Richter scale of measurement. Dr. Lee, Dr. Ku, and the revolutionary committee of the Institute of Geophysics allowed me to photograph a manuscript map of those 530 quakes. Seventeen of them were graded as greater than magnitude eight, and about eighty were between magnitudes seven and eight. . . .

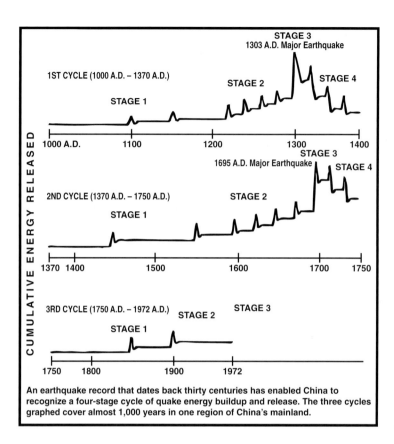

An earthquake record that dates back thirty centuries has enabled China to recognize a four-stage cycle of quake energy buildup and release. The three cycles graphed cover almost 1,000 years in one region of China's mainland.

Anyone with a grasp of modern geophysics could have seen at a glance that the earthquakes marked on the maps were concentrated along definite lines of tectonic weakness. One of these lines goes straight north from Kunming and crosses the Yellow River. Another extends down the Red River to Hanoi. A third cuts across the Yellow River near Siam to the neighborhood of Peking. Other quake epicenter lines pass along the Altai and Tien Shan mountains, encircle Tibet, underlie Taiwan, and skirt the Russian border at Vladivostok. . . .

With instruments derived from [those designed by Caltech professor Hugo Benioff], the Chinese have plotted the rate of energy release along the great faults revealed by their maps. Thanks to their long historical record, they have fixed the dates and magnitudes of earthquakes along each line over a range of time unequaled anywhere else in the world.

They have found that they can divide the earthquake history of each tectonically active area into cycles. Each cycle has four stages. During the first stage, which may last a century or more, the strain on the rocks accumulates. Only a few small earthquakes occur in this period. The second stage is one of transition, during which earthquakes become larger and more frequent. In the third stage the main energy is released. There is one huge earthquake, possibly followed by a few other large earthquakes within a short span of years. The fourth stage of the cycle is characterized by lesser aftershocks, completing the cycle.

In Shansi Province, west of Peking, for example, one cycle extended from A.D. 1000 to 1370. A second cycle ran from 1370 to 1750, culminating in a big shock near Peking that killed 100,000 people. A third cycle started in 1750. It is now in stage two and may be approaching stage three. If stage three is at hand, great shocks are expectable in the near future in the vicinity of Peking. The Chinese did not quite admit this to me, but such an anticipation would be an obvious explanation for the stress that is being laid on the predicting of earthquakes in China today.

7-6   Geologic Jeopardy – Richard H. Jahns

In this extract, Richard Henry Jahns (1915–1983), former Dean of the School of Earth Sciences at Stanford University, discusses the need for recognition of the geologic jeopardy in which society still stands. In spite of humanity's technological accomplishments and extractive skills, we are more, rather than less, vulnerable to the effects of natural hazards due to burgeoning population, endless wastes, and planning blunders. Jahns analyzes the various geologic hazards of life in California, including flood and landslides, as well as earthquakes, from which section of the article our extract is taken. Though informed legislation such as more stringent grading ordinances and building codes can reduce dangers from most of these hazards, far more comprehensive planning and prediction are required, and Jahns argues for a

new level of public stewardship. Trained at the California Institute of Technology and Northwestern, Jahns worked at the United States Geological Survey, and then held appointments at the California Institute of Technology, Pennsylvania State University, and Stanford University. Jahns made fundamental contributions in the fields of mineralogy and petrology, but also made significant scientific contributions in engineering geology, southern California geology, New England glacial geology, economic geology, and earthquake hazards. He received many prominent awards not only for his science, but also his role as an educator who helped shape the modern approach to Earth science education. This extract is taken from an article in The Texas Quarterly *(1968).*

Not long ago, man was often inclined to reflect with unqualified satisfaction upon his growing record of accomplishment in competing with nature. But as he has continued to reshape terrain, to modify much of its drainage, to extract useful materials from the subsurface, and to control various elements of his environment on larger and larger scales, it has become less and less clear that so pleasant a view is justified by the record. Today man is being more widely recognized as the kind of schizoid competitor he really is – imaginative, ingenious, resourceful, and remarkably courageous, but with distressing capacities for vastly increasing his own numbers, for enveloping himself with wastes of many kinds, and for making serious mistakes in dealing with his natural surroundings.

As human population has burgeoned and clustered during recent decades, unpleasant confrontations with geologic reality have become more frequent and more challenging. Among the so-called geologic hazards, or risks, those most commonly encountered are related to floods, earthquakes, and various kinds of unstable ground, and as such they are normal and widespread manifestations of natural processes operating upon and within the earth's crust. From time to time some of the risks are translated into disasters, either unavoidably or through the active cooperation of man; the nature, location, and even the magnitude of such disasters often can be predicted well in advance of their occurrence, but the advance dating of these occurrences is another matter. Intervals between successive natural catastrophes of the same kind ordinarily are so great that studies of the geologic and historic records rarely lead to forecasts sharply enough focused to be useful without supplementary information of other kinds.

A brief sampling of well-documented geologic risks and disasters from the state of California may provide some notion of the diversity of problems that have been recognized. By no means does California have a corner on such problems, but its great variety of geologic materials and features, combined with major concentrations of population, establishes it as an excellent testing ground for the basic wisdom of its occupants. And in few other places are large numbers of residents confronted by such stimulating assemblages of natural hazards as those in the Los Angeles and San Francisco regions, where contrasts in topography, climate, and geology are especially prominent. . . .

From California's subsurface come those recurring violent actions known as earthquakes, several of which have dealt rather harshly with man and his works during the period of historic record. These shocks have originated along faults, or breaks in the earth's crust that represent repeated slippage over very long spans of time. Thousands of faults are known within the state, and many of them can be classed as large in terms of their total displacements. Many of them also are geologically active in the sense of having moved within the past 10,000 years, and more than a few have been active in historic time.

The release of energy in the form of a large earthquake can be assumed to begin with sudden fault movement initiated at some depth in the earth's crust, and to continue as this slippage is rapidly propagated in all directions along the fault. Under appropriate conditions, the displacement may reach the earth's surface and appear as horizontal, vertical, or oblique offsets along the trace of the fault. Such surface faulting also may accompany a relatively small earthquake if the focus, or point of original rupture, is sufficiently shallow.

During the San Francisco earthquake of 1906, predominantly horizontal surface displacement occurred along perhaps as much as 270 miles of the San Andreas fault, California's widely known master break. Maximum observed offset of reference features such as roads and fences was nearly twenty feet. Displacements of similar magnitude may well have occurred farther south along the same fault during the great Fort Tejon earthquake of 1857. Among earthquakes originating in California and nearby parts of Nevada during the past century, at least twenty are known to have been attended by measurable surface faulting. The largest offsets observed in relatively recent years were nineteen feet (horizontal) with the Imperial Valley earthquake of 1940, and twelve feet (horizontal component) and fourteen feet (vertical component) with the Fairview Peak, Nevada, earthquake of 1954.

Although surface rupture is neither common nor widespread over periods of human generations, there is obvious risk in erecting homes or other structures athwart the most recent traces of past movements along active faults. Yet this is precisely what has been done, especially along the San Andreas fault southward from San Francisco and along the San Andreas and San Jacinto faults in the San Bernardino area of southern California. Moreover, this kind of "gamble-in-residence" is being taken more and more often, and in areas where the positions of active faults are well known; it is particularly distressing to note the number of schools represented among installations that some day might serve as reference features for large-scale shear. The gamble might seem safe enough in terms of the odds against losing, but any loss in this ill-advised game could be a major disaster. Fortunately, the nature of such risk is being increasingly noted and appraised for aqueducts, dams, and other special engineering works, if not for housing developments.

A more immediate threat to some installations that extend across fault traces is the slow creep, or progressive slippage in the apparent absence of earthquakes, that is being recognized as currently characteristic of several major breaks in California. These include the Hayward fault on the east side of San Francisco Bay, the Calaveras fault in the vicinity of Hollister, and parts of the San Andreas fault for a distance of about two hundred miles southeastward from San Francisco. Tunnels, railroad tracks, roads, fences, culverts, and buildings are being deformed and displaced where they straddle the narrow strands of creep, with cumulative offsets generally measured in inches over periods of years or decades. At the Almaden–Cienega winery southeast of Hollister, horizontal slippage along the San Andreas fault averages more than half an inch per year; the main building, which evidently lies astride the active fault trace, has been aptly described as experiencing "obvious structural distress."

Returning now to large earthquakes, it should be emphasized that by far their most widespread effects upon man are results of severe ground shaking that accompanies the sudden rupturing along faults. For a given moment of energy released, the severity of shaking can vary greatly from one locality to another according to the interplay of many factors. Perhaps most important among these is the nature of the rocks or other foundation materials at the site in question, as demonstrated by marked variations in the distribution of damage from individual historic shocks. In general, shaking is least intense in hard, firm rocks like granite and gneiss, and most intense where the energy is coupled into relatively soft and loose materials like alluvial silts, sands, and gravels, swamp and lake deposits, and hydraulic fill. The oft-used analogy of a block of jello on a vibrating platter is grossly simplified but nonetheless reasonable, and assuredly a structure that in effect is built upon the jello must be designed to accommodate greater dynamic stresses if it is to survive the shaking as effectively as a comparable structure built upon the platter.

This focuses principally upon California's metropolitan areas, large parts of which are underlain by relatively soft and poorly consolidated materials, and it further points up the need for prudent design of buildings to be erected on reclaimed marsh land or filled portions of lakes and bays. Much of value has been learned during recent decades in the important field of earthquake-resistant design for structures, and a great deal of this knowledge has been wisely applied in actual construction. Unhappily, however, it rarely is possible to offer more than broad generalizations concerning the nature of ground motion to be expected beneath these structures during future earthquakes. Relations between the release of energy at its source and the attendant ground response at a given surface locality are incompletely understood, and pertinent empirical data are still sketchy at best; thus the engineer with capability for earthquake-resistant design cannot be readily supplied with precise design criteria. . . .

It seems obvious that man cannot take for granted the ground he occupies, and that responsibility for troubles stemming from a careless attitude cannot be easily fixed upon someone else, legally or otherwise. Nor can a defeatist attitude survive under the growing pressure of population increase, with corollary expansion of settlement into areas where questions of ground stability must be faced and answered. Here some real progress already has been made, especially in the San Francisco and Los Angeles regions, as more geologists, engineers, land developers, and public officials appear to be asking themselves:

> Will posterity participate
> In chaos we create,
> Or will our heirs commemorate
> Mistakes we didn't make?

Granting man's limitations in controlling certain important elements of his geologic environment, in California and elsewhere, at least he is learning that when he imposes improperly upon nature, nature is likely to respond by imposing more seriously upon him. He has been modifying his approach by seeking better to understand natural processes and more effectively to apply this understanding in the primary struggle, which really lies more with himself than with nature. Nature now can be identified less as the antagonist than as the arena in which ever-increasing numbers of people are competing with one another for food, for air and water, for energy, and for space – there is no other readily available arena, hence this one needs a bit more respect and care.

Toward environmental understanding and improvement, geological scientists and engineers are vigorously investigating many kinds of so-called natural hazards. Active and potentially active faults are now being precisely mapped over large areas, and their respective styles of behavior during the geologic past are being deciphered via a remarkable variety of approaches. New data on creep along faults, the accumulating strain in ground adjacent to faults, and the behavior of the ground during recent earthquakes are revealing some instructive and unexpected relationships. Engineering seismology has fully emerged as a highly significant field of study, with major efforts now being devoted to determining the response of bedrock terranes and surficial deposits to earthquake shaking, the behavior of buildings and other structures during earthquakes, and the most satisfactory types of seismic design for many kinds of structures.

Prediction of earthquakes no longer seems to be an objective for some future century; indeed, at least one important break-through in this area can be expected within the coming decade. Soon perhaps we shall have warning of a few minutes to as much as an hour in advance of strong shocks along two or three of this country's most prominent faults, a contribution of incalculable value to people if not to their property. In the meantime, increasing coordination of empirical and basic data and of observation, experiment, and theoretical analysis can be expected further to improve our dealings with floods, landslides, ground subsidence, and other kinds of natural hazards. May we look forward to the days when all kinds of ground failure can be forestalled or reduced in impact, when existing landslides can be made safe for useful development, when our buildings and utilities can survive the severest earthquake, and when sound programs for disaster insurance can be predicted upon knowledge not yet available.

The human side of these dealings is even more complex than the problems posed by nature. The general public and numerous stewards at all levels of government have been rapidly awakening to the existence and scope of geologic jeopardy in its numerous forms, with reactions ranging from apathy to panic but tending properly to consolidate into deep concern. The growing record of damage and death, especially in some thickly populated areas, has become so compelling that direct responses are now extending beyond temporary reactions to individual disasters. Homeowners and public officials, scientists and engineers, universities and utilities, conservation groups and industrial organizations, government agencies and legislatures, and increasingly large numbers of individual citizens are discovering that they have a common stake in learning how better to live with their physical environment, regardless of their other interests. As more and more of them come to recognize the game we all have been playing in nature's arena for so many centuries, more and more of them will want to know what the score is. Geoscientists and engineers must be continuingly ready with the answer, however constrained or unpalatable it might be at any given place or time.

8. Controversies

"A professor" it has been said "is one who thinks otherwise." Thinking otherwise, questioning orthodoxy, and challenging conventional wisdom are essential components of the tool kit of science. For though science grows largely by personal curiosity, its refinement requires personal skepticism and individual doubt. Doubt arises when existing explanations prove inadequate to account for personal experience, and especially when new data and new experiences impose an additional strain upon existing explanations. Doubt is resolved either by new experience, and especially the contrived experience that we call experiment, or by a comparison of individual experiences and competing explanations, involving at times intense debate, which in turn allows a modification of detailed explanation or general theory. One aspect of verification in science is that experiments are always described and published in such a way that they may be repeated by any trained observer who chooses to duplicate them. This opening of private experience to public scrutiny is the ultimate basis for the large body of agreement that characterizes the scientific world, as well as its classic disputes.

Because much of this debate takes place in public, the history of science is inevitably marked by a good deal of spirited controversy. Some of this has been deeply personal, for many scientists find it difficult to avoid a paternalistic attitude toward their own discoveries and explanations. Other bitter controversies have involved the non-scientific implications of scientific discoveries, and there have been relatively frequent debates between scientists and those of other groups. Still other controversies have wracked the whole scientific world, and their resolution has produced significant turning points in the history of scientific thought. The controversies regarding the nature of the universe following publication of Copernicus's work and those concerning the organic world following publication of Darwin's *Origin of Species* are cases in point. Because the effect of such disputes as these is often to change the whole scientific framework in which explanation is offered, such major controversies are rarely settled quickly; indeed some of them may extend over decades. The notion of continental drift, for example, was first suggested on the basis of scientific evidence in a map published by Antonio Snyder in 1855, but it was at first ignored and then later ridiculed, especially during the middle years of the twentieth century. It was only the availability of new information in the form of paleomagnetic anomalies, and the development of the embracing concept of plate tectonics in the late 1960s, that provided general acceptance for a much modified theory of continental drift.

This example suggests that, even with the availability of new data, where two views compete for acceptance, it is relatively rare for either view to be ultimately fully proved or wholly disproved. What tends to happen is that one or the other explanation, or sometimes a wholly new one, slowly emerges as more general in its applicability and more economical or elegant in its explanation.

8-1   Apes, Angels, and Victorians – William Irvine

William Irvine (1906–1964) was a literary historian who served as Professor of English at Stanford. His writings deal chiefly with influential nineteenth-century figures, including Robert Browning, George Bernard Shaw, and Walter Bagshot. Our extract is taken from his book Apes, Angels and Victorians *(1955). Irvine's account of the "Oxford Meeting" describes an early round in a major controversy. Charles Darwin (1809–1882) published* The Origin of Species *on November 24, 1859. The first edition of 1,250 copies was sold out on the day of its publication, and the book soon reached an audience much wider than that of the scientific world to whom it was addressed. The book explained the use of artificial selection in domestic breeding and argued for the existence of a similar selection in nature, acting in an analogous way and favoring the survival of those organisms better adapted to the*

environment in which they lived. The book also dealt with what was then known of the laws of inheritance and variation, and the more general questions of evolution, or "descent with modification" as Darwin called it, grappling especially with difficulties in accepting the evolutionary origin of species.

The book produced a controversy which extended into almost every area of human understanding. The controversy reached its height with the meeting of the British Association for the Advancement of Science in Oxford in 1860, at which Samuel Wilberforce, Bishop of Oxford, engaged in debate with T.H. Huxley, one of the leading zoologists of the nineteenth century, and the foremost champion of evolutionary theory, who was sometimes referred to as "Darwin's bulldog." The account that follows gives something of the flavor of their meeting.

In June 1860, the British Association met at Oxford. Science was not very much at home there, and neither was Professor Huxley. Beneath those dreaming spires, he always felt as though he were walking about in the Middle Ages; and Professor Huxley did not approve of the Middle Ages. At Oxford, he feared, ideas were as ivy-covered as the buildings, and minds as empty and dreamy as the spires and quiet country air. Professor Huxley's laboratory was set squarely in the middle of the nineteenth century, in the narrow downtown London thoroughfare of Jermyn Street, which was as crowded and busy as Professor Huxley's own intellect.

Reciprocally, Oxford did not feel in the least at home with such people as Professor Huxley. In fact, she felt rather desperately at bay between a Tractarian past and a scientific future. Newman's conversion to Roman Catholicism had opened an abyss of conservatism on one side; now Mr. Darwin's patient and laborious heresy had opened an abyss of liberalism on the other. The ground of sanity seemed narrow indeed. But sanity can always be defended. After all, there was something obviously ridiculous in a heresy about monkeys.

Mr. Darwin himself was too ill to attend the meeting of the Association – as on a former even more important occasion, when Joseph Hooker and Sir Charles Lyell had acted in his place. A portentous absence from crucial events which deeply concerned him was already making Mr. Darwin a legend. It was just six months since his *Origin of Species* had appeared. Of course Darwinism was in everybody's mind. It was also on the program.

In Section D of the meeting, Dr. Daubeny of Oxford read a paper "On the final causes of sexuality in plants, with special references to Mr. Darwin's work on *The Origins of Species.*" Huxley was invited to comment by the president but avoided discussion of the vexed issue before "a general audience, in which sentiment would unduly interfere with intellect." Thereupon, Sir Richard Owen, the greatest anatomist of his time, rose and announced that he "wished to approach the subject in the spirit of the philosopher." In other words, he intended, as in his anonymous review of a few months before, to strike from under the cloak of scientific impartiality. "There were facts," he felt, "by which the public could come to some conclusion with regard to the probabilities of the truth of Mr. Darwin's theory." He then declared that the brain of the gorilla "presented more differences, as compared with the brain of man, than it did when compared with the brains of the very lowest and most problematical of the quadrumana."

Huxley rose, gave Owen's words "a direct and unqualified contradiction," pledged himself to "justify that unusual procedure elsewhere," and sat down. The effect was as though he had challenged Owen to a duel, and infinitely more dramatic than an immediate refutation, however convincing, would have been, though that duly appeared in the dignified pages of *The Natural History Review.*

Between Darwin and anti-Darwin the lines of battle were now drawn. The program of the Association encouraged peace on Friday, but the air was filled with rumors. A general clerical attack was to be made on Saturday, when a somewhat irrelevant American was to speak on the "Intellectual Development of Europe considered with reference to the views of Mr. Darwin." The Bishop of Oxford was arming in his tent and Owen was at his elbow, whispering the secret weaknesses of the enemy. Quite unaware that his larger destiny awaited him in a lecture room next day, Huxley had decided not to witness the onslaught. He was very tired, and eager to rejoin his wife at Reading. He knew that the Bishop was an able controversialist and felt that prevailing sentiment was strongly against the Darwinians. On Friday evening he met the much reviled evolutionist Robert Chambers in the street, and on remarking that he did not see the good of staying "to be episcopally pounded," was beset with such remonstrances and talk of desertion that he exclaimed, "Oh! if you are going to take it that way, I'll come and have my share of what is going on."

Perhaps revolutions often have their quiet beginnings in the classroom, but they seldom have their turbulent crises there. Saturday, June 30, 1860, was the exception. The Museum Lecture Room proved too small for the crowd, and the meeting was moved to a larger place, into which 700 people were packed. The ladies, in bright summer dresses and with fluttering handkerchiefs, lined the windows. The clergy, "shouting lustily for the Bishop," occupied the center of the room, and behind them a small group of undergraduates waited to cheer for the little known champions of "the monkey theory." On the platform, among others, sat the American Dr. Draper, the Bishop, Huxley, Hooker, Lubbock, and, as president of the section, Darwin's old teacher Henslow.

Dr. Draper had the compound misfortune to be at once a bore and the center of this exciting debate, having chosen to pick up the burning question of the day by its biggest and hottest handle. His American accent added a quaint remoteness to his metaphysical fulminations. "I can still hear," writes one witness, "the American accents of Dr. Draper's opening address when he asked 'Air we a fortuitous concourse of atoms?'" But had he luxuriated in the combined gifts of Webster and Emerson, he would still have seemed an irrelevance. The audience wanted British personalities, not Yankee ponderosities; and they had already smelled blood.

Dr. Draper droned away for an hour, and then the discussion began. It was evident that the audience had tolerated its last bore. Three men spoke and were shouted down in nine minutes. One had attempted to improve on Darwin with a mathematical demonstration. "Let this point A be man and that point B be the mawnkey." He was promptly overwhelmed with cries of "mawnkey." And now there were loud demands for the Bishop. He courteously deferred to Professor Beale and then, with the utmost good humor, rose to speak.

Bishop Wilberforce, widely known as "Soapy Sam," was one of those men whose moral and intellectual fibers have been permanently loosened by the early success and applause of a distinguished undergraduate career. He had thereafter taken to succeeding at easier and easier tasks, and was now, at fifty-four, a bluff, shallow, good-humored opportunist and a formidable speaker before an undiscriminating crowd. His chief qualification for pronouncing on a scientific subject derived, like nearly everything else that was solid in his career, from the undergraduate remoteness of a first in mathematics.

Huxley listened to the jovial, confident tones of the orator and observed the marked hostility of the audience toward Darwinians. How could he make an effective reply? He could hardly expound *The Origin of Species*, theory and evidence, in ten minutes. But Huxley was not the man to brood on disadvantages. He was encouraged to find that, though crammed to the teeth by Owen, the Bishop did not really know what he was talking about. Nevertheless, exploiting to the full the popular tendency to regard every novelty as an absurdity, he belabored Darwinism with such resources of obvious wit and sarcasm, saying nothing with so much gusto and ingenuity, that he was clearly taking even sober scientists along with him. Finally, overcome by success, he turned with mock politeness to Huxley and "begged to know, was it through his grandfather or his grandmother that he claimed his descent from a monkey?"

This was fatal. He had opened an avenue to his own vacuity. Huxley slapped his knee and astonished the grave scientist next to him by softly exclaiming, "The Lord hath delivered him into mine hands." The Bishop sat down amid a roar of applause and a sea of fluttering white handkerchiefs. Now there were calls for Huxley, and at the chairman's invitation he rose, a tall, slight, high-shouldered figure in a long black coat and an enormous high collar, which seemed to press the large, close-set features even more tightly together. His face was very pale, his eyes and hair were very black, and his wide lips were calculatingly, defiantly protruded. His manner, gauged with an actors' instinct, was as quiet and grave as the Bishop's had been loud and jovial. He said that he was there only in the interests of science, that he had heard nothing to prejudice his client's case. Mr. Darwin's theory was much more than an hypothesis. It was the best explanation of species yet advanced. He touched on the Bishop's obvious ignorance of the sciences involved; explained, clearly and briefly, Darwin's leading ideas; and then, in tones even more grave and quiet, said that he would not be ashamed to have a monkey for his ancestor; but he would be "ashamed to be connected with a man who used great gifts to obscure the truth."

The sensation was immense. A hostile audience accorded him nearly as much applause as the Bishop had received. One lady, employing an idiom now lost, expressed her sense of intellectual crisis by fainting. The Bishop had suffered a sudden and involuntary martyrdom, perishing in the diverted avalanches of his own blunt ridicule. Huxley had committed forensic murder with a wonderful artistic simplicity, grinding orthodoxy between the facts and the supreme Victorian value of truth-telling.

At length Joseph Hooker rose and botanized briefly on the grave of the Bishop's scientific reputation. Wilberforce did not reply. The meeting adjourned. Huxley was complimented, even by the clergy, with a frankness and fairness that surprised him. Walking back to lodgings with Hooker, he remarked that this experience had changed his opinion "as to the practical value of the art of public speaking," and that from this time forth he would "carefully cultivate it, and try to leave off hating it." Huxley had just enough of the sensitive romantic in him to imagine that he hated public speaking. How he actually felt at the time, he himself indicates in another sentence, "I was careful ... not to rise to reply till the meeting called for me – then I let myself go."

Huxley's destiny had thus been captured by another man's book, and discovering almost with astonishment how many talents for action he possessed, this young professor of paleontology became the acknowledged champion of science at one of the most dramatic moments in her history. He defended Darwinian evolution because it seemed to constitute, for terrestrial life, a scientific truth as significant and far-reaching as Newton's for the stellar universe – more particularly, because it seemed to promise that human life itself, by learning the laws of its being, might one day become scientifically rational and controlled.

8-2   The Great Piltdown Hoax – William L. Straus, Jr

In 1911 near Piltdown, on the Sussex Downs, a portion of a human skull was recovered from a gravel pit. Additional skull fragments, several humanlike teeth and other mammalian remains were found during the following three or four years. These fragments comprise the corpus delecti of a geological detective story. A controversy raged for four decades over the Piltdown discoveries and their significance, but late in 1953 the remains of the "earliest Englishman" were shown to be a hoax. Just how this scientific forgery was exposed is described by William L. Straus, Jr (1900–1981), who was Professor of Anthropology at Johns Hopkins University. This extract is from an article published in the journal Science in 1954.

When Drs. J.S. Weiner, K.P. Oakley, and W.E. Le Gros Clark recently announced that careful study had proven the famous Piltdown skull to be compounded of both recent and fossil bones, so that it is in part a deliberate fraud, one of the greatest of all anthropological controversies came to an end. Ever since its discovery, the skull of "Piltdown man" – termed by its enthusiastic supporters the "dawn man" and the "earliest Englishman" – has been a veritable bone of contention. To place this astounding and inexplicable hoax in its proper setting, some account of the facts surrounding the discovery of the skull and of the ensuing controversy seems in order.

Charles Dawson was a lawyer and an amateur antiquarian who lived in Lewes, Sussex. One day, in 1908, while walking along a farm road close to nearby Piltdown Common, he noticed that the road had been repaired with peculiar brown flints unusual to that region. These flints he subsequently learned had come from a gravel pit (that turned out to be a Pleistocene age) in a neighboring farm. Inquiring there for fossils, he enlisted the interest of the workmen, one of whom, some time later, handed Dawson a piece of an unusually thick human parietal bone. Continuing his search of the gravel pit, Dawson found, in the autumn of 1911, another and larger piece of the same skull, belonging to the frontal region. His discoveries aroused the interest of Sir Arthur Smith Woodward, the eminent paleontologist of the British Museum. Together, during the following spring (1912), the two men made a systematic search of the undisturbed gravel pit and the surrounding spoil heaps; their labors resulted in the discovery of additional pieces of bone, comprising – together with the fragments earlier recovered by Dawson – the larger part of a remarkably thick human cranium or brain-case and the right half of an apelike mandible or lower jaw with two molar teeth *in situ*. Continued search of the gravel pit yielded, during the summer of 1913, two human nasal bones and fragments of a turbinate bone (found by Dawson), and an apelike canine tooth (found by the distinguished archeologist, Father Teilhard de Chardin). All these remains constitute the find that is known as Piltdown I.

Dawson died in 1916. Early in 1917, Smith Woodward announced the discovery of two pieces of a second human skull and a molar tooth. These form the so-called Piltdown II skull. The cranial fragments are a piece of thick frontal bone representing an area absent in the first specimen and a part of a somewhat thinner occipital bone that duplicates an area recovered in the first find. According to Smith Woodward's account, these fragments were discovered by Dawson early in 1915 in a field about two miles from the site of the original discovery.

The first description of the Piltdown remains, by Smith Woodward at a meeting of the Geological Society of London on December 18, 1912, evoked a controversy that is probably without equal in the history of paleontological science and which raged, without promise of a satisfactory solution, until the studies of Weiner, Oakley, and Clark abruptly ended it. With the announcement of the discovery, scientists rapidly divided themselves into two main camps representing two distinctly different points of view (with variations that need not be discussed here).

Smith Woodward regarded the cranium and jaw as belonging to one and the same individual, for which he created a new genus, *Eoanthropus*. In this monistic view toward the fragments he found ready and strong support. In addition to the close association within the same gravel pit of cranial fragments and jaw, there was advanced in support of this interpretation the evidence of the molar teeth in the jaw (which were flatly worn down in a manner said to be quite peculiar to man and quite unlike the type of wear ever found in apes) and, later, above all, the evidence of a second, similar individual in the second set of skull fragments and molar tooth (the latter similar to those imbedded in the jaw and worn away in the same unapelike manner). A few individuals (Dixon, Kleinschmidt, Weinert), moreover, have even thought that proper reconstruction of the jaw would reveal it to be essentially human, rather than simian. Reconstructions of the skull by adherents to the monistic view produced a brain-case of relatively small cranial capacity, and certain workers even fancied that they had found evidences of primitive features in the brain from examination of the reconstructed endocranial cast – a notoriously unreliable procedure; but subsequent alterations of reconstruction raised the capacity upward to about 1400 cc – close to the approximate average for living men.

A number of scientists, however, refused to accept the cranium and jaw as belonging to one and the same kind of individual. Instead, they regarded the brain-case as that of a fossil but modern type of man and the jaw (and canine tooth) as that of a fossil anthropoid ape which had come by chance to be associated in the same deposit. The supporters of the monistic view, however, stressed the improbability of the presence of a hitherto unknown ape in England during the Pleistocene epoch, particularly since no remains of fossil apes had been found in Europe later than the Lower Pliocene. An anatomist, David Waterston, seems to have been the first to have recognized the extreme morphological incongruity between the cranium and the jaw. From the announcement of the discovery he voiced his disbelief in their anatomical association. The following year (1913) he demonstrated that superimposed tracings taken from radiograms of the Piltdown mandible and the mandible of a chimpanzee were "practically identical"; at the same time he noted that the Piltdown molar teeth not only "approach the ape form, but in several respects are identical with them." He concluded that since "the cranial fragments of the Piltdown skull, on the other hand, are in practically all their details essentially human ... it seems to me to be as inconsequent to refer the mandible and the cranium to the same individual as it would be to articulate a chimpanzee foot with the bones of an essentially human thigh and leg." ...

A third and in a sense neutral point of view held that the whole business was so ambiguous that the Piltdown discovery had best be put on the shelf, so to speak, until further evidence, through new discoveries, might become available. I have not attempted anything resembling a thorough poll of the literature, but I have the distinct impression that this point of view has become increasingly common in recent years, as will be further discussed. Certainly, those best qualified to have an opinion, especially those possessing a sound knowledge of human and primate anatomy, have held largely – with a few notable exceptions – either to a dualistic or to a neutral interpretation of the remains, and hence have rejected the monistic interpretation that led to the reconstruction of a "dawn man." Most assuredly, and contrary to the impression that has been generally spread by the popular press when reporting the hoax, "Eoanthropus" has remained far short of being universally accepted into polite anthropological society.

An important part of the Piltdown controversy related to the geological age of the "Eoanthropus" fossils. As we shall see, it was this aspect of the controversy that eventually proved to be the undoing of the synthetic Sussex "dawn man." Associated with the primate remains were those of various other mammals, including mastodon, elephant, horse, rhinoceros, hippopotamus, deer, and beaver. The Piltdown gravel, being stream-deposited material, could well contain fossils of different ages. The general opinion, however, seems to have been that it was of the Lower Pleistocene (some earlier opinions even allocated it to the Upper Pliocene), based on those of its fossils that could be definitely assigned such a date. The age of the remains of "Piltdown man" thus was generally regarded as Lower Pleistocene, variously estimated to be from 200,000 to 1,000,000 years. To the proponents of the monistic, "dawn-man" theory, this early dating sufficed to explain the apparent morphological incongruity between cranium and lower jaw.

In 1892, Carnot, a French mineralogist, reported that the amount of fluorine in fossil bones increases with their geological age – a report that seems to have received scant attention from paleontologists. Recently, K.P. Oakley, happening to come across Carnot's paper, recognized the possibilities of the fluorine test for establishing the relative ages of bones found within a single deposit. He realized, furthermore, that herein might lie the solution of the vexed Piltdown problem. Consequently, together with C.R. Hoskins, he applied the fluorine test to the "Eoanthropus" and other mammalian remains found at Piltdown. The results led to the conclusion that "all the remains of *Eoanthropus* ... are contemporaneous"; and that they are, "at the earliest, middle Pleistocene." However, they were strongly indicated as being of late or Upper Pleistocene age, although "probably at least 50,000 years" old. Their fluorine content was the same as that of the beaver remains but significantly less than that of the geologically older, early Pleistocene mammals of the Piltdown fauna. This seemed to increase the probability that cranium and jaw belonged to one individual. But at the same time, it raised the enigma of the existence in the late Pleistocene of a human-skulled, large-brained individual possessed of apelike jaws and teeth – which would leave "Eoanthropus" an anomaly among Upper Pleistocene men. To complete the dilemma, if cranium and jaw were attributed to two different animals – one a man, the other an ape – the presence of an anthropoid ape in England near the end of the Pleistocene appeared equally incredible. Thus the abolition of a Lower Pleistocene dating did not solve the Piltdown problem. It merely produced a new problem that was even more disturbing.

As the solution of this dilemma, Dr. J.S. Weiner advanced the proposition to Drs. Oakley and Clark that the lower jaw and canine tooth are actually those of a modern anthropoid ape, deliberately altered so as to resemble fossil specimens. He demonstrated experimentally, moreover, that the teeth of a chimpanzee could be so altered by a combination of artificial abrasion and appropriate staining as to appear astonishingly similar to the molars and canine tooth ascribed to "Piltdown man." This led to a new study of all the "Eoanthropus" material that "demonstrated quite clearly that the mandible and canine are indeed deliberate fakes." It was discovered that the "wear" of the teeth, both molar and canine, had been produced by an artificial planing down, resulting in occlusal surfaces unlike those developed by normal wear. Examination under a microscope revealed fine scratches such as would be caused by an abrasive. X-ray examination of the canine showed that there was no deposit of secondary dentine, as would be expected if the abrasion had been due to natural attrition before the death of the individual.

An improved method of fluorine analysis, of greater accuracy when applied to small samples, had been developed since Oakley and Hoskins made their report in 1950. This was applied to the Piltdown specimens. The results clearly indicate that whereas the Piltdown I cranium is probably Upper Pleistocene in age, as claimed by Oakley and Hoskins, the attributed mandible and canine tooth are "quite modern." As for Piltdown II, the frontal fragment appears to be Upper Pleistocene (it probably belonged originally to Piltdown I cranium), but the occipital fragment and the isolated molar tooth are of recent or modern age. . . .

Weiner, Oakley, and Clark also discovered that the mandible and canine tooth of Piltdown I and the occipital bone and molar tooth of Piltdown II had been artificially stained to match the naturally colored Piltdown I cranium and Piltdown II frontal. Whereas these latter cranial bones are all deeply stained, the dark color of the faked pieces is quite superficial. The artificial color is due to chromate and iron. This aspect of the hoax is complicated by the fact that, as recorded by Smith Woodward, "the colour of the pieces which were first discovered was altered a little by Mr. Dawson when he dipped them in a solution of bichromate of potash in the mistaken idea that this would harden them." The details of the staining, which confirm the conclusions arrived at by microscopy, fluorine analysis, and nitrogen estimation, need not be entered into here.

In conclusion, therefore, the *disjecta membra* of the Piltdown "dawn man" may now be allocated as follows: (1) the Piltdown I cranial fragments (to which should probably be added Piltdown II frontal) represent a modern type of human brain-case that is in no way remarkable save for its unusual thickness and which is, at most, late Pleistocene in age; (2) Piltdown I mandible and canine tooth and Piltdown II molar tooth are those of a modern anthropoid ape (either a chimpanzee or an orangutan) that have been artificially altered in structure and artificially colored so as to resemble the naturally colored cranial pieces – moreover, it is almost certain that the isolated molar of Piltdown II comes from the original mandible, thus confirming Hrdlicka's earlier suspicion; and (3) Piltdown II occipital is of recent human origin, with similar counterfeit coloration.

Weiner, Oakley, and Clark conclude that "the distinguished palaeontologists and archaeologists who took part in the excavations at Piltdown were the victims of a most elaborate and carefully prepared hoax" that was "so extraordinarily skilful" and which "appears to have been so entirely unscrupulous and inexplicable, as to find no parallel in the history of palaeontological discovery."

8-3 Fossils and Free Enterprisers – Howard S. Miller

Edward Drinker Cope (1840–1897), wealthy, talented, ambitious, prolific author, and brilliant scientist, studied at Princeton. He was one of the leading students of his time in herpetology and ichthyology and subsequently devoted his attention to fossil vertebrates. In 1870, he worked under F.V. Hayden with the United States Geological Survey, and discovered many new groups of fish and reptiles. He was deeply concerned not only with systematic paleontology but also with the value of fossil vertebrates in correlation, and with the evolutionary mechanisms involved in the fossil record. He rejected Darwin's theory of natural selection and embraced a neo-Lamarckian mechanism.

Othniel Charles Marsh (1831–1899) was born in Lockport, New York, and was a student at Andover and Yale. He subsequently traveled to Europe and, on his return to the United States, was appointed Professor of Paleontology at Yale. In 1870, he organized the first

of four Yale scientific expeditions to the West and published extensively on the fossils that were collected on the expeditions. His millionaire uncle, George Peabody, founded the Peabody Museum at Yale, which, under Marsh's direction, became one of the leading museums in the world. Though of independent means, Marsh became vertebrate paleontologist at the United States Geological Survey in 1892, and it was during this period that his rivalry with E.D. Cope reached a climax.

In the extract below, taken from his Dollars for Research: science and its patrons in nineteenth century America *(1970), Howard S. Miller writes about the conflict between Cope and Marsh. Miller was born in 1936 in Pontiac, Illinois, educated at Bradley University and the University of Wisconsin and held academic appointments in history at the University of Southern California and the University of Missouri, St. Louis.*

One summer day in 1868 the Union Pacific made an unscheduled stop at Antelope Station, a tiny dot on the rolling Nebraska prairies, just east of the Wyoming line. A dapper, obviously Eastern gentleman leapt from a parlor car and began searching through a mound of dirt beside a recently dug well. Othniel C. Marsh, a Yale paleontologist, had come in response to newspaper reports of fossil human bones unearthed by a well digger. He failed to find fragments of early man, but what he did find excited him hardly less. For scattered over the ground were remains of many animals, among them a diminutive horse which Marsh subsequently identified as an important link in the evolutionary chain. He left the scene only after the impatient conductor had flagged the train ahead, leaving the preoccupied scientist to run after the last car. "I could only wonder," Marsh later recalled, "if such scientific truths as I had now obtained were concealed in a single well, what untold treasures must there be in the whole Rocky Mountain region. This thought promised rich rewards to the enthusiastic explorer in this new field, and thus my own life work seemed laid out before me."

During the next three decades enthusiastic explorers, notably Joseph Leidy, Marsh himself, and Edward Drinker Cope, unearthed an extinct fauna of unimagined richness and variety. Before the three entered the field there were fewer than one hundred species of fossil vertebrates known in America. By the end of the century they had dug more than two thousand new species from the fossil cemeteries of the Great Plains. In the process they assembled impressive support for Darwinism, laid the foundations of modern paleontology in the United States, and became embroiled in one of the bitterest scientific and political controversies of the late nineteenth century. ...

But unveiling the past in the geological record was easier said than done, as Charles Darwin was the first to admit. Thus far fossil strata simply had not displayed the "finely graduated organic chain" demanded by Darwin's theory. Instead of orderly evolution, in fact, the rocks showed complex organisms suddenly appearing and disappearing without a trace of either ancestor or descendant. The evidence seemed to favor Progressionism, the respectable doctrine of Genesis and geology championed by Louis Agassiz. Undaunted, Darwin and the Darwinists plunged ahead, blaming the missing links on the imperfection of the geological record. Fossilization only occurred under rare circumstances, they argued, and in any case only a fraction of the globe had been explored. The geological record was, in short, "a history of the world imperfectly kept, and written in a changing dialect," a book of which whole chapters had been lost, or never imprinted in the sediments of primeval seas.

The fossil strata of the Great Plains held many of Darwin's missing chapters. Fortunately for the scientific community the territory was just now becoming accessible, as westward expansion gained momentum after the Civil War. Soldiers and miners, cattlemen and homesteaders all quickly discovered the practical utility of a careful scientific reconnaissance of the public domain. Their practical needs plus natural curiosity added up to political pressure, so that by the late 1860s the federal government had launched elaborate scientific surveys of the western territories. ...

From the 1870s on O.C. Marsh and E.D. Cope contested for supremacy in vertebrate paleontology. Had the two men lived in different eras, each would undoubtedly have had a less controversial career. While Joseph Leidy seemed to value peace at any price, Marsh and Cope were prepared for war at any cost. Both were independently wealthy, scientific robber barons, fitting contemporaries of Daniel Drew and Commodore Vanderbilt. Joseph Leidy was forced to surrender, a casualty in the Cope–Marsh war. "Professors Marsh and Cope, with long purses, offer money for what used to come to me for nothing, and in that respect I cannot compete with them. So now ... I have gone back to my microscope and my Rhizopods and make myself busy and happy with them."

Marsh owed his long purse – indeed much of his scientific career – to his uncle, George Peabody, a self-made man who had amassed a fortune in Anglo–American trade and finance. ...

Peabody was a doting bachelor uncle, who paid special attention to the education and well-being of his nieces and nephews. Routine family philanthropy prompted him to see O.C. Marsh through preparatory school and Yale, through two years graduate work in the Sheffield Scientific School, plus three more years of study abroad. Marsh set his sights on a professorship of paleontology at Yale. His mentors, James Dwight Dana and the younger Silliman, had also schooled him in the logistics of scientific enterprise, so that by 1862 Marsh was prepared to combine his request for further personal subsidy with a general plea for science at Yale. Marsh reported that Peabody had strong commitments to Harvard, and that he had already promised Edward Everett $100,000 for astronomy, or perhaps art. Still, there was hope. "As I have at present no interest in any institution except Yale," he assured Silliman, "I shall use all my influence with Mr. P. in her favor."

Silliman and Dana had long since envisioned a grand natural history museum and research center to fill out the Scientific School. Their protégé was to be the living bond between the plan and the patron. By May 1863 Marsh had secured a $100,000 Peabody legacy for Yale science. Three more years of complicated revision converted the legacy into a lifetime donation, and increased the amount to $150,000. In the meantime Marsh had also helped the Cambridge scientific corps redirect Peabody's proposed Harvard gift from a school of design to a museum of archeology and ethnology, and in the interest of institutional parity, had convinced his uncle to make equal gifts to the two universities.

George Peabody drew the formal trust deed in October 1866, establishing a museum specializing in zoology, geology, and mineralogy. It was, for all practical purposes, his nephew's private research institute. Several months before the grateful Yale Corporation had invited Marsh to fill a full professorship of paleontology created for the occasion. He had agreed to serve without salary (Uncle George provided a substantial allowance, and at his death in 1869 would leave him $100,000), so that the university could not demand that he teach under-graduate courses. Moreover, as a legal trustee of the museum as well as its director, Marsh could dictate research policy, control the collections, and publish his research at institutional expense. It was an enviable position for a fledgling scientist of thirty-five.

Under Marsh's direction the Peabody Museum quickly gained prominence. In the summer of 1870 he launched the first of a series of annual wild west fossil hunts. The Yale expeditions were technically private affairs, funded by the museum, but their familiarity with the government surveys, and their constant use of army facilities and personnel, clothed them at least partially in the public interest. The expeditions provided field training, augmented Marsh's collections, won him national publicity, and, at least to his own satisfaction, established squatter sovereignty over the fossil fields of the Great Plains.

In 1871 Edward Drinker Cope trespassed on Marsh's self-proclaimed preserve. "I am now in a territory which interests me greatly," he reported from Kansas. "The prospects are that I will be able to do something in my favorite line of Vertebrate Paleontology. ... Marsh has been doing a great deal I find, but has left more for me." Cope had only recently committed himself to a full-time career in science. His wealthy Quaker father had intended that he become a farmer, and in the interests of scientific agriculture had provided Cope with a comprehensive if unorthodox scientific education at home and abroad. But Cope hated farming, and found the teaching of zoology at Haverford College little better. He had attached himself to the Hayden (and later the Wheeler) Survey, and began with a certain malicious enthusiasm to compete with Marsh for priority of discovery and publication. If Marsh had entrée to *Silliman's Journal* in New Haven, Cope had the Government Printing Office. "Hayden has fathered my Paleontological Bulletins. ... A disagreeable pill for the Yale College People."

For the next twenty years the Cope–Marsh conflict sputtered intermittently, much to the embarrassment of the scientific community at large. It proved in the long run to be a war of attrition in which Marsh countered Cope's brilliant but erratic forays with a less spectacular but relentless marshaling of financial, personal, and institutional support. The turning point came in 1878, when Marsh succeeded Joseph Henry as President of the National Academy of Sciences. Marsh's commanding position in the academy carried prestige and power. But even more significant in 1878 was the fact that Congress had recently asked the academy, as official science advisor to the government, to recommend a unified scheme for the scientific surveys which would "secure the best results for the least possible cost." Fate had placed Marsh in a position to cut off his rival's source of support. . . .

The practical outcome was quickly told. Marsh's National Academy report (adopted with the sole dissent of E.D. Cope, who understood that reform would in effect freeze him out), urged a scheme which would have provided for geodesy, the systematic disposition of public lands, and theoretical and practical geology. Congress, in keeping with the general tenor of the Gilded Age, ignored all but the latter when it authorized the creation of a new United States Geological Survey. After all, said Congressman Hewitt, "the science of geology and the science of wealth are indissolubly linked." Nations became great only as the developed "a genius for grasping the forces and materials of nature within their reach and converting them into a steady flowing stream of wealth and comfort." The Hayden and Powell surveys, once work in progress had been completed, were to be either absorbed or abolished. With Marsh's help Clarence King emerged as director of the new survey, only to be replaced by John Wesley Powell, when King left government service the following year. Largely in return for past favors, in 1882 Powell appointed Marsh official paleontologist of the Geological Survey. E.D. Cope, cut off from federal patronage, could only bewail his dwindling fortune, appeal for congressional and private support, and rage at his enemies, who he charged had created "a gigantic politico-scientific monopoly next in importance to Tammany Hall."

More was at stake than a personal feud and bureaucratic empire building. The survey reorganization of 1878–79 was of lasting significance as the beginning of a ten year debate over the proper relationship between science and the federal government. Except in special circumstances, notably Joseph Henry's Smithsonian battle of the 1850s, heretofore government science had been neither costly nor controversial enough to attract searching congressional criticism. But by the late 1870s it had grown too big to ignore, and politicians began to understand what years of salutory neglect had wrought.

8-4   The K-T Extinction – Charles Officer and Jake Page

Charles Officer is a Columbia PhD who is Research Professor in Earth Sciences and Engineering at Dartmouth. Jake Page, a professional writer, is founder and past director of Smithsonian Books and the editorial director of Natural History. The present extract, taken from their book Tales of the Earth: paroxysms and perturbations of the blue planet *(1993), discusses the demise of the dinosaurs.*

Many of us have a picture in mind of dinosaurs as monstrously large creatures that once roamed a swampy Earth but then were suddenly gone. The question of why they vanished so suddenly has fascinated humans of all ages and persuasions since a fossil dinosaur bone was first recognized for what it was. This was the last major extinction crisis on Earth, occurring about sixty-six million years ago, bringing down with the dinosaurs a host of other kinds of creatures, mostly marine plankton and shellfish. And it all happened in what amounts to a geologic instant. In fact, the extinctions took place over a period of time we can barely imagine – thousands to hundreds of thousands of years. Writing originated 7,000 years ago and our human experience of changes on Earth can be measured in tens, perhaps hundreds, of years. From our own limited experience of environmental change, it is nearly impossible to appreciate, even to grasp, the kinds of changes that have taken place over eons. Imagine, then, the excitement of finding a distinct line in nature that corresponds to the distinct line in the geologic timetable: the moment the dinosaurs vanished, the transition between Cretaceous times and Tertiary times, known specifically as the Cretaceous/Tertiary transition (in shorthand, the K/T boundary). Imagine the excitement of finding a single event that could have brought down that vast race of terrible lizards, a single cause for such a great dying.

Scientists have long speculated about the extinction of the dinosaurs, sometimes coming up with explanations as marvelous as the animals themselves. At one point, it was suggested that the dinosaurs might have died out from constipation because, as flowering plants (angiosperms) became dominant over evergreens (gymnosperms), the oil in angiosperm seeds would have blocked dinosaurian digestive systems. Such interest on the part of paleontologists and others had tottered along, a minor theme in science, the majority satisfied generally with the notion that some complicated changes in climate and environment would have led to the dinosaurs' end. Then, in 1980, a physicist named Luis Alvarez came forward with the hypothesis that this last great extinction was caused by the impact of a gigantic asteroid. The impact would have created a huge dust cloud that shrouded the Earth, curtailing if not cutting off photosynthesis and leading to the demise of those animals dependent on plants – such as herbivorous dinosaurs, followed promptly by the carnivores.

Alvarez was not given to wild speculations. He was a Nobel Prize winner in physics and a highly respected scientist. When he spoke, people listened. He could point to a smoking gun in the geologic record. His son geologist Walter Alvarez and others had discovered in rock strata located in Italy, Denmark, and New Zealand and dated precisely to the Cretaceous/Tertiary boundary a thin layer bearing anomalous amounts of a rare element, iridium. Iridium, a metallic element akin to platinum, is extremely rare in the Earth's crust. But in the widespread, presumably worldwide layer Alvarez pointed to, it occurred in the range of five to ten parts per billion (ppb) – an iridium bonanza that could only be explained as having come from an extraterrestrial source, since iridium is far more common in meteorites. . . .

The Alvarez hypothesis was simple, straightforward, and dramatic, and it was argued by an eminent scientist. As Australian journalist Ian Warden put it in the May 20, 1984, issue of *The Canberra Times*: "To connect the dinosaurs, creatures of interest to but the veriest dullard, with a spectacular extraterrestrial event like the deluge of meteors ... seems a little like one of those plots that a clever publisher might concoct to guarantee enormous sales. All the Alvarez–Raup theories lack is some sex and the involvement of the Royal family and the whole world would be paying attention to them."

Indeed, practically the whole world did. Dinosaurs made the cover of *Time*. And in the interval, it is fair to say, virtually all interested lay persons, are satisfied that the answer to the dinosaurs' extinction is now known: a catastrophic global nuclear-type winter brought on by extraterrestrial impact, or impacts. But geologists had over the past century painfully rid themselves of the previous school of catastrophism, and to them this idea smacked of neocatastrophism. Furthermore, the impact theory left many geologic and paleontogical matters unaddressed. And so the battle was joined.

The vertebrate paleontologists said the geologic record showed that the demise of the various dinosaur species occurred sequentially over 100,000 to 500,000 years, not in a relatively brief cataclysmic event that left the planet suddenly littered with dinosaur carcasses. The final disappearance in the fossil record, they said, of the nearly ubiquitous *Triceratops*, the last of the dinosaurs, occurred stratigraphically *below* (meaning earlier) the level of iridium-bearing clays. A few such cautionary facts would have been sufficient to scotch the asteroid-impact theory, but it had its own momentum by then. Many scientists had "invested" in it, the public loved it, even some elements of the "scientific press" loved it. (For example, *Science* magazine, the journal of the American Association for the Advancement of Science, appeared reluctant to publish any findings that gainsaid the impact theory.) As has happened from time to time in the brief history of geology, matters became rancorous. Paraphrasing another, earlier physicist, Lord Kelvin, Luis Alvarez lashed out against paleontologists. In *The New York Times* of January 19, 1988, Alvarez said, "They're really not very good scientists. They're more like stamp collectors."

But a continuing collection of what we might call "geophilatelic insights" nonetheless raised questions about the impact theory. The theory calls for the creation of a dust cloud 1,000 times greater than that produced by the eruption of Krakatoa. But Dennis Kent of Columbia University pointed out that 75,000 years ago the still larger volcano Toba sent up a dust cloud estimated to be at least 400 times that of Krakatoa. Yet no abnormal rate of extinction occurred after the Toba eruption.

While dinosaurs have naturally received the greatest attention during the controversy, particularly among the public (which may not even be aware that a controversy exists, so foregone a conclusion has the impact theory become in the press), the dinosaurs are not the most satisfactory group of animals to examine when trying to determine the specifics of the extinction process. Only a relative few species of them existed at any given time during their 160 million years on the planet, and relatively few specimens have been preserved in the geologic record as fossils – and these are only found in isolated locales. The theory must explain why so many forms of life persevered: other reptiles, including big ones like crocodiles; the mammals; and those dinosaurian derivatives, the birds, to name but a few. It must explain as well the other extinctions of the age – the plankton and the shellfish. Here we are dealing with a large number of species, and a large number of specimens.

Paleontologist Gerta Keller looked at the extinction record of ocean plankton at the Brazos River in Texas and at El Kef in Tunisia, and found that the plankton species disappeared in a sequential, stepwise manner over a period of a few hundred thousand years. Other researchers have found the same extended, sequential extinction record for shellfish species in Antarctica during this general time period. The onset of the shellfish extinction series also preceded that of the ocean plankton. Clearly, extinctions are not simple, instantaneous events. They are complex and occur over periods of geologic, not human, time (until recently). Such paleontological findings should be enough, in and of themselves, to rule out an asteroid impact as their cause. In a system as complex as the Earth and its living forms, it may be useful to keep in mind a remark H.L. Mencken made in another context: "For every complex problem, there is a solution that is simple, neat and wrong."

The proponents of the impact hypothesis were not dissuaded, and the controversy has gone on, an ancient detective story with many dead bodies and few clues to the cause of death.

8-5   The Founders of Geology – Sir Archibald Geikie

Sir Archibald Geikie (1835–1924) – whom we first encountered on page 64 – wrote in 1897 The Founders of Geology, from which our present extract is taken. This is a rare account of one of the decisive discoveries that ultimately led to the resolution of the great eighteenth and early nineteenth century debate between the Neptunists (those who believed that all rocks had been precipitated from aqueous solution) and the Plutonists (who regarded some of them as igneous in origin). In 1788, Nicolas Desmarest (1725–1815) was appointed by the King of France as Inspector-General and Director of the Manufactures of France. His influence continued even after the Revolution. His studies of French industry were of enormous importance, but he was interested, too, in geology. His biographer records "... he made his journeys on foot, with a little cheese as all his sustenance. No path seemed impracticable to him, no rock inaccessible. He never sought the country mansions, he did not even halt at the inns. To pass the night on the hard ground of some herdsman's hut, was to him only an amusement." During these journeys, he found himself in Auvergne in 1763, and was fascinated by the hexagonal pillars of basalt which characterize those lava fields.

Shortly after the middle of the eighteenth century, the Governments of Europe, wearied with ruinous and profitless wars, began to turn their attention towards the improvement of the industries of their peoples. The French Government especially distinguished itself for the enlightened views which it took in this new line of national activity. It sought to spread throughout the kingdom a knowledge of the best processes of manufacture, and to introduce whatever was found to be superior in the methods of foreign countries. Desmarest was employed on this mission from 1757 onwards. At one time he would be sent to investigate the cloth-making processes of the country: at another to study the various methods adopted in different districts in the manufacture of cheese. Besides being deputed to examine into the condition of the industries of different provinces of France, he undertook two journeys to Holland to study the paper-making system of that country. He prepared elaborate reports of the results of his investigations, which were published in the *Mémoires* of the Academie des Sciences, or in the *Encyclopédie Méthodique*. At the last in 1788 he was named by the King Inspector-General and Director of the Manufactures of France.

He continued to hold this office until the time of the Revolution, when his political friends – Trudaine, Malesherbes, La Rochefoucault, and others – perished on the scaffold or by the knife of the assassin. He himself was thrown into prison, and only by a miracle escaped the slaughter of the 2nd September. After the troubles were over, he was once more called to assist the Government of the day with his experience and judgment in all matters connected with the industrial development of the country. It may be said of Desmarest that "for three quarters of a century it was under his eyes, and very often under his influence, that French industry attained so great a development."

Such was his main business in life, and the manner in which he performed it would of itself entitle him to the grateful recollection of his fellow-countrymen. But these occupations did not wholly engross his time or his thoughts. Having early imbibed a taste for scientific investigation, he continued to interest himself in questions that afforded him occupation and solace, even when his fortunes were at the lowest ebb.

"Resuming the rustic habits of his boyhood," says his biographer, "he made his journeys on foot, with a little cheese as all his sustenance. No path seemed impracticable to him, no rock inaccessible. He never sought the country mansions, he did not even halt at the inns. To pass the night on the hard ground in some herdsman's hut, was to him only an amusement. He would talk with quarrymen and miners, with blacksmiths and masons, more readily than with men of science. It was thus that he gained that detailed personal acquaintance with the surface of France with which he enriched his writings."

During these journeyings, he was led into Auvergne in the year 1763, where, eleven years after Guettard's description had been presented to the Academy, he found himself in the same tract of Central France, wandering over the same lava-fields, from Volvic to the heights of Mont Dore. Among the many puzzles reported by the mineralogists of his day, none seems to have excited his interest more than that presented by the black columnar stone which was found in various parts of Europe, and for which Agricola, writing in the middle of the sixteenth century, had revived Pliny's old name of "basalt." The wonderful symmetry, combined with the infinite variety of the pillars, the vast size to which they reached, the colossal cliffs along which they were ranged in admirable regularity, had vividly aroused the curiosity of those who concerned themselves with the nature and origins of minerals and rocks. Desmarest had read all that he could find about this mysterious stone. He cast longing eyes towards the foreign countries where it was developed. In particular, he pictured to himself the marvels of the Giant's Causeway of the north of Ireland, as one of the most remarkable natural monuments of the world, where Nature had traced her operations with a bold hand, but had left the explanation of them still concealed from mortal ken. How fain would he have directed his steps to that distant shore. Little did he dream that the solution of the problems presented by basalt was not to be sought in Ireland, but in the heart of his own country, and that it was reserved for him to find. . . .

Much had been learnt as to the diffusion of basalt in Europe, and many excellent drawings had been published of the remarkable prismatic structure of this rock. But no serious attempt seems to have been made to grapple with the problem of its origin. Some absurd notions had indeed been entertained on this subject. The long regular pillars of basalt, it was gravely suggested, were jointed bamboos of a former period, which had somehow been converted into stone. The similarity of the prisms to those of certain minerals led some mineralogists to regard basalt as a kind of schorl, which had taken its geometrical forms in the process of crystallization. Romé de Lisle is even said to have maintained that each basalt prism ought to have a pyramidal termination, like the schorls and other small crystals of the same nature.

Guettard, as we have seen, drew a distinction between basalt and lava, and this opinion was general in his time. The basalts of Central and Western Europe were usually found on hill tops, and displayed no cones or craters, or other familiar signs of volcanic action. On the contrary, they were not infrequently found to lie upon, and even to alternate with, undoubted sedimentary strata. They were, therefore, not unnaturally grouped with these strata, and the whole association of rocks was looked upon as having had one common aqueous origin. It was also a prevalent idea that a rock which had been molten must retain obvious traces of that condition in a glassy structure. There was no such conspicuous vitreous element in basalt, so that this rock, it was assumed, could never have been volcanic. As Desmarest afterward contended, those who made such objections could have but little knowledge of volcanic products.

We may now proceed to trace how the patient and sagacious Inspector of French industries made his memorable contribution to geological theory. It was while traversing a part of Auvergne in the year 1763 that he detected for the first time columnar rocks in association with the remains of former volcanoes. On the way from Clermont to the Puy de Dôme, climbing the steep slope that leads up to the plateau of Prudelle, with its isolated outlier of a lava-stream that flowed long before the valley below it had been excavated, he came upon some loose columns of a dark compact stone which had fallen from the edge of the overlying sheet of lava. He found similar columns standing vertically all along the mural front of the lava, and observed that they were planted on a bed of scoriæ and burnt soil, beneath which lay the old granite that forms the foundation rock of the region. He noticed still more perfect prisms a little further on, belonging to the same thin cake of dark stone that covered the plain which leads up to the foot of the great central puy.

Every year geological pilgrims now make their way to Auvergne, and wander over its marvellous display of cones, craters and lava-rivers. Each one of them climbs to the plateau of Prudelle, and from its level surface gazes in admiration across the vast fertile plain of the Limagne on the one side, and up to the chain of the puys on the other. Yet how few of them connect that scene with one of the great triumphs of their science, or know that it was there that Desmarest began the observations which directly led to the fierce contest over the origin of basalt!

That cautious observer tells us that amidst the infinite variety of objects around him, he drew no inference from this first occurrence of columns, but that his attention was aroused. He was kept no long time in suspense on the subject. "On the way back from the Puy de Dôme," he tells us, "I followed the thin sheet of black stone and recognised in it the characters of a compact lava. Considering further the thinness of this crust of rock, with its underlying bed of scoriæ, and the way in which it extended from the base of hills that were obviously once volcanoes, and spread out over the granite, I saw in it a true lava-stream which had issued from one of the neighbouring volcanoes. With this idea in my mind, I traced out the limits of the lava, and found again everywhere in its thickness the faces and angles of the columns, and on the top their cross-section, quite distinct from each other. I was thus led to believe that prismatic basalt belonged to the class of volcanic products, and that its constant and regular form was the result of its ancient state of fusion. I only thought then of multiplying my observations, with the view of establishing the true nature of the phenomenon, and its conformity with what is to be found in Antrim – a conformity which would involve other points of resemblance."

He narrates the course of his discoveries as he journeyed into the Mont Dore, detecting in many places fresh confirmation of the conclusion he had formed. But not only did he convince himself that the prismatic basalts of Auvergne were old lava-streams, he carried his induction much further and felt assured that the Irish basalts must also have had a volcanic origin. "I could not doubt," he says, "after these varied and repeated observations, that the groups of prismatic columns in Auvergne belonged to the same conformation as those of Antrim, and that the constant and regular form of the columns must have resulted from the same cause in both regions.

What convinced me of the truth of this opinion was the examination of the material constituting the Auvergne columns with that from the Giant's Causeway, which I found to agree in texture, colour and hardness, and further, the sight of two engravings of the Irish locality which at once recalled the scenery of parts of Mont Dore. I draw, from this recognised resemblance and the facts that establish it, a deduction which appears to be justified by the strength of the analogy – namely, that in the Giant's Causeway, and in all the prismatic masses which present themselves along the cliffs of the Irish coast, in short even among the truncated summits of the interior, we see the operations of one or more volcanoes which are extinct, like those of Auvergne."

8-6   To a Rocky Moon – Don E. Wilhelms

Don Wilhelms, a geologist with the United States Geological Survey, is one of the pioneers of the nation's space program in planetary geology. The book from which this extract is taken – To a Rocky Moon: a geologist's history of lunar exploration (1993) – describes the growing interest in lunar exploration, from ancient times through the various Apollo missions, for each of which the planning, the crews, and the scientific results are described.

Our extract is concerned with earlier lunar observations of G.K. Gilbert (one of the greatest geologists of all time) and especially the once-vexed question of whether lunar craters were formed by impact or by vulcanism. Not only is the impact theory of origin now generally accepted – though volcanic lunar features are known in many mare regions – but the moon itself is thought to have originated in the impact of a Mars-size body with Earth.

People seem to need heroes, so let us consider Grove Karl Gilbert (1843–1918) of the United States Geological Survey. Gilbert was surely one of the greatest geologists who ever lived, and his genius touched almost all aspects of the science: geomorphology, glaciology, sedimentation, structure, hydrology, and geophysics. He was in Berkeley in April 1906 when he awoke early one morning "with unalloyed pleasure" at realizing that a vigorous earthquake was in progress, and he caught the first available ferry to San Francisco. He carefully recorded how long the subsequent fire took to consume the wooden buildings on Russian Hill (where I now live) and contributed major parts of the subsequent official report. His personality seems to have been mild and subdued, even "saintlike," in an era of rough-hewn and feisty pioneers of western geology. His recent biographer Stephen Pyne has applied to him the same term Gilbert applied to the Geological Survey: a great engine of research.

In 1891, while chief geologist of the Survey (the insider's term for the USGS), Gilbert was attracted by reports of large amounts of meteoric iron, the Canyon Diablo meteorites, around a crater in Arizona then called Coon Mountain or Coon Butte. Apparently he had already been thinking about the possible impact origin of lunar craters, and he alone realized that the Coon crater might itself be a "scar produced on the earth by the collision of a star." If so, a large iron meteorite might lie buried underneath the crater. He reasoned that such a body should (1) show up magnetically at the surface and (2) displace such a large volume that the ejecta of the crater should be more voluminous than its interior.

He tested both ideas and got negative results. In October 1891 he and his assistants carefully surveyed the volumes of ejecta and of the crater and found them to be identical at 82 million cubic yards (63 million m³). Their magnetic instruments showed no deflections whatsoever between the rim and the interior. Gilbert reluctantly concluded that the crater was formed by a steam explosion; that is, it was a *maar*. There the matter appeared to rest for a while.

But he was not ready to give up on impacts. Calling himself temporarily a selenologist, he observed the Moon visually for 18 nights in August, September, and October 1892 with the 67-cm refracting telescope the Naval Observatory in Washington. A member of Congress assessed this activity and Gilbert's parent organization as follows: "So useless has the Survey become that one of its most distinguished members has no better way to employ his time than to sit up all night gaping at the Moon." But those 18 nights left a tremendous legacy. The use to which Gilbert put them shows that the quality of scientific research depends first and foremost on the quality of the scientist's mind. It was not lack of data that led others of the time to so many erroneous conclusions.

Gilbert presented his conclusions in a paper titled "The Moon's Face," the first in the history of lunar geoscience with a modern ring. He knew he was not the first to suggest an impact origin for lunar craters; he mentioned Proctor, A. Meydenbauer, and "Asterios," the pseudonym for two Germans. Apparently, however, he was the first to adduce solid scientific arguments favoring impact for almost all lunar craters from the smallest to the largest – "phases of a single type" as he put it. Most earlier observers had seen the trees but not the forest: the subtypes but not the overall unity of form. Gilbert's contemporary, Nathaniel Southgate Shaler (1841–1906) recognized the unity of origin but got the origin wrong. Now, almost everything fit. Gilbert's sketches, descriptions, and interpretations could be used in a modern textbook. He knew that the inner terraces of craters formed by landslip. He wrote that the depression of lunar crater floors below the level of the surrounding "outer plain" made them totally unlike most terrestrial volcanoes. He noted that central peaks are common in craters of medium size but not in those smaller than about 20 km across and rarely in those larger than 150 km; but this is a regular relation and does not destroy the basic unity of form. The peaks lie below the crater rim and even mostly below the outer plain, unlike cones of terrestrial volcanoes of the Vesuvius type. The volcanic-collapse craters (*calderas*) of Hawaii were a somewhat better match, as others had said, but Gilbert listed enough dissimilarities to damn this comparison as well. He pointed out that the largest lunar craters (including those we call basins) far exceed the largest terrestrial craters in size. In his words, "volcanoes appear to have a definite size limit, while lunar craters do not. Form differences effectually bar from consideration all volcanic action involving the extensive eruption of lavas."

What Gilbert called "meteoric" theories fit the craters' sizes and forms much better. Impacts could have created the raised, complexly structured rim-flank deposits that he called "wreaths," the low floors, and even the central peaks, which he surmised were formed when material responded to the impact by flowing toward the center from all sides. He realized that impacts would weaken lunar materials to the point of plasticity (hence the peaks) and could melt them (hence the flat floors). His conclusions were based partly on simple experiments with projectiles and targets composed of everyday materials. He so completely accepted the origin of the "white" *rays* which radiate from many craters as splashes from impacts that "it is difficult to understand why the idea that they really are splashes has not sooner found its way into the moon's literature."

8-7 Properties and Composition of Lunar Materials: Earth Analogies – Edward Schreiber and Orson L. Anderson

This tongue-in-cheek article appeared in the prestigious weekly journal Science *in 1970 and reviews compressional velocities of lunar* rocks with those of Earth rocks and other "terrestrial materials."

Abstract. *The sound velocity data for the lunar rocks were compared to numerous terrestrial rock types and were found to deviate widely from them. A group of terrestrial materials were found which have velocities comparable to those of the lunar rocks, but they do obey velocity-density relations proposed for earth rocks.*

Certain data from Apollo 11 and Apollo 12 missions present some difficulties in that they require explanations for the signals received by the lunar seismograph as a result of the impact of the lunar module (LEM) on the lunar surface. In particular, the observed signal does not resemble one due to an impulsive source, but exhibits a generally slow build-up of energy with time. In spite of the appearance of the returned lunar samples, the lunar seismic signal continued to ring for a remarkably long time – a characteristic of very high Q material. The lunar rocks, when studied in the laboratory, exhibited a low Q. Perhaps most startling of all, however, was the very low sound velocity indicated for the outer lunar layer deduced from the LEM impact signal. The data obtained on the lunar rocks and finds agree well with the results of the Apollo 12 seismic experiment. These rock velocities are startlingly low. The measured velocities on a vesicular medium grained, igneous rock (10017) having a bulk density of $3.2 \text{g}/\text{cm}^3$ were $v_p = 1.84$, and $v_s = 1.05 \text{ km}/\text{sec}$. The results for a microbreccia (10046) with a bulk density of $2.2 \text{ g}/\text{cm}^3$ were $v_p = 1.25$ and $v_s = 0.74 \text{ km}/\text{sec}$ for the compressional (v_p) and shear (v_s) velocities.

It was of some interest to consider the behavior of these lunar rocks in terms of the expected behavior based on measurements of earth materials. Birch first proposed a simple linear relation between compressional velocity and density for rocks. This relation was examined further by Anderson who showed that this was a first approximation to a more general relation, derivable from a dependence of the elastic moduli with the density through a power function. Comparison of the results obtained from the returned lunar rocks with the predictions of these relationships expresses graphically the manner they deviate from the behavior of rocks found on earth. The velocities are remarkably lower than what would be predicted from either the Birch or Anderson relationships.

To account for this very low velocity, we decided to consider materials other than those listed initially by Birch or more detailed compilation of Anderson and Liebermann. The search was aided by considerations of much earlier speculations concerning the nature of the moon, and a significant group of materials was found which have velocities that cluster about those actually observed for lunar rocks.

These materials are summarized in Table 1, where, for emphasis, common rock types found on earth are listed for comparison. The materials studied were chosen so as to represent a broad geographic distribution in order to preclude any bias that might be introduced by regional sampling. It is seen that these materials exhibit

Table I. Comparison of compressional velocities of lunar rocks and various earth materials.

Earth rocks and minerals	v_p (km/sec)	Lunar rocks and cheeses	v_p (km/sec)
Corundum	10.80	Sapsego (Swiss)	2.12
Garnet	8.53	Lunar Rock 10017	1.84
Hematite	7.90	Gjetost (Norway)	1.83
Eclogite	6.89	Provolone (Italy)	1.75
Gabbro	6.80	Romano (Italy)	1.75
Diabase	6.33	Cheddar (Vermont)	1.72
Greywacke	6.06	Emmenthal (Swiss)	1.65
Quartz	6.05	Muenster (Wisconsin)	1.57
Granite	5.90	Lunar Rock 10046	1.25
Schist	5.10		
Limestone	5.06		
Sandstone	4.90		
Gneiss	4.90		

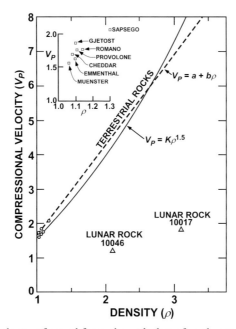

Fig. I. Comparison of the velocity of sound for rocks with that of earth materials.

compressional velocities that are in consonance with those measured for the lunar rocks – which leads us to suspect that perhaps old hypotheses are best, after all, and should not be lightly discarded.

A comparison of these low velocity materials with the predictions of Birch and of Anderson is shown in Figure 1. It is at once apparent that these materials do yield values of velocity that are predicted by these relations for their densities. Thus the curve of Birch for the rock types diabase, gabbros, and eclogites fit the cheeses surprisingly well. This apparent inconsistency, in that the cheeses do obey these relationships by having a velocity appropriate to their density, in contrast to the lunar rocks with which they compare so well, may readily be accounted for when one considers how much better aged the lunar materials are.

8-8 CFCs – Joel L. Swerdlow

Joel L. Swerdlow was born in Washington, DC, spent part of his childhood in Burma, graduated in political science at Syracuse and has a PhD from Cornell. He is a professional journalist and has written widely for radio, TV, and print on topics ranging from Watergate to the brain to Walt Whitman. He now serves as Assistant Editor of National Geographic, *from which our present article (1998), dealing with the controversial topic of global warming, is taken.*

Discoveries here on earth are changing much of our understanding of the physical world. Until recently no one thought that human actions could affect the world on a global scale. Now some scientists believe that for the first time in recorded history such changes are occurring. In 1974 scientists first observed that chlorofluorocarbons (CFCs) – synthetic compounds invented about 70 years ago and used today in refrigeration and the electronics industry – threaten the ozone layer of the earth's atmosphere. Ozone depletion also occurs when CFCs interact with aerosols from volcanic eruptions, although volcanoes alone do not cause ozone depletion.

For roughly a billion years the ozone layer has shielded earth from the sun's ultraviolet rays. But CFCs and other man-made compounds have punched a hole twice the size of Europe over Antarctica. The subsequent increase in ultraviolet radiation reduces some crop yields, has been linked to a higher incidence of skin cancers and cataracts in humans, and may suppress our immune systems. As of 1997 more than 160 countries had signed the Montreal Protocol, committing to cut use of CFCs in half by 2000. Some countries, including the U.S., have promised more aggressive reductions. Concentrations of ozone-depleting chemicals have started to decrease, in part because cars with air conditioners are spewing less of the CFC dichlorodifluoromethane. Fortunately for people who like to stay cool, engineers have devised an air-conditioning system that relies on a non-ozone-destroying compound. All new vehicles sold in the U.S. after 1994 must use this new system.

As the supply of dichlorodifluoromethane dwindles, the price of comfort has gone up for people like me who own older vehicles. The installation of the ozone-destroying Freon that had cost me $30 when my car was new now costs $130.

Mike Cunningham, who owns an auto shop in Bethesda, Maryland, showed me how my car air conditioner's condenser connects to the accumulator.

"Your car holds 2.12 pounds of Freon, and you've got about one and a half pounds in here," Mike said.

"What's happened to the rest," I asked.

"Into the atmosphere," Mike replied. "We'll inject dye, and see if we can find the leak."

"You mean I've been destroying the ozone layer?" I said, not hiding concern. One chlorine atom from a CFC can destroy more than 100,000 ozone molecules. The leak in my air-conditioning will destroy millions, if not billions, of ozone molecules.

"Don't feel bad," Mike said. "About half of all car ACs leak, and we never find the source." He injected the dye but with no success.

That ozone depletion is reversible provides some reassurance. With curtailed use of chlorofluorocarbons, experts say, the ozone layer will likely return to its original state around the year 2050.

But my automobile sends a lot more into the air than CFCs. Burning gasoline also releases carbon dioxide, which traps heat in the atmosphere. Since the industrial revolution such greenhouse gases, including methane and nitrous oxide, have been raising earth's global temperature, half a degree this century alone.

Some experts argue that global warming is part of the planet's natural rhythm, unrelated to human activity. Why curtail the use of fossil fuels, they ask, when the causes of global warming are not clearly understood?

Whatever its causes, global warming may have contributed to more extreme weather, such as droughts and severe floods. Another indicator of climate change may be the gradual shrinking of ice shelves in Antarctica. Several ice shelves have retreated and collapsed in the past 50 years.

Global warming may also be raising sea levels through the thermal expansion of water and the melting of glaciers around the world. With a three-foot rise in sea level (a possibility by the end of the [twenty-first] century), many coastal populations would be at risk. Nearly one-fifth of Bangladesh would disappear.

The problem of global warming may be harder to solve than ozone depletion. Output of greenhouse gases would increase significantly if the world's population, now about six billion, reaches – as experts suggest – more than nine billion by 2050.

Part 3

Language of the Earth

We share this small planet with two million or more other species. Yet, so far as we know, we are the only talking animals; no other species communicates as we do, with the array of symbols – spoken and written – on which daily existence and our complex social structures have come to depend. Other animals communicate: some, like bees, with complex "dances"; a few can understand limited human speech; some others can mimic human sounds; but we alone classify our cries and modulate our utterances to share our experience. And that sharing provides us with a huge potential range, not only of personal expression, but also of cumulative information and social coordination that is unavailable to other species.

Utterances about Earth must have been early components of the human vocabulary. Cries of warning at the eruption of a volcano, cries of fear at an approaching tsunami, cries for help as the ground trembled, cries of alarm at the varying moods of Earth. And other noises – grunts, hisses, expostulations, whoops, exclamations – at the use of fire, at the versatility of flint, at the discovery of a pool of clear water – all these must have been part of the proto-human repertoire.

So language, the ability to think aloud, became both an expression of our dependence on Earth and the means of our growing independence from its vagaries. It is language that supports human society, in all its many expressions.

We have chosen to interpret "language" broadly. We include, not only poetry and prose, for example, but also painting, sculpture, and architecture, for we regard these as means of expression and communication which, while themselves nonverbal in character, may be analyzed, understood and, in one sense, recreated by the use of language.

9. Prose

The dictionary tells us that prose is the ordinary language of writing or speaking, as opposed to verse or poetry. Samuel Coleridge (1772–1834) in his lectures on Shakespeare and Milton distinguished between prose and poetry when he observed, "I wish our clever young poets would remember my homely definitions of prose and poetry; that is, prose – words in their best order; poetry – the best words in their best order." In this section we include a group of prose selections, all concerned in one way or another with the Earth, by writers who show how to use words in their best order.

The examples we include are diverse, ranging from the humor of Samuel Clemens (Mark Twain) to Ernest Hemingway, who writes of hunting in Africa. We also include excerpts from Antoine de St. Exupéry, author of *The Little Prince*, who describes the spectacular scenery of Patagonia viewed from the air, and T.E. Lawrence who takes the reader across the rugged terrain of Arabian deserts and imparts a feeling of mystery to these arid lands. The roll call of prose writers who write of geology and geologists is a lengthy one. In our brief collection, we include Nobel and Pulitzer Prize-winning novelists, together with best-selling contemporary authors and writers of earlier times – from Goethe to Conan Doyle.

9-1   Out of Africa – Isak Dinesen

Karen Christentze Dinesen Blixen (1885–1962) was born into a wealthy family in Denmark and wrote in Danish under the name Karen Blixen and in English under the pseudonym Isak Dinesen. She studied English at Oxford and painting in Copenhagen, Paris, and Rome, and she explained that she could best describe landscape by seeing it through the eyes of painters. After marrying Baron Bror Blixen-Finecke in 1914, she moved to a six thousand acre coffee farm near Nairobi in Kenya. The marriage was deeply unhappy and she contracted syphilis from her husband, whom she later divorced. The farm proved to be a financial disaster and was ultimately sold. It was during these distressing times that Karen Blixen resumed her earlier interest in writing, telling the stories she composed to Denys Finch Hatton, an English pilot and big game hunter. Their relationship was portrayed by Meryl Streep and Robert Redford in the 1985 movie, Out of Africa. *Plagued throughout the rest of her life by ill health, she wrote a series of successful books, from one of which – Out of Africa (1937) – our extract is* taken; this book was described by one reviewer as "perhaps the finest book ever written about Africa." *Twice nominated for the Nobel Prize in Literature, she is celebrated as one of Denmark's greatest writers.*

I had a farm in Africa, at the foot of the Ngong Hills. The Equator runs across these highlands, a hundred miles to the North, and the farm lay at an altitude of over six thousand feet. In the day-time you felt that you had got high up, near to the sun, but the early mornings and evenings were limpid and restful, and the nights were cold.

The geographical position, and the height of the land combined to create a landscape that had not its like in all the world. There was no fat on it and no luxuriance anywhere; it was Africa distilled up through six thousand feet, like the strong and refined essence of a continent. The colours were dry and burnt, like the colours in pottery. The trees had a light delicate foliage, the structure of which was different from that of the trees in Europe; it did not grow in bows or cupolas, but in horizontal layers, and the formation gave to the tall solitary trees a likeness to the palms, or a heroic and romantic air like fullrigged ships with their sails clewed up, and to the edge of a wood a strange appearance as if the whole wood were faintly vibrating. Upon the grass of the great plains the crooked bare old thorn-trees were scattered, and the grass was spiced like thyme and bog-myrtle; in some places the scent was so strong, that it smarted in the nostrils. All the flowers that you found on the plains, or upon the creepers and liana in the native forest, were diminutive like flowers of the downs, – only just in the beginning of the long rains a number of big, massive heavy-scented lilies sprang out on the plains. The views were immensely wide. Everything that you saw made for greatness and freedom, and unequalled nobility.

The chief feature of the landscape, and of your life in it, was the air. Looking back on a sojourn in the African highlands, you are struck by your feeling of having lived for a time up in the air. The sky was rarely more than pale blue or violet, with a profusion of mighty, weightless, ever-changing clouds towering up and sailing on it, but it has a blue vigour in it, and at a short distance it painted the ranges of hills and the woods a fresh deep blue. In the middle of the day the air was alive over the land, like a flame burning; it scintillated, waved and shone like running water, mirrored and doubled all objects, and created great Fata Morgana. Up in this high air you breathed easily, drawing in a vital assurance and lightness of heart. In the highlands you woke up in the morning and thought: Here I am, where I ought to be.

The Mountain of Ngong stretches in a long ridge from North to South, and is crowned with four noble peaks like immovable darker blue waves against the sky. It rises eight thousand feet above the Sea, and to the East two thousand feet above the surrounding country; but to the West the drop is deeper and more precipitous, – the hills fall vertically down towards the Great Rift Valley.

The wind in the highlands blows steadily from the North–North-East. It is the same wind that, down at the coasts of Africa and Arabia, they name the Monsoon, the East Wind, which was King Solomon's favourite horse. Up here it is felt as just the resistance of the air, as the Earth throws herself forward into space. The wind runs straight against the Ngong Hills, and the slopes of the hills would be the ideal place for setting up a glider, that would be lifted upwards by the currents, over the mountain top. The clouds, which were traveling with the wind, struck the side of the hill and hung round it, or were caught on the summit and broke into rain. But those that took a higher course and sailed clear of the reef, dissolved to the West of it, over the burning desert of the Rift Valley. Many times I have from my house followed these mighty processions advancing, and have wondered to see their proud floating masses, as soon as they had got over the hills, vanish in the blue air and be gone.

The hills from the farm changed their character many times in the course of the day, and sometimes looked quite close, and at other times very far away. In the evening, when it was getting dark, it would first look, as you gazed at them, as if in the sky a thin silver line was drawn all along the silhouette of the dark mountain; then, as night fell, the four peaks seemed to be flattened and smoothed out, as if the mountain was stretching and spreading itself.

From the Ngong Hills you have a unique view, you see to the South the vast plains of the great game-country that stretches all the way to Kilimanjaro; to the East and North the park-like country of the foot-hills with the forest behind them, and the undulating land of the Kikuyu-Reserve, which extends to Mount Kenya a hundred miles away, – a mosaic of little square maize-fields, banana-groves and grass-land, with here and there the blue smoke from a native village, a small cluster of peaked mole-casts. But towards the West, deep down, lies the dry, moon-like landscape of the African low country. The brown desert is irregularly dotted with the little marks of the thornbushes, the winding river-beds are drawn up with crooked dark-green trails; those are the woods of the mighty, wide-branching Mimosa-trees, with thorns like spikes; the cactus grows here, and here is the home of the Giraffe and the Rhino.

The hill-country itself, when you get into it, is tremendously big, picturesque and mysterious; varied with long valleys, thickets, green slopes and rocky crags. High up, under one of the peaks, there is even a bamboo-grove. There are springs and wells in the hills I have camped up here by them.

In my day, the Buffalo, the Eland and the Rhino lived in the Ngong Hills, – the very old Natives remembered a time when there were Elephants there, – and I was always sorry that the whole Ngong Mountain was not enclosed in the Game Reserve. Only a small part of it was Game Reserve, and the beacon on the Southern peak marked the boundary of it. When the Colony prospers and Nairobi, the capital, grows into a big city, the Ngong Hills might have made a matchless game park for it. But during my last years in Africa many young Nairobi shop-people ran out into the hills on Sundays, on their motor-cycles, and shot at anything they saw, and I believe that the big game will have wandered away from the hills, through the thorn-thickets and the stony ground further South.

Up on the very ridge of the hills and on the four peaks themselves it was easy to walk; the grass was short as on a lawn, with the grey stone in places breaking through the sward. Along the ridge, up and down the peaks, like a gentle switch-back, there ran a narrow game-path. One morning, at the time that I was camped in the hills, I came up here and walked along the path, and I found on it fresh tracks and dung of a herd of Eland. The big peaceful animals must have been up on the ridge at sunrise, walking in a long row, and you cannot imagine that they had come for any other reason than just to look, deep down on both sides, at the land below.

9-2 Seven Pillars of Wisdom – T.E. Lawrence

That Thomas Edward Lawrence (1888–1935), archaeologist, scholar, soldier, adventurer, writer, and lover of the desert, should have a poetic sense for landscape is not unexpected. Lawrence first visited the Middle East in 1909 – a year before he graduated from Oxford – when he went to Syria and Palestine to study the architecture of Crusader castles. He returned as an officer in the British Army during World War I, where he organized Arab groups in the desert to fight against Turkey. Although known to millions as a war hero through the 1962 movie Lawrence of Arabia, *Lawrence himself hoped to be remembered "rather as a man of letters than a man of action." Lawrence published an account of his wartime experiences,* Seven Pillars of Wisdom *(privately printed in 1922 and 1926, first published in 1935), which is his greatest work. The account is laden with descriptions of dunes, desert pavement, oases, wadis, lava fields, watersheds, and escarpments. We include a passage from Lawrence, in which he relates the crossing of a rough lava plateau.*

Dawn found us crossing a steep short pass out of Wadi Kitan into the main drainage valley of these succeeding hills. We turned aside into Wadi Reimi, a tributary, to get water. There was no proper well, only a seepage hole in the stony bed of the valley; and we found it partly by our noses: though the taste, while as foul, was curiously unlike the smell. We refilled our waterskins. Arslan baked bread, and we rested for two hours. Then we went on through Wadi Amk, an easy green valley which made comfortable marching for the camels.

When the Amk turned westward we crossed it, going up between piles of the warped grey granite (like cold toffee) which was common up-country in the Hejaz. The defile culminated at the foot of a natural ramp and staircase: badly broken, twisting, and difficult for camels, but short. Afterwards we were in an open valley for an hour, with low hills to the right and mountains to the left. There were water pools in the crags, and Merawin tents under the fine trees which studded the flat. The fertility of the slopes was great: on them grazed flocks of sheep and goats. We got milk from the Arabs: the first milk my Ageyl had been given in the two years of drought.

The track out of the valley when we reached its head was execrable, and the descent beyond into Wadi Marrakh almost dangerous; but the view from the crest compensated us. Wadi Marrakh, a broad, peaceful avenue, ran between two regular straight walls of hills to a circus four miles off where valleys from left, right and front seemed to meet. Artificial heaps of uncut stone were piled about the approach. As we entered it, we saw that the grey hill-walls swept back on each side in a half-circle. Before us, to the south, the curve was barred across by a straight wall or step of blue-black lava, standing over a little grove of thorn trees. We made for these and lay down in their thin shade, grateful in such sultry air for any pretence of coolness.

The day, now at its zenith, was very hot; and my weakness had so increased that my head hardly held up against it. The puffs of feverish wind pressed like scorching hands against our faces, burning our eyes. My pain made me breathe in gasps through the mouth; the wind cracked my lips and seared my throat till I was too dry to talk, and drinking became sore; yet I always needed to drink, as my thirst would not let me lie still and get the peace I longed for. The flies were a plague.

The bed of the valley was of fine quartz gravel and white sand. Its glitter thrust itself between our eyelids; and the level of the ground seemed to dance as the wind moved the white tips of stubble grass to and fro. The camels loved this grass, which grew in tufts, about sixteen inches high, on slate-green stalks. They gulped down great quantities of it until the men drove them in and couched them by me. At the moment I hated the beasts, for too much food made their breath stink; and they rumblingly belched up a new mouthful from their stomachs each time they had chewed and swallowed the last, till a green slaver flooded out between their loose lips over the side teeth, and dripped down their sagging chins.

Lying angrily there, I threw a stone at the nearest, which got up and wavered about behind my head; finally it straddled its back legs and staled in wide, bitter jets; and I was so far gone with the heat and weakness and pain that I just lay there and cried about it unhelping. The men had gone to make a fire and cook a gazelle one of them had fortunately shot; and I realized that on another day this halt would have been pleasant for me; for the hills were very strange and their colours vivid. The base had the warm grey of old stored sunlight; while about their crests ran narrow veins of granite-coloured stone, generally in pairs, following the contour of the skyline like the rusted metals of an abandoned scenic railway. Arslan said the hills were combed like cocks, a sharper observation.

After the men had fed we re-mounted, and easily climbed the first wave of the lava flood. It was short, as was the second, on the top of which lay a broad terrace with an alluvial plot of sand and gravel in its midst. The lava was a nearly clean floor of iron-red rock-cinders, over which were scattered fields of loose stone. The third and other steps ascended to the south of us: but we turned east, up Wadi Gara.

Gara had, perhaps, been a granite valley down whose middle the lava had flowed, slowly filling it, and arching itself up in a central heap. On each side were deep troughs, between the lava and the hill-side. Rain water flooded these as often as storms burst in the hills. The lava flow, as it coagulated, had been twisted like a rope, cracked, and bent back irregularly upon itself. The surface was loose with fragments through which many generations of camel parties had worn an inadequate and painful track.

We struggled along for hours, going slowly, our camels wincing at every stride as the sharp edges slipped beneath their tender feet. The paths were only to be seen by the droppings along them, and by the slightly bluer surfaces of the rubbed stones. The Arabs declared them impassable after dark, which was to be believed, for we risked laming our beasts each time our impatience made us urge them on. Just before five in the afternoon, however, the way got easier. We seemed to be near the head of the valley, which grew narrow. Before us on the right, an exact cone-crater, with tidy furrows scoring it from lip to foot, promised good going; for it was made of black ash, clean as though sifted, with here and there a bank of harder soil, and cinders. Beyond it was another lava-field, older perhaps than the valleys, for its stones were smoothed, and between them were straths of flat earth, rank with weeds. In among these open spaces were Beduin tents, whose owners ran to us when they saw us coming; and, taking our headstalls with hospitable force, led us in.

9-3   Green Hills of Africa – Ernest Hemingway

Ernest Hemingway (1899–1961) is one of the greatest and most controversial of all twentieth century American literary figures. He grew up in Oak Park, Illinois, later serving as an ambulance driver in Italy towards the end of World War I, where he was wounded in action.

Returning to Europe in 1921, his work gained immediate attention, both for his books and for his short stories. His prose style is taut and spare, and his topics are often concerned with a violence and belligerence that, in part, reflects his own tough, troubled and turbulent character. Hemingway – bullfighter, big-game hunter, traveler – was a skilled and observant outdoorsman. In all his short stories and novels set in the out-of-doors, Hemingway accurately describes landscape and almost invariably some aspect of the terrain is integral to the story as it unfolds. In the segment presented here from The Green Hills of Africa *(1935), Hemingway describes a hunting party in search of kudu and sable.*

In the morning Karl and his outfit started for the salt-lick and Garrick, Abdullah, M'Cola and I crossed the road, angled behind the village up to a dry watercourse and started climbing the mountains in a mist. We headed up a pebbly, boulder-filled, dry stream bed. ...

We kept along the face of this hill on a pleasant sort of jutting plateau and then came out to the edge of the hill where there was a valley and a long open meadow with timber on the far side and a circle of hills at its upper end where another valley went off to the left. We stood in the edge of the timber on the face of this hill looking across the meadow valley which extended to the open out in a steep sort of grassy basin at the upper end where it was backed by the hills. To our left there were steep, rounded, wooded hills, with outcropping of limestone rock that ran, from where we stood, up to the very head of the valley and there formed part of the other range of hills that headed it. Below us, to the right, the country was rough and broken in hills and stretches of meadow and then a steep fall of timber that ran to the blue hills we had seen to the westward beyond the huts where the Roman and his family lived. I judged camp to be straight down below us and about five miles to the northwest through the timber.

The husband was standing, talking to the brother and gesturing and pointing out that he had seen the sable feeding on the opposite side of the meadow valley and that they must have fed either up or down the valley. We sat in the shelter of the trees and sent the Wanderobo–Masai down into the valley to look for tracks. He came back and reported there were no tracks leading down the valley below us and to the westward, so we knew they had fed on up the meadow valley.

Now the problem was to so use the terrain that we might locate them, and get up and into range of them without being seen.

9-4 Wind, Sand and Stars – Antoine de Saint-Exupéry

In the decade before World War II, Antoine de Saint-Exupéry (1900–1944) was a best-selling author in France, England, and the United States. An aviator, entrepreneur, peacemaker, reporter, and author, Exupéry was also a keen observer of the geologic features of the areas over which he flew. Born in Lyons, Saint-Exupéry studied architecture for 15 months at the Ecole des Beaux Arts but these studies were terminated in 1921 when he joined the French Air Force. From his later experience as a commercial pilot, he wrote in 1929 Southern Mail, *followed by* Night Flight; Wind, Sand and Stars; *and* Flight to Arras. *The latter was written in 1942,* two years before his death, missing in action on a reconnaissance flight over occupied France. As a writer, one of his biographers noted his unique blend of graphic narrative, rich imagery, and of poetic meditation on human values.

The excerpt presented is from a chapter entitled "The Plane and the Planet" in Wind, Sand and Stars (1939), translated from the French by Lewis Galantière, and illustrates these talents, as well as a superb ability to describe landscape.

The pilot flying towards the Straits of Magellan sees below him, a little to the south of the Gallegos River, an ancient lava flow, an erupted waste of a thickness of sixty feet that crushes down the plain on which it has congealed. Farther south he meets a second flow, then a third; and thereafter every hump on the globe, every mound a few hundred feet high, carries a crater in its flank. No Vesuvius rises up to reign in the clouds; merely, flat on the plain, a succession of gaping howitzer mouths.

This day, as I fly, the lava world is calm. There is something surprising in the tranquillity of this deserted landscape where once a thousand volcanoes boomed to each other in their great subterranean organs and spat forth their fire. I fly over a world mute and abandoned, strewn with black glaciers.

South of these glaciers there are yet older volcanoes veiled with the passing of time in a golden sward. Here and there a tree rises out of a crevice like a plant out of a cracked pot. In the soft and yellow light the plain appears as luxuriant as a garden; the short grass seems to civilize it, and round its giant throats there is scarcely a swelling to be seen. A hare scampers off; a bird wheels in the air; life has taken possession of a new planet where the decent loam of our earth has at last spread over the surface of the star.

Finally, crossing the line into Chile, a little north of Punta Arenas, you come to the last of the craters, and here the mouths have been stopped with earth. A silky turf lies snug over the curves of the volcanoes, and all is suavity in the scene. Each fissure in the crust is sutured up by this tender flax. The earth is smooth, the slopes are gentle; one forgets the travail that gave them birth. This turf effaces from the flanks of the hillocks the sombre sign of their origin.

We have reached the most southerly habitation of the world, a town born of the chance presence of a little mud between the timeless lava and the austral ice. So near the black scoria, how thrilling it is to feel the miraculous nature of man! What a strange encounter! *Who knows how, or why, man visits these gardens ready to hand, habitable for so short a time – a geologic age – for a single day blessed among days?*

I landed in the peace of evening. Punta Arenas! I leaned against a fountain and looked at the girls in the square. Standing there within a couple of feet of their grace, I felt more poignantly than ever the human mystery. . . .

But by the grace of the airplane I have known a more extraordinary experience than this, and have been made to ponder with even more bewilderment the fact that this earth that is our home is yet in truth a wandering star.

A minor accident had forced me down in the Río de Oro region, in Spanish Africa. Landing on one of those table-lands of the Sahara which fall away steeply at the sides, I found myself on the flat top of the frustrum of a cone, an isolated vestige of a plateau that had crumbled round the edges. In this part of the Sahara such truncated cones are visible from the air every hundred miles or so, their smooth surfaces always at about the same altitude above the desert and their geologic substance always identical. The surface sand is composed of minute and distinct shells; but progressively as you dig along a vertical section, the shells become more fragmentary, tend to cohere, and at the base of the cone form a pure calcareous deposit.

Without question I was the first human being ever to wander over this … this iceberg: its sides were remarkably steep, no Arab could have climbed them, and no European had as yet ventured into this wild region.

I was thrilled by the virginity of a soil which no step of man or beast had sullied. I lingered there, startled by this silence that never had been broken. The first star began to shine, and I said to myself that this pure surface had lain here thousands of years in sight only of the stars.

But suddenly my musings on this white sheet and these shining stars were endowed with a singular significance. I had kicked against a hard, black stone, the size of a man's fist, a sort of moulded rock of lava incredibly present on the surface of a bed of shells a thousand feet deep. A sheet spread beneath an apple-tree can receive only apples; a sheet spread beneath the stars can receive only star-dust. Never had a stone fallen from the skies made known its origin so unmistakably.

And very naturally, raising my eyes, I said to myself that from the height of this celestial apple-tree there must have dropped other fruits, and that I should find them exactly where they fell, since never from the beginning of time had anything been present to displace them.

Excited by my adventure, I picked up one and then a second and then a third of these stones, finding them at about the rate of one stone to the acre. And here is where my adventure became magical, for in a striking foreshortening of time that embraced thousands of years, I had become the witness of this miserly rain from the stars. The marvel of marvels was that there on the rounded back of the planet, between this magnetic sheet and those stars, a human consciousness was present in which as in a mirror that rain could be reflected.

9-5   The French Lieutenant's Woman – John Fowles

John Fowles (1926–2005) was born in Essex, England, educated at Oxford, served in the Royal Marines, and taught in various colleges and schools in France, Greece, and London. He has been widely celebrated, not only as a novelist, but also as a writer of poetry, essays, short stories, guidebooks, commentary, and translations. A runaway best-seller in the early 1970s, The French Lieutenant's Woman (1969) is a novel of mid-Victorian England. The life and mores of that era have rarely been recreated in better fashion than by John Fowles. That the hero, Charles, is a self-trained paleontologist is not by chance nor are the references to Darwinism. Gentlemen of that era were attracted to nature, so much so that geology was a popular avocation, and Darwinism was a topic for discussion and argumentation by most educated people. Here then is a novel built around a paleontologist and fossil collecting and including sea urchin (Echinocorys and Micraster) tests, ammonites, and The Old Fossil Shop, based on the shop of Mary Anning (1799–1847), "the greatest fossilist the world ever knew," who was involved in the first discovery of Jurassic marine reptiles.

Two days passed during which Charles's hammers lay idle in his rucksack. He banned from his mind the thoughts of the tests lying waiting to be discovered: and thoughts, now associated with them, of women lying asleep on sunlit ledges. But then,

Ernestina having a migraine, he found himself unexpectedly with another free afternoon. He hesitated a while; but the events that passed before his eyes as he stood at the bay window of his room were so few, so dull. The inn sign – a white lion with the face of an unfed Pekinese and a distinct resemblance, already remarked on by Charles, to Mrs. Poulteney – stared glumly up at him. There was little wind, little sunlight . . . a high gray canopy of cloud, too high to threaten rain. He had intended to write letters, but he found himself not in the mood.

To tell the truth he was not really in the mood for anything; strangely there had come ragingly upon him the old travel-lust that he had believed himself to have grown out of those last years. He wished he might be in Cadiz, Naples, the Morea, in some blazing Mediterranean spring not only for the Mediterranean spring itself, but to be free, to have endless weeks of travel ahead of him, sailed-towards islands, mountains, the blue shadows of the unknown.

Half an hour later he was passing the Dairy and entering the woods of Ware Commons. He could have walked in some other direction? Yes, indeed he could. But he had sternly forbidden himself to go anywhere near the cliff-meadow; if he met Miss Woodruff, he would do, politely but firmly, what he ought to have done at that last meeting – that is, refuse to enter into conversation with her. In any case, it was evident that she resorted always to the same place. He felt sure that he would not meet her if he kept well clear of it.

Accordingly, long before he came there he turned northward, up the general slope of the land and through a vast grove of ivyclad ash trees. They were enormous, these trees, among the largest of the species in England, with exotic-looking colonies of polypody in their massive forks. It had been their size that had decided the encroaching gentleman to found his arboretum in the Undercliff; and Charles felt dwarfed, pleasantly dwarfed as he made his way among them towards the almost vertical chalk faces he could see higher up the slope. He began to feel in a better humor, especially when the first beds of flint began to erupt from the dog's mercury and arum that carpeted the ground. Almost at once he picked up a test of *Echinocorys scutata*. It was badly worn away . . . a mere trace remained of one of the five sets of converging pinpricked lines that decorate the perfect shell. But it was better than nothing and thus encouraged, Charles began his bending, stopping search.

Gradually he worked his way up to the foot of the bluffs where the fallen flints were thickest, and the tests less likely to be corroded and abraded. He kept at this level, moving westward. In places the ivy was dense – growing up the cliff face and the branches of the nearest trees indiscriminately, hanging in great ragged curtains over Charles's head. In one place he had to push his way through a kind of tunnel of such foliage; at the far end there was a clearing, where there had been a recent fall of flints. Such a place was most likely to yield tests; and Charles set himself to quarter the area, bounded on all sides by dense bramble thickets, methodically. He had been at his task perhaps ten minutes, with no sound but the lowing of a calf from some distant field above and inland, the clapped wings and cooings of the wood pigeons; and the barely perceptible wash of the tranquil sea far through the trees below. He heard then a sound as of a falling stone. He looked, and saw nothing, and presumed that a flint had indeed dropped from the chalk face above. He searched on for another minute or two; and then, by one of those inexplicable intuitions, perhaps the last remnant of some faculty from our paleolithic past, knew he was not alone. He glanced sharply around.

She stood above him, where the tunnel of ivy ended, some forty yards away. He did not know how long she had been there; but he remembered that sound of two minutes before. For a moment he was almost frightened; it seemed uncanny that she should appear so silently. She was not wearing nailed boots, but she must even so have moved with great caution. To surprise him; therefore she had deliberately followed him.

"Miss Woodruff!" He raised his hat. "How come you here?"

"I saw you pass."

He moved a little closer up the scree towards here. Again her bonnet was in her hand. Her hair, he noticed, was loose, as if she had been in wind; but there had been no wind. It gave here a kind of wildness, which the fixity of her stare at him aggravated. He wondered why he had ever thought she was not indeed slightly crazed.

"You have something . . . to communicate to me?"

Again that fixed stare, but not through him, very much down at him. Sarah had one of those peculiar female faces that vary very much in their attractiveness; in accordance with some subtle chemistry of angle, light, mood. She was dramatically helped at this moment by an oblique shaft of wan sunlight that had found its way through a small rift in the clouds, as not infrequently happens in a late English afternoon. It lit her face, her figure standing before the entombing greenery behind her; and her face was suddenly very beautiful, truly beautiful, exquisitely grace and yet full of an inner, as well as outer, light. Charles recalled that it was just so that a peasant near Gavarnie, in the Pyrenees, had claimed to have seen the Virgin Mary standing on a *déboulis* beside his road . . . only a few weeks before Charles once passed that way. He was taken to the place; it had been most insignificant. But if such a figure as this had stood before him!

However, this figure evidently had a more banal mission. She delved into the pockets of her coat and presented to him, one in each hand, two excellent *Micraster* tests. He climbed close enough to distinguish them for what they were. Then he looked up in surprise at her unsmiling face. He remembered – he had talked briefly of paleontology, of the importance of sea urchins, at Mrs. Poulteney's that morning. Now he stared again at the two small objects in her hands.

"Will you not take them?"

She wore no gloves, and their fingers touched. He examined the two tests; but he thought only of the touch of those cold fingers.

"I am most grateful. They are in excellent condition."

"They are what you seek?"

"Yes indeed."

"They were once marine shells?"

He hesitated, then pointed to the features of the better of the two tests: the mouth, the ambulacra, the anus. As he talked, and was listened to with a grave interest, his disapproval evaporated. The girl's appearance was strange; but her mind – as two or three questions she asked showed – was very far from deranged. Finally he put the two tests carefully in his own pocket.

"It is most kind of you to have looked for them."

"I had nothing better to do."

"I was about to return. May I help you back to the path?"

But she did not move. "I wished also, Mr. Smithson, to thank you . . . for your offer of assistance."

"Since you refused it, you leave me the more grateful."

There was a little pause. He moved up past her and parted the wall of ivy with his stick, for her to pass back. But she stood still, and still facing down the clearing.

"I should not have followed you."

He wished he could see her face, but he could not.

"I think it is better if I leave."

9-6 Trip to the Middle and North Forks of San Joaquin River – John Muir

No anthology of the Earth could be considered complete without an extract from the writings of John Muir, mountain man, environmentalist (before they became known as such), fighter for National Parks, friend of Presidents, founder of the Sierra Club – a man who understood the beauty of solitude. Born in Scotland, Muir (1838–1914) attended the University of Wisconsin but never graduated. With a degree of financial independence, Muir spent most of his life in the mountains of California. In addition to his many journeys afoot and mostly alone in the then little known Sierras of California, John Muir also visited Alaska and in 1880 the Arctic.

Our selection is an excerpt from The Yosemite *(1912) in which the Ritter glacier is described. John Muir not only documents the large and spectacular landforms, glaciers, and rugged mountain crests, but he exhibits the naturalist's eye for smaller features such as rocks, plants, clouds, and birds as well. His powers of description and his fundamental knowledge of geology are clearly shown in the passage that is presented here.*

Camp at head of main Joaquin Canyon. Altitude 9800 feet.

Here the river divides, one fork coming in from the north and the other from Ritter. The Ritter fork comes down the mountain-side here in a network of cascades, wonderfully woven, as are all slate cascades of great size near summits, when the slate has a cleavage well pronounced. The mountains rise in a circle, showing their grand dark bosses and delicate spires on the starry sky. Down the canyon a company of sturdy, long-limbed mountain pines show nobly. All the rest of the horizon is treeless, because moraineless. How fully are all the forms and languages of waters, winds, trees, flowers, birds, rocks, subordinated to the primary structure of the mountains ere they were ice-sculptured! When all was planned and ready, snow-flowers were dusted over them, forming a film of ice over the mountain plate. And so all this development – the photography of God.

Linnæus says Nature never leaps, which means that God never shouts or spouts or speaks incoherently. The rocks and sublime canyons, and waters and winds, and all life structures – animals and ouzels, meadows and groves, and all the silver stars – are words of God, and they flow smooth and ripe from his lips.

The branches of no pine have so long a downward swooping curve as those of *Pinus monticola*. The ends are tufted and conspicuous, the bark redder the higher you go. And it is nobler in form, the colder and balder the rocks and mountains about it.

Never saw so many large gentians as today (*Gentiana frigida*). I counted thirty-three flowers on a patch not much larger than my hand – some flowers one and a quarter inches in diameter.

Ursa Major nearly horizontal, and has gone to rest. So must I. I am in a small grove of mountain hemlock. Their feathery boughs are extended above my head like hands of gentle spirits. Good night to God and His stars and mountains.

August 21 (?).

Clouds, dense and black, come from the southwest, over the black crests and peaks of Ritter, the lowest torn and raked to shreds. The cold wind is tuned to the opaque, somber sky. Now a little rain, snow, and hail.

I wish to cross to the east side of Ritter today, visiting the main north glacier on my way. I have never crossed these summits, and fear to try on so dark a day. . . . At noon, the clouds breaking, I decide to make the attempt, climbing up the fissured and flowery edge of the rock near Ritter Cascade. At the top of the Cascade I find a grand amphitheater, where glaciers from Ritter and peaks to the north once congathered – now meadow and lake and red and black walls with wild undressed falls and rapids and a few hardy flowers. From the rim of this I have a glorious view of high fountain glaciers and névés, with pointed spears and tapering towers and spires innumerable. Here, too, I find a white stemless thistle, larger than the purple one of Mono.

Yet higher, following the wild streams and climbing around inaccessible gorges, at length I am in sight of the lofty top crest of Ritter and know I am exactly right. The clouds grow dark again and send hail, but I know the rest of my way too well to fear. I can push down to the tree-line if need be, in any kind of storm. A glacier which I had not before seen heaves in sight. It is on the north side of the highest of the many spires that run off from Ritter Peak as a center. It is one of the best I've seen in this wild region, occupying a most delicately curved basin and discharging into a lake.

I come in sight of the main North Ritter Glacier, its snout projecting from a narrow opening gap into a lake about three hundred yards in diameter. Its waters are intensely blue. The basin was excavated by the Ritter Glacier and those of Banner Peak to the north at the time of their greater extension. The glacier enters on the east side. On the west is a splendid frame of pure white névé, abounding in deep caves with arched openings. One of these has a span of forty feet, and a height of thirty. This frame of névé comes down to the water's edge and in places reaches out over the lake. Most of its face is made precipitous by sections breaking off into the lake after the manner of icebergs. These are snowbergs. The strength of the glacier itself is so nearly spent ere it reaches the lake, it does not break off in bergs, but melts with many a rill, giving a network of sweet music.

Wishing to reach the head of this glacier, I follow up the left lateral moraine that extends from the flank of the mountain to the lake until I reach the top. Then I try to cross the glacier itself where it is not so steep. But I find it bare, unwalkable ice. The least slip would send me down a slope of thirty-five degrees, a distance of three hundred yards, into the lake. I begin to cut steps, but the snout of the glacier is so hard, and every step has to be so perfect, that I make slow and wearisome progress with blankets on my back. I soon realize it will be dark ere I succeed in crossing, therefore I resolve to retrace my steps and attempt crossing at the foot of the glacier close to the lake where it is narrower, and if I find that too dangerous from crevasses, to go around the lake, which last alternative, from the roughness of its shore, would be no easy thing. I find the ice on the end of the glacier softer and cross, then push rapidly up the right moraine to where the glacier becomes much more easily accessible on account of the lowness of its slope and because its surface is roughened with ridges and rocks.

In going up between the edge of the glacier and the moraine I discover the main surface stream of the glacier has cut a large cave, which I enter and find is made of clear-veined ice – a very inhuman place, with strong water gurgles and tinkles.

The lower steep portion of the glacier is alive with swift-running rills. A larger stream begins its course at the head of the glacier, and runs in a westerly direction along near the right side. It has cut for itself a strangely curved and scalloped channel in the solid green ice, in which it glides with motions and tones and gestures I never before knew water to make. . . .

When my observations on this most interesting glacier (the main North Ritter Glacier) were finished, it was near sunset, and I had to make haste down to the tree-line. Yet I lingered reveling in the grandeur of the landscape. I was on the summit of the pass, looking upon Ritter Lake with its snowy crags and banks, and many a wide glacial fountain beyond, rimmed with peaks, the wind making stern music among their thousand spires, the sky with grand openings in the huge black clouds – openings jagged, walled, and steep like the passes of the mountains beneath them. Eastward lay Islet Lake with its countless little rocky isles; to the left, the splendid architecture of Mammoth Mountain, and in the distance range on range of mountains yet unnamed, with Mono plains and the magnificent lake and volcanic cones, sunlit and warm, between. To the eastward over the Great Basin swelled a range of alabaster cumuli, presenting a series of precipices deeply cleft with shadowy canyons, the whole fringed about their bases with a grand talus of the same alabaster material. Here and there occurred black masses with clearly defined edges like metamorphic slate in granite. Beneath these noble cloud mountains were horizontal bars and feathery touches of rose and crimson with clear sky between, of that exquisite spiritual kind that is connected in some way with our other life, and never fails, wherever we chance to be, to produce a hush of all cares and a longing, longing, longing. . . .

9-7 Roughing It – Mark Twain

Mark Twain (Samuel Longhorne Clemens) (1835–1910) is one of the best-known and best-loved figures in American literature. Born in Missouri, he grew to love the Mississippi, working as typesetter, journalist and river boat pilot, and serving in the Confederate Army. After the end of the war, Twain traveled west with his brother, who had been appointed territorial secretary of Nevada. Twain wrote for the Virginia City *Territorial Enterprise*, fled to San Francisco to avoid a duel, wrote for the local paper and fled again – this time to the Sierras, where he panned for gold – to avoid a lawsuit from the San Francisco police department. His first book, *The Celebrated Jumping Frog of Calaveras County, and Other Sketches*, was published in 1867. In one of his most famous books, *Roughing It* (1872), a chapter or two is devoted to a visit to Mono Lake, California. In the excerpt chosen, Twain considers with humor and aplomb what he termed "this lonely tenant of the loneliest spot on Earth."

Mono Lake lies in a lifeless, treeless, hideous desert, eight thousand feet above the level of the sea, and is guarded by mountains two thousand feet higher, whose summits are always clothed in clouds. This solemn, silent, sailless sea – this lonely tenant of the loneliest spot on earth – is little graced with the picturesque. It is an unpretending expanse of grayish water, about a hundred miles in circumference, with two islands in its center, mere upheavals of rent and scorched and blistered lava, snowed over with gray banks and drifts of pumice-stone and ashes, the winding sheet of the dead volcano, whose vast crater the lake has seized upon and occupied.

The lake is two hundred feet deep, and its sluggish waters are so strong with alkali that if you only dip the most hopelessly soiled garment into them once or twice, and wring it out, it will be found as clean as if it had been through the ablest of washer-women's hands. While we camped there our laundry work was easy. We tied the week's washing astern of our boat, and sailed a quarter of a mile, and the job was complete, all to the wringing out. . . .

Mono Lake is a hundred miles in a straight line from the ocean – and between it and the ocean are one or two ranges of mountains – yet thousands of sea-gulls go there every season to lay their eggs and rear their young. One would as soon expect to find sea-gulls in Kansas. And in this connection let us observe another instance of Nature's wisdom. the islands in the lake being merely huge masses of lava, coated over with ashes and pumice-stone, and utterly innocent of vegetation or anything that would burn; and sea-gulls' eggs being entirely useless to anybody unless they be cooked, Nature has provided an unfailing spring of boiling water on the largest island, and you can put your eggs in there, and in four minutes you can boil them as hard as any statement I have made during the past fifteen years. Within ten feet of the boiling spring is a spring of pure cold water, sweet and wholesome. So, in that island you get your board and washing free of charge – and if nature had gone further and furnished a nice American hotel clerk who was crusty and disobliging, and didn't know anything about the time tables, or the railroad routes – or – anything – and was proud of it – I would not wish for a more desirable boarding-house.

Half a dozen little mountain brooks flow into Mono Lake, but *not a stream of any kind flows out of it*. It neither rises nor falls, apparently, and what it does with its surplus water is a dark and bloody mystery.

There are only two seasons in the region round about Mono Lake – and these are, the breaking up of one winter and the beginning of the next. . . .

About seven o'clock one blistering hot morning – for it was now dead summertime – Higbie and I took the boat and started on a voyage of discovery to the two islands. We had often longed to do this, but had been deterred by the fear of storms; for they were frequent, and severe enough to capsize an ordinary rowboat like ours without great difficulty – and once capsized, death would ensue in spite of the bravest swimming, for that venomous water would eat a man's eyes out like fire, and burn him out inside, too, if he shipped a sea. It was called twelve miles, straight out to the islands – a long pull and a warm one – but the morning was so quiet and sunny, and the lake so smooth and glassy and dead, that we could not resist the temptation. So we filled two large tin canteens with water (since we were not acquainted with the locality of the spring said to exist on the large island), and started. Higbie's brawny muscles gave the boat good speed, but by the time we reached our destination we judged that we had pulled nearer fifteen miles than twelve.

We landed on the big island and went ashore. We tried the water in the canteens, now, and found that the sun had spoiled it; it was so brackish that we could not drink it; so we poured it out and began a search for the spring – for thirst augments fast as soon as it is apparent that one has no means at hand of quenching it. The island was a long, moderately high hill of ashes – nothing but gray ashes and pumice-stone, in which we sunk to our knees at every step – and all around the top was a forbidding wall of scorched and blasted rocks. When we reached the top and got within the wall, we found simply a shallow, far-reaching basin, carpeted with ashes, and here and there a patch of fine sand. In places, picturesque jets of steam shot up out of crevices, giving evidence that although this ancient crater had gone out of active business, there was still some fire left in its furnaces. Close to one of these jets of steam stood the only tree on the island – a small pine of most graceful shape and most faultless symmetry; its color was a brilliant green, for the steam drifted unceasingly through its branches and kept them always moist. It contrasted strangely enough, did this vigorous and beautiful outcast, with its dead and dismal surroundings. It was like a cheerful spirit in a mourning household.

We hunted for the spring everywhere, traversing the full length of the island (two or three miles), and crossing it twice – climbing ash-hills patiently, and the sliding down the other side in a sitting posture, plowing up smothering volumes of gray dust, but we found nothing but solitude, ashes, and a heart-breaking silence. Finally we noticed that the wind had risen, and we forgot our thirst in a solicitude of greater importance; for, the lake being quiet, we had not taken pains about securing the boat. We hurried back to a point overlooking our landing place, and then – but mere words cannot describe our dismay – the boat was gone! The chances were that there was not another boat on the entire lake. The situation was not comfortable – in truth, to speak plainly, it was frightful. We were prisoners on a desolate island, in aggravating proximity to friends who were for the present helpless to aid us; and what was still more uncomfortable was the reflection that we had neither food nor water. But presently we sighted the boat. It was drifting along, leisurely, about fifty yards from shore, tossing in a foamy sea. It drifted, and continued to drift, but at the same safe distance from land, and we walked along abreast it and waited for fortune to favor us. At the end of an hour it approached a jutting cape, and Higbie ran ahead and posted himself on the utmost verge and prepared for the assault. If we failed there, there was no hope for us. It was driving gradually shoreward all the time, now; but whether it was driving fast enough to make the connection or not was the momentous question. When it got within thirty steps of Higbie I was so excited that I fancied I could hear my own heart beat. When, a little later, it dragged slowly along and seemed about to go by, only one little yard out of reach, it seemed as if my heart stood still; and when it was exactly abreast him and began to widen away, and he still standing like a watching statue, I knew my heart did stop. but when he gave a great spring, the next instant, and lit fairly in the stern, I discharged a war-whoop that awoke the solitudes!

But it dulled my enthusiasm, presently, when he told me he had not been caring whether the boat came within jumping distance or not, so that it passed within eight or ten yards of him, for he had made up his mind to shut his eyes and mouth and swim that trifling distance. Imbecile that I was, I had not thought of that. It was only a long swim that could be fatal.

9-8 A Place on the Glacial Till – Thomas Fairchild Sherman

Thomas Fairchild Sherman grew up in the Finger Lakes region of New York, graduated from Oberlin and served as a wilderness guide. Now retired, he was a Rhodes scholar, a research fellow at Harvard, Yale, and Hopkins, and a long time faculty member in biology at Oberlin. Our extract is taken from his book A Place on the Glacial Till: time, land, and nature within an American town (1997), a narrative connecting the geology, flora, fauna and human settlers of his long time home, Oberlin, Ohio.

The form of the land is shaped by water, and so is the form of life. I walked along country roads one sunny day in early March. The red-winged blackbirds, recently back from the South, were chattering in the trees above the Black River, and everywhere snow was melting along the edges of the lanes. The fields were alive with the glitter of light on patches of ice and snow, and little freshets of water ran everywhere across the land, forming their own lacy routes to the ditches beside the roads and to the meandering tributaries of Elk Creek and the Black River. The day was witness to a world of creation, as the crystals of winter liberated again the flowing waters of life. Here in a countryside of roads and agricultural fields, the wild forces of nature were everywhere awake, generating new patterns across the greening land.

The most common things of our world are also the most miraculous. All living architecture is based on water, for water is more prevalent in organisms than all other substances combined. Without this most common of substances, all the miracles of living activity cease. Spirit and body instinctively sense the crucial link between life and water. An injured animal will seek shelter by a pool, drawn to its healing powers. A lonely spirit may seek a flowing brook, model and metaphor of our nature. Like a river, all life is water moving beneath the energy of the sun, a union of the rain, the uplands, and the ocean.

The magic of water is often hidden from our view. Seldom do we stop to think that it is the most universal of solvents, attuned to holding the myriad chemicals of life in its gently electric matrix. Its cohesiveness within and strength along its surface, and its thermal stability combined with conductivity all contribute to the organization and activity of life. And it dwells at just the right thermal interfaces between liquid, solid, and gas. Water, as the ancient philosophers and poets foresaw, can hold earth and air and gentle fire within it.

When water freezes, it can suddenly reveal – to our watery eyes and nerves and brain – the miracles it holds inside. On a winter's morning, the first rays of the sun disperse their colors through fractal fronds of frost on the windows. How many hours would a craftsman labor to make such fine crystal patterns from the nighttime air? Yet from formless gas, by the mere chill on the world, come the frost crystals on the window or the snowflakes in the air. From simple actions and common matter, the complex beauty of nature emerges.

That the landscape is shaped by water is evident to all who live in a country of soft shales and clay. When I walk the streambeds of the Vermilion River, even in the driest of summer seasons, water seeps out at many points along the high walls of the gorge above me, and with every trickle of water, the flakes of shale come loose and fall, sometimes forming great conical heaps of soft stone beside the river. As winter ice along the steep banks gives way to the rays of the springtime sun, the shale itself seems to melt from the canyon walls, and a chorus of falling stone accompanies the songs of dripping water. In such a country, landforms give way to the forces of water before our very eyes. At the junction of Chance Creek and the Vermilion River, the little promontory on which I like to sit is greatly altered from the form it had before the torrential rains of July 4–5, 1969, when the rivers flowed 12 feet or more above their customary shallow levels and swept away much of the point. Another time will see a different world.

The sedimentary rocks beneath us were created by the forces of water on ancient landscapes, as rain and streams brought other highlands down to other lowlands. In the mud of our rivers flows the land into a new time and different form. In soft country like this, everyday observations lead easily to the idea that large floods may have helped to shape the land we see before us.

Early geologists were well instructed as children in the biblical flood, which required Noah to build an ark, and for many the account was accepted as earth history rather than human allegory. Fossils of seashells found high in the sedimentary rocks of the Alps seemed to be evidence that the biblical flood had indeed been extraordinary. If the idea seems preposterous that water once covered central Europe to a depth of 14,000 feet, how likely does it seem, from the perspective of lives on apparently solid ground, that the Alps have been raised that distance from a position previously below the ocean? Marine fossils on mountaintops show that land and water have undergone remarkable processes of one kind or another.

Other geologic features of Switzerland – and of northern Ohio – baffled nineteenth-century observers. The story that science has invented to explain them may seem far less probable than Noah's flood. The scientific verdict, almost universally accepted today, is that millions of square miles of Europe, Asia, and North America were recently covered not with 14,000 feet of water, but with an equivalent thickness of ice.

Northwest of the Alps in Switzerland and France, across the plain of Geneva, lie the Jura Mountains (for which the Jurassic period was named). High in the Jura Mountains are large rocks unlike those of the Juras, but identical to those of the Alps. They are analogous to the large rocks on Tappan Square in Oberlin: They do not belong where they are, but one does not have to go too many miles away to find where they might have come from. Then there are scratches and grooves and polish on the limestone of the Jura Mountains (as at Sandusky Bay or Kelleys Island in Lake Erie), as if giants had once carved and shined the rocks. Darwin and Sedgwick walked past such sculptured rocks in a narrow valley of North Wales as they hunted for fossils, but such features of the land seem to defy any plausible explanation.

No one living in England or in northern Ohio was likely to imagine any natural process to account for such things. But climbing in the Swiss Alps above Zermatt, I once looked down many hundreds of feet on the Monte Rosa glacier and saw its massive ice, laden with rocks, on the summer land. It is not too hard to imagine that the glacial ice might be slowly falling down the mountain.

When young Louis Agassiz went to Les Diablerets glaciers in the summer of 1836, alpine farmers and woodcutters had long known that mountain glaciers moved, that they carried rocks long distances, and that glaciers had once extended much farther down the mountains. Agassiz became convinced that mountain glaciers had once completely overrun the Jura Mountains from the Alps, carrying and dropping those erratic rocks, and carving and polishing the limestone of the Juras. From his perspective in the mountains, Agassiz could also imagine what was unimaginable to anyone in the lowlands: that ice might have come down from the Arctic and stretched across the whole northern world. Was this a vision of intellectual genius, or an aesthetic response to the grandeur of his native mountains? Or just an expansive leap of the unfettered mind in an air made thin of oxygen? At high elevations, the mind is free to create bold thoughts. Agassiz was seized by this novel idea and soon was convincing colleagues in Britain and America that the Highlands of Scotland and Wales, the White Mountains of New Hampshire, and our own Great Lakes had been sculptured by ice. All this came to Agassiz's mind long before European science discovered that Antarctica and Greenland are even today covered by vast sheets of ice.

9-9 Basin and Range – John McPhee

The writings of John McPhee, whom we met on page 4, cover an astonishing range of topics, from Russian art to shipping to medicine to *tennis to oranges to biography to nuclear energy. Our present extract is taken from* Basin and Range *(1981).*

Geologists mention at times something they call the Picture. In an absolutely unidiomatic way, they have often said to me, "You don't get the Picture." The oolites and dolomite – tuff and granite, the Pequop siltstones and shales – are pieces of the Picture. The stories that go with them – the creatures and the chemistry, the motions of the crust, the paleoenvironmental scenes – may well, as stories, stand on their own, but all are fragments of the Picture.

The foremost problem with the Picture is that ninety-nine percent of it is missing – melted or dissolved, torn down, washed away, broken to bits, to become something else in the Picture. The geologist discovers lingering remains, and connects them with dotted lines. The Picture is enhanced by filling in the lines – in many instances with stratigraphy: the rock types and ages of strata, the scenes at the times of deposition. The lines themselves to geologists represent structure – folds, faults, flat-lying planes. Ultimately, they will infer why, how, and when a structure came to be – for example, why, how, and when certain strata were folded – and that they call tectonics. Stratigraphy, structure, tectonics. "First you read ze Kafka," I overheard someone say once in a library elevator. "Ond zen you read ze Turgenev. Ond zen ond only zen are – you – ready – for – ze Tolstoy."

And when you have memorized Tolstoy, you may be ready to take on the Picture. Multidimensional, worldwide in scope and in motion through time, it is sometimes called the Big Picture. The Megapicture. You are cautioned not to worry if at first you do not wholly see it. Geologists don't see it, either. Not all of it. The modest ones will sometimes scuff a boot and describe themselves and their colleagues as scientific versions of the characters in John Godfrey Saxe's version of the Hindu fable of the blind men and the elephant. "We are blind men feeling the elephant," David Love, of the Geological Survey, has said to me at least fifty times. It is not unknown for a geological textbook to include snatches of the poem.

> It was six men of Indostan
> To learning much inclined,
> Who went to see the Elephant
> (Though all of them were blind).
> That each by observation
> Might satisfy his mind.

The first man of Indostan touches the animal's side and thinks it must be some sort of living wall. The second touches a tusk and thinks an elephant is like a spear. The others, in turn, touch the trunk, an ear, the tail, a knee – "snake," "fan," "rope," "tree."

> And so these men of Indostan
> Disputed loud and long,
> Each in his own opinion
> Exceeding stiff and strong,
> Though each was partly in the right,
> And all were in the wrong!

The blind men and the elephant are kept close at hand mainly to slow down what some graduate students refer to as "arm-waving" – the delivery, with pumping elbows, of hypotheses so breathtakingly original that the science seems for the moment more imaginative than descriptive. Where it is solid, it is imaginative enough. Geologists are famous for picking up two or three bones and sketching an entire and previously unheard-of creature into a landscape long established in the

Picture. They look at mud and see mountains, in mountains oceans, in oceans mountains to be. They go up to some rock and figure out a story, another rock, another story, and as the stories compile through time they connect – and long case histories are constructed and written from interpreted patterns of clues. This is detective work on a scale unimaginable to most detectives, with the notable exception of Sherlock Holmes, who was, with his discoveries and interpretations of little bits of grit from Blackheath or Hampstead, the first forensic geologist, acknowledged as such by geologists to this day. Holmes was a fiction, but he started a branch of a science; and the science, with careful inference, carries fact beyond the competence of invention. Geologists, in their all but closed conversation, inhabit scenes that no one ever saw, scenes of global sweep, gone and gone again, including seas, mountains, rivers, forests, and archipelagoes of aching beauty rising in volcanic violence to settle down quietly and then forever disappear – *almost* disappear. If some fragment has remained in the crust somewhere and something has lifted the fragment to view, the geologist in his tweed cap goes out with his hammer and his sandwich, his magnifying glass and his imagination, and rebuilds the archipelago.

I once dreamed about a great fire that broke out at night at Nasser Aftab's House of Carpets. In Aftab's showroom under the queen-post trusses were layer upon layer and pile after pile of shags and broadlooms, hooks and throws, para-Persians and polyesters. The intense and shriveling heat consumed or melted most of what was there. The roof gave way. It was a night of cyclonic winds, stabs of unseasonal lightning. Flaming debris fell on the carpets. Layers of ash descended, alighted, swirled in the wind, and drifted. Molten polyester hardened on the cellar stairs. Almost simultaneously there occurred a major accident in the ice-cream factory next door. As yet no people had arrived. Dead of night. Distant city. And before long the west wall of the House of Carpets fell in under the pressure and weight of a broad, braided ooze of six admixing flavors, which slowly entered Nasser Aftab's showroom and folded and double-folded and covered what was left of his carpets, moving them, as well, some distance across the room. Snow began to fall. It turned to sleet, and soon to freezing rain. In heavy winds under clearing skies, the temperature fell to six below zero. Celsius. Representatives of two warring insurance companies showed up just in front of the fire engines. The insurance companies needed to know precisely what had happened, and in what order, and to what extent it was Aftab's fault. If not a hundred per cent, then to what extent was it the ice cream factory's fault? And how much fault must be – regrettably – assigned to God? The problem was obviously too tough for the Chicken Valley Police Department, or, for that matter, for any ordinary detective. It was a problem, naturally, for a field geologist. One shuffled in eventually. Scratched-up boots. A puzzled look. He picked up bits of wall and ceiling, looked under the carpets, tasted the ice cream. He felt the risers of the cellar stairs. Looking up, he told Hartford everything it wanted to know. For him this was so simple it was a five-minute job.

9-10 Neanderthal – John Darnton

John Darnton, born in New York in 1941, is a Pulitzer prize-winning journalist, with a long association with The New York Times, *of which he is now Cultural News Editor. A graduate of Wisconsin and Columbia, he previously served as* Times *correspondent in Nigeria and*

Kenya, and bureau chief in Warsaw, Madrid, and London. The novel Neanderthal (1996), from which our extract is taken, describes the discovery of two surviving isolated Neanderthal groups and the exploits of those sent to study them.

Van stopped before a heavy oak door, knocked, waited for a reply, and entered. The room was unexpectedly dim; it took a few seconds to focus and find the angular figure seated behind a desk along the wall, away from the window, which was blocked by closed blinds. The man was smoking. A cloud hung over his head.

"Ah, come in . . . welcome." The voice was nasal but seductive, authoritative.

They stepped closer. The man behind the desk did not rise but offered a bent hand across the spotless blotter.

"Dr. Arnot, Dr. Mattison. I am Harold Eagleton. Welcome to the Institute for Prehistoric Research."

His tone suggested that he was accustomed to having his name recognized. He held his cigarette in the Eastern European manner, between the thumb and forefinger of his left hand, the other fingers splayed like a fan.

As Matt shook his hand, towering over him, Susan studied Eagleton. He was an extraordinary sight, hunched over, all askew: pale skin, cocked head, tilted steel-rimmed glasses. There was a glint of metal under the desk, the rounded steel and black rubber of a wheelchair. So that was why he seemed sprawled out, caved in like a soufflé. She smelled a peculiar odor she couldn't place. Disinfectant, perhaps.

Eagleton turned to face her. "We're grateful, my dear, that you could come so quickly. Kellicut needs your help, as do we."

"There didn't seem to be a lot of choice," said Matt. "What's it all about?"

Eagleton looked him over. "Well, let's not stand on ceremony, shall we?" He puffed another cloud. "The Institute . . . you've heard of us, yes? Good." It was hard to tell if he was genuinely pleased. "We're involved in many aspects of prehistoric research – many areas. Areas that other institutions might not look into. We have ample funding and we place a strong value on good fieldwork. We have projects around the world and we want nothing but the best. People like Dr. Kellicut. We need them."

Matt was struck by the wording: need? "Why do you need them? What do you need them for?"

"Whatever," said Eagleton, waving him off.

Matt looked at Susan, who was staring at Eagleton, fascinated. Van sat wordlessly on a sofa. The walls were covered with maps and what appeared to be satellite reconnaissance photographs. A small Degas was in one corner. Matt spotted some framed degrees and plotted the ascent they represented: University of Tennessee, Columbia, Harvard, Edinburgh, St. John's Oxford.

Eagleton followed his gaze. He didn't miss much. "Ah, the old paper trail," he said. "So meaningless, isn't it?" He paused reflectively.

"Where was I?" A cloud of smoke went up. "Well, we've sponsored quite a few expeditions recently and some have been more . . . orthodox than others. Lately we've decided to specialize on the Neanderthal – or rather, we've had the decision thrust upon us. We're very keen on it. Interesting stuff. Not all of us had that kind of background, you understand, though we've managed to assemble a neat little stable of experts, as I'm sure you'll – "

"I'm afraid I don't understand," Matt cut in. "Why did you get involved in Neanderthal research? What exactly do you hope to gain?"

Eagleton's tone changed. Now there was a hard edge to it.

"Why . . . why, it could change everything. It could change the whole field, don't you see? Actually, it was your friend Dr. Kellicut who got us involved. He was quite enthusiastic, so we sponsored him. To the Caucasus. It sounded a bit crackpot, really, but one never knows, does one?" He stubbed out the cigarette, reached under his desk, and flipped a switch. The smoke disappeared through a vent in the ceiling and a slight mist fell. "Antibacteriological agent," he explained. "Hope you don't mind."

"Go on, please," said Susan. "Tell us, where is Dr. Kellicut now?"

"Well, that's just it. We don't know. We know generally, of course, but we don't know specifically. That's where you come in. That's why we need you, to find him. It takes a paleoanthropologist to find a paleoanthropologist and all that, you know."

Eagleton appeared agitated. His right hand made an arc through the air and came to rest on his forehead, fingers pointing down. He raked it backward through his hair and rocked slightly. Matt began to wonder if this ditzy professorial air wasn't all just an act.

"I mean, he is where we sent him – or rather, where he wanted to go – with our blessing. The thing is, we had not heard from him for a long time – until recently, that is. Until he sent for you."

"For us!"

"Yes."

"Both of us?"

"Yes. Here is how it came." Eagleton opened his desk drawer and pulled out a flattened piece of brown wrapping paper which had Kellicut's familiar scrawl:

Dr. S. Arnot / Dr. M. Mattison
care of: Institute for Prehistoric Research
1290 Brandywine Lane
Bethesda, MD 09763
USA

"Where's the message?" asked Susan. "Where's the note?"

"It's not a note," said Eagleton, "but I guess you could say it *is* a message. It's what was inside the package." He nodded toward Van, who went to a closet and came back with a square, battered, wooden box about a foot high. He set it in the center of Eagleton's desk, lifted off the top, reached in, and pulled out an object covered with a dirty white cloth.

"This is just the way it came," said Egaleton. He reached over and snapped the cloth away.

Underneath, gleaming and surprisingly white, was a skull. It seemed to grin at them from the center of Eagleton's desk. Van held it up, Hamlet-like. A rush of recognition: that long sloping forehead, the clipped chin, and of course the thick impenetrable band of beetle-shaped bone above the eyes.

"It's perfect!" exclaimed Susan, reaching for it excitedly. She held it in both hands, like a Christmas present. "A perfect specimen. I've never seen one so complete, so well-preserved. It's the find of the century!"

Eagleton grunted. "That it is," he said.

"It's almost too perfect," put in Matt. "It looks unreal. Did you date it?"

"Of course," replied Eagleton. He lit another cigarette.

"And?"

"That's the strange part."

"What? How old is it?"

Eagleton puffed. "Twenty-five."

"Twenty-five?" said Matt, incredulous.

"That's impossible," said Susan. Matt shot her a look. "Neanderthal was not alive twenty-five thousand years ago."

"Not twenty-five thousand years," said Eagleton. He was caught up in a sudden fit of coughing so that he could barely wheeze out the words. "Twenty-five years."

He waved at the air, and the smoke cloud over his head undulated.

9-11 Antarctica – Kim Stanley Robinson

Kim Stanley Robinson is one of the most widely read contemporary authors of science fiction. Born in Waukegan, Illinois, in 1952 and educated at the University of California, San Diego and Boston University, Robinson is also a literary critic and historian. Showing extra-ordinary versatility and creativity, Robinson's books have ranged from The Memory of Whiteness *(1985), in which music becomes the universal language for multi-planetary societies, to his best-selling* Mars Trilogy *(1993,* Red Mars*; 1994,* Green Mars*; 1996,* Blue Mars*), a seventeen hundred page "his-tory" of the adaptation of humans to Mars. Our present extract is from his book* Antarctica *(1998), based upon his meticulous first-hand observations of the White Continent during a six-week stay there.*

Graham walked some distance behind Geoffrey and Harry and Misha, over broken dolerite. They were checking a beautiful long band of Sirius sandstone plastered above them against one of the dolerite cliffs of the Apocalypse Peaks. The band was a succession horizontally stratified at scales both large and small. Diamictons contain-ing different mixes of boulders, cobbles and laminated silts, and bounded above and below by distinct disconformities, horizontal to slightly inclined. Planar to slightly undulating small-scale relief. One line was traceable for more than thirty meters as Graham walked it off. Certainly this succession had been deposited *in situ*, and then plowed away by the next grounded glacier to pass over, leaving only this sandy blond band against the rock, where the force of the passing ice was lessened just enough to leave a trace against the wall.

He leaned down and tapped at white diatomite with his geological hammer, then scratched at it. The D-7 disconformity, about a centimeter wide. Below it all the diamictons were marine; above it they were subaerial. Which did not mean that this line had been sea level, but that the rise of the Transantarctics had lifted this region out of the seas for good at about the time this disconformity had been laid. As they were now at some fifteen hundred meters above sea level, it implied an uplift rate of about five hundred meters per million years, if you accepted the Pliocene dating of the Sirius, as Graham did. That was a fairly rapid uplift rate, and one of the ways that the stabilists criticized the dynamicists' conclusions; but faster rates were certainly known, and it was hard to argue the evidence, displayed here on the cliffside like a classroom diagram. Graham would have very much liked to take his first thesis advisor by the scruff of the neck and haul him to this very point and shove his face into such a display, clearer even than the Cloudmaker Formation. See that! he would have said. How can you deny the facts!

But of course it was not actually a cliff of facts, but of stone. Interpretations were open to argument, at least until the matter was firmly pinned down and black-boxed, as Geoffrey put it, meaning become something that all the scientists working in that field took for granted, going on to further questions. Some scientific controversies resolved themselves fairly quickly, and others didn't; and this was proving to be one of the slow ones. As they had not black-boxed this particular question, it was still sediment only and not yet fact.

This process was something that Graham had not understood early in his career, and it had gotten him into considerable trouble. He had started his graduate work in geology at Cambridge, working with Professor Martin, unaware of the fact that Martin's own work dating ash deposits in the Transantarctics allied him with the stabilists in the Sirius controversy. Graham had merely wanted to work in the Antarctic and knew that that was Martin's area of research. He had been very naïve, having been educated mostly in physics, with only a late switch to geology because of the field work, and the tangibility of rock – a switch just begun at that time in his life, and by little more than his entry into graduate level work in Martin's group. Then he had been so immersed in catching up on the basics of geology that he had not been fully aware of the Sirius controversy and how Martin fit into it, and so he had not understood why Martin had been so cool to his geomorphological research into the question of why the Transantarctics were there at all. As part of that study Graham had examined the question of how quickly the range was uplifting, and had come to the conclusion that although the range was quite old, dating from about eighty million years ago, when the East Antarctic craton began to show intracration rifting, still it looked like it had been rising at a fairly rapid clip in the most recent period, perhaps (he had ventured rashly) because of the lithostatic pressure of the ice cap. And Martin had been cool, and had never devoted any of his time to critiquing Graham's papers on the subject, or contributing any of what he needed to contribute, as second author and principal investigator, to make the papers publishable. In a fit of angry frustration Graham had sent one paper to a journal without Martin's approval, as the approval seemed likely never to come; and the paper had been rejected, Martin informed of the submission, and Graham basically dismissed from the program, as he was not invited to return to Antarctica in Martin's group the following season.

This experience had made him bitter. He had gone back to New Zealand, and there one night in a pub one of his old teachers from the university in Christchurch had shaken his head and explained some things to him. Martin had cast his lot in with the stabilists in the Sirius controversy because his findings in volcanic ash convinced him that the Transantarctics and Antarctica generally had been in the deep freeze for at least twelve million years. One of the many other aspects of the controversy had to do with the rate of uplift of the Transantarctics, with the stabilists maintaining that there was no reason for the range to be rising anywhere near as fast as the dynamicists claimed, so that the Sirius formations had to be older than they claimed in order to be found now at such high elevations. And so naturally, his old teacher told him, Graham's conclusions had not been welcome.

This had outraged Graham. A perversion of science! he cried. But his old professor had chided him. No no, he had said, it's your own fault; you should have known better. Perhaps it's even my fault; I should have taught you better than I did how science works, obviously.

There was nothing particularly untoward in Martin's response, Graham's teacher explained to him, with no outrage or indignation whatsoever. Indeed, he said, if Graham had joined the program of one of the dynamicists, and begun to produce work indicating that the ice had lain heavy on Antarctica for millions and millions of years, he would not have prospered there either. It was not a matter of evil-doing either way; the simple truth was that science was a matter of making alliances to help you to show what you wanted to show, and to make clear also that what you were showing was important. And your own graduate students and post-docs were necessarily your closest allies in that struggle to pull together all the strings of an argument.

All this became even more true when there was a controversy ongoing, when there were people on the other side publishing articles with titles like "Unstable Ice or Unstable Ideas?" and so on, so that the animus had grown a bit higher than normal.

So, Graham had been forced to conclude, thinking over his talk with his old teacher in the days after: it was not that Martin was evil, but that he Graham had been naïve, and, yes, even stupid. Thick, anyway. Bitterness was not really appropriate. Science was not a matter of automatons seeking Truth, but of people struggling to black-box some facts.

So his education began again, in effect, after two years wasted in Cambridge. Which was not so very great a length of time in scientific terms. Many scientists had taken far longer to learn how their disciplines worked. And so Graham had become reconciled to the experience and had shelved it, and gotten into a program in glaciology at the University of Sydney, and gone on from there.

That had all happened a long time ago. And yet still the Sirius controversy raged on, with both sides finding new allies and students, and producing papers published in peer-reviewed journals. Graham thought he began to see a tilt on the part of outsiders toward the dynamicist interpretation; but as he was a dynamicist himself now, he supposed he could not really tell for sure. Anyway, the case was beginning to look stronger to him as the years passed and evidence from other parts of Antarctica was collected. Over in the Prince Charles Mountains, for instance, on the other side of Antarctica, the Aussies were making a good case that there had been Pliocene-era seas as far as five hundred kilometers inland from the current shore. The Beardmore Glacier had been pretty conclusively shown to be a paleofjord, and there were unallied scientists referring to a "Beardmore paleofjord" in other contexts, also to *Nothofagus beardmorensis*, the beech type found in the Cloudmaker Formation and named by dynamicists to underline its location of discovery. And evidence of beech forests was showing up elsewhere in Sirius formations, with seeds and beetles and other plant material all pointing to the Pliocene or late Miocene. No, the case was coming together at last, after all these long years; all of Geoff Michelson's career, effectively, and passed on to him from his advisor Brown, who had also spent a career working away at it; and now, come to think of it, a good fair fraction of Graham's career had been devoted to it as well. All to build the walls that would box this part of the story up for good, building it brick by brick over years and generations. Because the stones did not speak, not really. They had to be translated.

9-12   The Lost World – Sir Arthur Conan Doyle

Born and educated in Edinburgh, Sir Arthur Conan Doyle (1859–1930) practiced medicine, serving as a ship's surgeon on an Arctic whaling voyage, as a British army surgeon in South Africa during the Boer War, and in various hospitals in southern England. His first novel, A Study in Scarlet, *appeared in 1887 and was followed by scores of detective stories, novels, plays, historical and military works, short story collections, poems, and more than a dozen books on spiritualism. His fame today is chiefly based on Sherlock Holmes and Dr. Watson, the activities at 221B Baker Street and the sinister Professor Moriarty, but Doyle was also well known as one of the first science-fiction writers. Two examples are the* Maracot Deep *(1929) and the more familiar* The Lost World *(1912), from which our extract is taken. Here Doyle's imagination soars as he describes a pterodactyl rookery on an isolated basalt plateau in South America.*

Creeping to his side, we looked over the rocks. The place into which we gazed was a pit, and may, in the early days, have been one of the smaller volcanic blowholes of the plateau. It was bowl-shaped, and at the bottom, some hundreds of yards from where we lay, were pools of green-scummed, stagnant water, fringed with bulrushes. It was a weird place in itself, but its occupants made it seem like a scene from the Seven Circles of Dante. The place was a rookery of pterodactyls. There were hundreds of them congregated within view. All the bottom area round the water-edge was alive with their young ones, and with hideous mothers brooding upon their leathery, yellowish eggs. From this crawling, flapping mass of obscene reptilian life came the shocking clamour which filled the air and the mephitic, horrible, musty odour which turned us sick. But above, perched each upon its own stone, tall, grey, and withered, more like dead and dried specimens than actual living creatures, sat the horrible males, absolutely motionless save for the rolling of their red eyes or an occasional snap of their rat-trap beaks as a dragonfly went past them. Their huge, membranous wings were closed by folding their forearms, so that they sat like gigantic old women, wrapped in hideous web-coloured shawls, and with their ferocious heads protruding above them. Large and small, not less than a thousand of these filthy creatures lay in the hollow before us.

Our professors would gladly have stayed there all day, so entranced were they by this opportunity of studying the life of a prehistoric age. They pointed out the fish and dead birds lying about among the rocks as proving the nature of the food of these creatures, and I heard them congratulating each other on having cleared up the point why the bones of this flying dragon are found in such great numbers in certain well-defined areas, as in the Cambridge Green-sand, since it was now seen that, like penguins, they lived in gregarious fashion.

Finally, however, Challenger, bent upon proving some point which Summerlee had contested, thrust his head over the rock and nearly brought destruction upon us all. In an instant the nearest male gave a shrill, whistling cry, and flapped its twenty-foot span of leathery wings as it soared up into the air. The females and young ones huddled together beside the water, while the whole circle of sentinels rose one after the other and sailed off into the sky. It was a wonderful sight to see at least a hundred creatures of such enormous size and hideous appearance all swooping like swallows with swift, shearing wing strokes above us; but soon we realized that it was not one on which we could afford to linger. At first the great brutes flew round in a huge ring, as if to make sure what the exact extent of the danger might be. Then, the flight grew lower and the circle narrower, until they were whizzing round and round us, the dry, rustling flap of their huge slate-coloured wings filling the air with a volume of sound that made me think of Hendon aerodrome upon a race day.

"Make for the wood and keep together," cried Lord John, clubbing his rifle. "The brutes mean mischief."

The moment we attempted to retreat the circle closed in upon us, until the tips of the wings of those nearest to us nearly touched our faces. We beat at them with the stocks of our guns, but there was nothing solid or vulnerable to strike. Then suddenly out of the whizzing, slate-coloured circle a long neck shot out, and a fierce beak made a thrust at us. Another and another followed. Summerlee gave a cry and put his hand to his face, from which the blood was streaming. I felt a prod at the back of my neck, and turned, dizzy with the shock. Challenger fell, and as I stooped to pick him up I was again struck from behind and dropped on the top of him. At the same instant I heard the crash of Lord John's elephant gun, and, looking up, saw one of the creatures with a broken wing struggling upon the ground, spitting and gurgling at us with a wide-opened beak and bloodshot, goggled eyes, like some devil in a mediæval picture. Its comrades had flown higher at the sudden sound, and were circling above our heads.

"Now," cried Lord John, "now for our lives!"

We staggered through the brushwood, and even as we reached the trees the harpies were on us again. Summerlee was knocked down, but we tore him up and rushed among the trunks. Once there we were safe, for those huge wings had no space for their sweep beneath the branches. As we limped homewards, sadly mauled and discomfited, we saw them for a long time flying at a great height against the deep blue sky above our heads, soaring round and round, no bigger than wood pigeons, with their eyes no doubt still following our progress. At last, however, as we reached the thicker woods they gave up the chase, and we saw them no more.

"A most interesting and convincing experience," said Challenger, as we halted beside the brook and he bathed a swollen knee. "We are exceptionally well informed, Summerlee, as to the habits of the enraged pterodactyl."

10. Poetry

This chapter concerns geology and poetry. What is poetry? "Poetry," Wordsworth declared, "is the image of man and nature. ... [It] takes its origin from emotion recollected in tranquility." To Johnson poetry was "the art of uniting pleasure with truth by calling imagination to the help of reason." "Musical thought" responded Carlyle. "The art of doing by means of words what the painter does by means of colours" affirmed Macaulay. To Coleridge, poetry was "that species of composition which is opposed to the works of science, by proposing for its immediate object, pleasure, not truth."

But poetry in one sense is complementary to science, rather than competitive. C.A. Coulson has illustrated this well by asking (though in a quite different context) "What is a primrose?" To the casual observer:

> "A primrose by the river's brim
> A yellow primrose was to him
> And it was nothing more!"

To the classical botanist, a primrose is a member of the species *Primula vulgaris*, short plants with large, wrinkled, light green leaves, slightly notched corolla lobes, and radical peduncles, each bearing a single large flower. But to the saint, the primrose is "God's promise of spring." All descriptions apply to the same small yellow flower, yet they have little in common, and the validity of one in no way invalidates the others. Even another botanist may not use the same definitions of our classical plant taxonomist, but may emphasize the delicate biochemical mechanisms and interchanges "requiring potash, phosphates, nitrogen and water in definite proportions."

What characterizes the poet's description is its insight, its imaginative artistry, and the emotive meter of its language.

Because our experience is earthbound, it is scarcely surprising that the Earth, in all its moods, features prominently in poetry, whether narrative, lyric, or dramatic. Our extracts provide a varied selection.

10-1 Landscape and Literature – Sir Archibald Geikie

On June 1, 1898, Sir Archibald Geikie (1835–1924) delivered the Romanes Lecture in the Sheldonian Theatre at Oxford University. His subject was "Types of Scenery and Their Influence on Literature." He classified the scenery of Britain into Lowlands, Uplands, and Highlands, and showed how each of these landscapes had influenced the work of a number of poets, including John Milton, William Cowper, James Thompson, Robert Burns, James Macpherson, Sir Walter Scott, and William Wordsworth. The complete essay is recommended to anyone interested in poetry and geology. In our third sample of Geikie's writing (see pages 64 and 185 for the others), we have chosen a passage from his essay "Landscape and Literature" in Landscape in History and Other Essays (1905). This extract is about one of the many writers described by Geikie – a poet who was born, worked, and lived in the English Lake District and who wrote of it with affection – William Wordsworth (1770–1850).

One mountainous district in Britain – that of the English Lakes – claims our attention for its influence on the progress of the national literature. Of all the isolated tracts of higher ground in these islands, that of the Lake District is the most eminently highland in character. It is divisible into two entirely distinct portions by a line drawn in a north-easterly direction from Duddon Sands to Shap Fells. South of that line the hills are comparatively low and featureless, though they enclose the largest of the lakes. They are there built up of ancient sedimentary strata, like those that form so much of the similar scenery in the uplands of Wales and the South of Scotland. But to the north of the line, most of the rocks are of a different nature, and have given rise to a totally distinct character of landscape. They consist of various volcanic materials which in early Palaeozoic time were piled up around submarine vents, and accumulated over the sea-floor to a thickness of many thousand feet. They were subsequently buried under the sediments that lie to the south, but, in after ages uplifted into land, their now diversified topography has been carved out of them by the meteoric agents of denudation. Thus pike and fell, crag and scar, mere and dale, owe their several forms to the varied degrees of resistance to the general waste offered by the ancient lavas and ashes. The upheaval of the district seems to have produced a dome-shaped elevation, culminating in a summit that lay somewhere between Helvellyn and Grasmere. At least from that centre the several dales diverge, like the ribs from the top of a half-opened umbrella.

The mountainous tract of the Lakes, though it measures only some thirty-two miles from west to east by twenty-three from north to south, rises to heights of more than 3,000 feet, and as it springs almost directly from the margin of the Irish Sea, it loses none of the full effect of its elevation. Its fells present a thoroughly highland type of scenery, and have much of the dignity of far loftier mountains. Their sky-line often displays notched crests and rocky peaks, while their craggy sides have been carved into dark cliff-girt recesses, often filled with tarns, and into precipitous scars, which send long trails of purple scree down the grassy slopes.

Moreover, a mild climate and copious rainfall have tempered this natural asperity of surface by spreading over the lower parts of the fells and the bottoms of the dales a greener mantle than is to be seen among the mountains further north. Though the naked rock abundantly shows itself, it has been so widely draped with herbage and woodland as to combine the luxuriance of the lowlands with the near neighbourhood of bare cliff and craggy scar.

Such was the scenery amidst which William Wordsworth was born and spent most of his long life. Thence he drew the inspiration which did so much to quicken the English poetry of the nineteenth century, and which has given to his dales and hills so cherished a place in our literature. The scenes familiar to him from infancy were loved by him to the end with an ardent and grateful affection which he never wearied of publishing to the world. No mountain-landscapes had ever before been drawn so fully, so accurately, and in such felicitous language. Every lineament of his hills and dales is depicted as luminously and faithfully in his verse as it is reflected on the placid surface of his beloved meres, but suffused by him with an ethereal glow of human sympathy. He drew from his mountain-landscape everything that

> 'Can give an inward help, can purify
> And elevate, and harmonize and soothe.'

It brought to him 'authentic tidings of invisible things'; and filled him with

> 'The sense
> Of majesty and beauty and repose,
> A blended holiness of earth and sky.'

For his obligations to that native scenery he found continual expression.

> 'Ye mountains and ye lakes,
> And sounding cataracts, ye mists and winds
> That dwell among the hills where I was born,
> If in my youth I have been pure in heart,
> If, mingling with the world, I am content
> With my own modest pleasures, and have lived
> With God and Nature communing, removed
> From little enmities and low desires –
> The gift is yours.'

Not only did his observant eye catch each variety of form, each passing tint of colour on his hills and valleys, he felt, as no poet before his time had done, the might and majesty of the forces by which, in the mountain-world, we are shown how the surface of the world is continually modified.

> 'To him was given
> Full many a glimpse of Nature's processes
> Upon the exalted hills.'

The thought of these glimpses led to one of the noblest outbursts in the whole range of his poetry, where he gives way to the exuberance of his delight in feeling himself, to use Byron's expression, 'a portion of the tempest' –

> 'To roam at large among unpeopled glens
> And mountainous retirements, only trod
> By devious footsteps; regions consecrate
> To oldest time; and reckless of the storm,

> . . . . . . while the mists
> Flying, and rainy vapours, call out shapes
> And phantoms from the crags and solid earth,
> . . . . . and while the streams
> Descending from the region of the clouds,
> And starting from the hollows of the earth,
> More multitudinous every moment, rend
> Their way before them – what a joy to roam
> An equal among mightiest energies!'

In this passage Wordsworth seems to have had what he would have called 'a foretaste, a dim earnest' of that marvellous enlargement of the charm and interest of scenery due to the progress of modern science. When he speaks of 'regions consecrate to oldest time,' he has a vague feeling that somehow his glens and mountains belonged to a hoary antiquity, such as could be claimed by none of the verdant plains around. Had he written half a century later he would have enjoyed a clearer perception of the vastness of that antiquity and of the long succession of events with which it was crowded.

10-2   The Excursion – William Wordsworth

William Wordsworth (1770–1850), the greatest of the English Romantic poets, was born in Cumberland into a comfortable and cultured family, disrupted by the early death of his mother, when he was eight, and his father five years later. Wordsworth was educated at the local grammar school (Hawkshead) and Cambridge, and after an unsettled period, living in France at the time of the revolutionary turmoil, Wordsworth inherited nine hundred pounds from a friend, who had hoped this might enable him to establish himself as a poet. Wordsworth settled with his sister in Dorset; he later married Mary Hutchinson and spent the last fifty years of his life in the English Lake District, with which his work is always associated. His work is full of descriptions of landscape and of the evocative implications and intimations of nature. Our extract from the third book of his longest poem, The Excursion (1814), gives a satirical view of the field geologist. Wordsworth was a friend of Adam Sedgwick, Woodwardian Professor of Geology at Cambridge and founder of the Cambrian system. There may, perhaps, be a touch of Sedgwick in these lines.

> Nor is that Fellow-wanderer, so deem I,
> Less to be envied, (you may trace him oft
> By scars which his activity has left
> Beside our roads and pathways, though, thank Heaven!
> This covert nook reports not of his hand)
> He who with pocket-hammer strikes the edge
> Of luckless rock or prominent stone, disguised
> In weather – stains or crusted o'er by Nature
> With her first growths, detaching by the stroke
> A chip or splinter – to resolve his doubts;
> And with that ready answer satisfied,

The substance classes by some barbarous name,
And hurries on; or from the fragments picks
His specimen, if but haply interveined
With sparkling mineral, or should crystal cube
Lurk in its cells – and thinks himself enriched,
Wealthier, and doubtless wiser, than before!

10-3   The Lisbon Earthquake – Voltaire

In 1733 the English poet Alexander Pope (1688–1744) published Essay on Man *which, for both its art and its ideas, became one of the most influential poems of the eighteenth century. One fundamental idea in this poem was* Philosophical Optimism, *originally grounded in the work of the German philosopher Gottfried Leibniz (1646–1716). Philosophical optimism was controversial because it tried to explain the existence of evil in the world. Leibniz basic-ally said that evil was unavoidably implied by the nature of God's own perfection, and that such evil as there was in the world was unavoidable. In his poem, Pope followed much the same line. A stanza in Pope's* Essay of Man *which many found worrying read:*

All nature is but art, unknown to thee
All chance, direction, which thou cans't not see
All discord, harmony, not understood
All partial evil, universal good
And, spite of pride, in erring reason's spite
One truth is clear, Whatever is, is right.

Pope seemed to be saying that everything had its place in the greater scheme of things and that individual evil was actually part of a greater pattern of good. Voltaire (1694–1778, see page 12), originally admired Pope's poem when it was first published in 1733, but turned against it following the 1755 Lisbon Earthquake, which claimed many thousands of lives. It just was not possible, Voltaire argued, to look at the Portuguese corpses and still maintain that it was all part of a greater plan and should not worry us too much. On hearing of the November 1, 1755 Lisbon Earthquake Voltaire wrote The Lisbon Earthquake, *from which our brief extract is taken, and as a preface, "An Inquiry into the Maxim, 'Whatever is, is right.'".*

Say what advantage can result to all,
From wretched Lisbon's lamentable fall?
Are you then sure, the power which could create
The universe and fix the laws of fate,
Could not have found for man a proper place,
But earthquakes must destroy the human race?
Will you thus limit the eternal mind?
Should not our God to mercy be inclined?
Cannot then God direct all nature's course?
Can power almighty be without resource?
Humbly the great Creator I entreat,

This gulf with sulphur and with fire replete,
Might on the deserts spend its raging flame,
God my respect, my love weak mortals claim;
When man groans under such a load of woe,
He is not proud, he only feels the blow.
Would words like these to peace of mind restore
The natives sad of that disastrous shore?
Grieve not, that others' bliss may overflow,
Your sumptuous palaces are laid thus low;
Your toppled towers shall other hands rebuild;
With multitudes your walls one day be filled;
Your ruin on the North shall wealth bestow,
For general good from partial ills must flow;
You seem as abject to the sovereign power,
As worms which shall your carcasses devour.
No comfort could such shocking words impart,
But deeper wound the sad, afflicted heart.
When I lament my present wretched state,
Allege not the unchanging laws of fate;
Urge not the links of the eternal chain,
'Tis false philosophy and wisdom vain.
The God who holds the chain can't be enchained;
By His blest will are all events ordained:
He's just, nor easily to wrath gives way,
Why suffer we beneath so mild a sway:
This is the fatal knot you should untie,
Our evils do you cure when you deny?
Men ever strove into the source to pry,
Of evil, whose existence you deny.
If he whose hand the elements can wield,
To the winds' force makes rocky mountains yield;
If thunder lays oaks level with the plain,
From the bolts' strokes they never suffer pain.
But I can feel, my heart oppressed demands
Aid of that God who formed me with His hands.
Sons of the God supreme to suffer all
Fated alike; we on our Father call.

10-4   The Fountains of the Earth – C.S. Rafinesque

Constantine Samuel Rafinesque (1783–1840) was born of French parents in what was then Constantinople, but the family returned the next year to Marseilles. At nineteen he came to the United States and stayed two-and-a-half years before returning to Europe, where he served as secretary to the American consul in Palermo. Within a few years he was able to devote his full time to natural history and he returned to the United States in 1815, spending the rest of his life there. Rafinesque was a naturalist with comprehensive interests, publishing – usually privately – descriptions and memoirs on plants (he named 6,700 species), molluscs, fish, the prehistory of indigenous peoples, philology, and the mutability of species. Controversial and assertive, he was disregarded by most of his contemporaries. The extract on mineral springs which we include is from Part IV "The Earth and Moon, Water, Fire and Land" from his 248 page poem The World, or Instability (1836). These lines reflect on the influence of "fountains of the earth," with Robert M. Hazen, in his book Poetry of Geology (1982), naming this section "Mineral Springs."

The fountains of the earth are earthy pores,
The sweat and moisture of this globe exuding.
How various and unsteady in their sizes,
Contents and functions? Few are always pure,
But liquid fluids of many kinds they throw,
Sweet or impure, both cold and tepid, warm
Or hot; that gently rise, or bubbling boil,
Nay spout on high. Now nearly dry becoming,
Or full their basins filling to the brim.
Not only water flows from earthly springs,
But mineral fluids, holding sulphur, iron,
Acids and gasses, lime, and many salts.
Naphtha and oils from fountains seldom flow;
Yet there are such, even liquid pitch
In bubbles bursting underground, in lakes
Expanding; thro' volcanic regions, prone
To offer firy [sic] springs, in heat evolving:
While spungy gound [sic], or marshy soil conceal
Of lurid swamps the deadly hues and mire.
Where none arise, where liquid outlets scarce,
Or if the soil they shun, a desert dry
The earth becomes; and if no fluid could moisten
This globe, it would have been a dreary wild,
Unfit for life, where life should be extinct.

10-5   To a Trilobite – Timothy A. Conrad

Timothy Abbot Conrad (1803–1877) was a member of a Philadelphia Quaker family devoted to scientific studies. A superb artist, he illustrated many of his own publications on Cretaceous and Cenozoic paleontology and stratigraphy. Conrad was associated with James Hall for a time, while serving as geologist to the State of New York. Although bitterly opposed to Darwin's evolutionary theory, Conrad was a prolific author. He was, however, an archetype absent-minded paleontologist, and was even found to have described some of his own new species twice, using different names. An extract from his ode To A Trilobite (1840) is included here.

Methinks I see thee gazing from the stone
　　With those great eyes, and smiling as in scorn
Of notions and of systems which have grown
　　From relics of the time when thou wert born.

10-6   A Shropshire Lad – A.E. Housman

After leaving Oxford, Alfred Edward Housman (1859–1936) worked for eleven years in the Civil Service in London, before being appointed Professor of Latin at University College, London, in 1892. He became the Kennedy Professor of Latin at Cambridge in 1911, where he was a Fellow of Trinity. Apart from his meticulous scholarly publications in classics – one biographer described him, at his death in 1936, as "perhaps the most learned Latin scholar in the world" – he is remembered also as a poet. Housman was reserved, passionate, complex, and sensitive, writing his poetry, he claimed, "as a morbid secretion." Written chiefly in times of illness or depression, his poetry is reflective on the loss of youth, the pains of friendship, and the loss of rustic simplicity. The present extract from A Shropshire Lad (1896) contains a description of Wenlock Edge, known to countless geologists as the type area of the Middle Silurian, which is here seen through the eyes of a poet. The Wrekin is a nearby hill of Precambrian rhyolitic lava, Uriconium was an adjacent Roman city, and the Severn, the gentle river that threads its way through that enfolding landscape.

On Wenlock Edge the wood's in trouble,
   His forest fleece the Wrekin heaves;
The gale, it plies the saplings double,
   And thick on Severn snow the leaves.

'Twould blow like this through holt and hanger
   When Uricon the city stood:
'Tis the old wind in the old anger,
   But then it threshed another wood.

Then, 'twas before my time, the Roman
   At yonder heaving hill would stare:
The blood that warms an English yeoman,
   The thoughts that hurt him, they were there.

There, like the wind through woods in riot,
   Through him the gale of life blew high;
The tree of man was never quiet:
   Then 'twas the Roman, now 'tis I.

The gale, it plies the saplings double
   It blows so hard, 'twill soon be gone:
To-day the Roman and his trouble
   Are ashes under Uricon.

10-7   Mente et Malleo – Andrew C. Lawson

Andrew Cowper Lawson (1861–1952) was born in Scotland, but his family moved to Canada while he was still a child. He was educated at the University of Toronto and Johns Hopkins, and was a pioneer worker on the geology of the Canadian shield. Strong and contradictory in his character, he spent 55 years as a professor at the University of California, Berkeley, and was a revered teacher of generations of students, who referred to him as "King." Lawson was an amateur carpenter, legendary raconteur, and art connoisseur (from his collection, he donated Rembrandts and Gainsboroughs to the University of California), and was also something of a poet. Mente et Malleo, written in 1888, was one of his earliest, though not one of his better, efforts, and was dedicated to the Logan Club, made up largely of members of the Geological Survey of Canada.

By thought and dint of hammering
Is the good work done whereof I sing,
And a jollier lot you'll rarely find,
Than the men who chip at earth's old rind,
And often wear a patched behind,
By thought of dint of hammering.

All summer through we're on the wing,
Kept moving by the skeeter's sting;
From Alaska unto Halifax,
With our compass and our little axe,
We make our way and pay our tax,
By thought and dint of hammering.

We crack the rocks and make them ring,
And many a heavy pack we sling;
We run our lines and tie them in,
We measure strata thick and thin,
And Sunday work is never sin,
By thought and dint of hammering.

Across the waters our paddles swing,
O'er wind and rapids triumphing;
Thro' mountain passes our slow mules trudge
As if they owed us a heavy grudge,
And often can't be got to budge
By thought and dint of hammering.

To the stars at night our thoughts we bring
But no maiden fair to our arm doth cling;
She, at Ottawa, with smiling lips,
The other fellow's ice cream sips;
You can't prevent these feminine slips
By thought and dint of hammering.

To array the "chiels that waunna ding"
Is our winter's work far into spring;
Some people think us wondrous wise;
Some maintain we're otherwise;
We're simply piercing Nature's guise
By thought and dint of hammering.

10-8   Selected Poems – John Stuart Blackie

John Stuart Blackie (1809–1895) was Professor of Humanities at Marischall College and later Professor of Greek at the University of Edinburgh, where he taught for 30 years. Friend of a number of contemporary Scottish geologists, the following are extracts from same of his poems, meant to be lighthearted verses to be sung at student gatherings. These and other examples of "Paleontology in Literature" are included in Archie Lamont's paper by that name in The Quarry Manager's Journal *(1947). Our extracts concern the creatures of Silurian, Pennsylvanian, Triassic, Jurassic, and Pleistocene times.*

Stratigraphical Palaeontology

The waters, now big with a novel sensation,
Brought corals and buckies and bivalves to view,
Who dwell in shell houses, a softbodied nation;
But fishes and fins were yet none, or few.
Buckies and bivalves, a numberless nation!
Buckies, and bivalves, and trilobites too!
These you will find in Silurian station,
When Ramsay and Murchison sharpen your view.

Coal Measures

God bless the fishes! – but now on the dry land,
In days when the sun shone benign on the poles,
Forests of ferns in the low and the high land
Spread their huge fans, soon to change into coals!
Forests of ferns – a wonderful verity!
Rising like palm trees beneath the North Pole;
And all to prepare for the golden prosperity
Of John Bull reposing on iron and coal!

Triassic Period

Now Nature the eye of the gazer entrances
With wonder on wonder from teaming abodes;
From the gills of the fish to true lungs she advances,
And bursts into blossoms of tadpoles and toads.
People Batrachian, strange, Triassic all,
Like Hippopotamus huge on the roads!
Call them ungainly, uncouth, and unclassical,
But great in the reign of the Trias were Toads!

Ichthyosaurus

Behold a strange monster our wonder engages,
If dolphin or lizard your wit may defy,
Some thirty feet long on the shore of Lyme-Regis
With saw for a jaw, and a big-staring eye.

A fish or a lizard? An ichthyosaurus,
With big goggle-eyes, and a very small brain,
And paddles like mill-wheels in clattering chorus
Smiting tremendous the dreadsounding main!

Frozen Mammoths

Mammoth, Mammoth! mighty old Mammoth!
Strike with your hatchet and cut a good slice;
The bones you will find, and the hide of the mammoth,
Packed in stiff cakes of Siberian ice.

10-9 Lyell's Hypothesis Again – Kenneth Rexroth

Kenneth Rexroth (1905–1982) had a varied and colorful career, not only as a poet, but also as an essayist, translator, playwright, artist, fruit picker and packer, jazz scholar, forest patrolman, factory hand, and attendant at a mental institution. Largely self-educated, Rexroth was a complex character, whose knowledge was vast. He played a major role in the Bay Area's literary community and amongst Beat generation writers. Widely read and admired, he is represented here by a short poem from his collection The Signature of All Things *(1950). The poem is based on the subtitle of Lyell's* Principles *and reflects on the supremacy of a present relationship between two people, insulated for a moment from the constraints of the "vast impersonal vindictiveness of the ruined and ruining world." Fossils, lava flows, waterfalls, and ice ages – all converge here in an intimate relationship, and "... ideograms printed on the immortal hydrocarbons of flesh and stone."*

The mountain road ends here,
Broken away in the chasm where
The bridge washed out years ago.
The first scarlet larkspur glitters
In the first patch of April
Morning sunlight. The engorged creek
Roars and rustles like a military
Ball. Here by the waterfall,
Insuperable life, flushed
With the equinox, sentient
And sentimental, falls away
To the sea and death. The tissue
Of sympathy and agony
That binds the flesh in its Nessus' shirt;
The clotted cobweb of unself
And self; sheds itself and flecks
The sun's bed with darts of blossom
Like flagellant blood above
The water bursting in the vibrant
Air. This ego, bound by personal

Tragedy and the vast
Impersonal vindictiveness
Of the ruined and ruining world,
Pauses in this immortality,
As passionate as apathetic,
As the lava flow that burned here once;
And stopped here; and said, "This far
And no further." And spoke thereafter
In the simple diction of stone.

Naked in the warm April air,
We lie under the redwoods,
In the sunny lee of a cliff.
As you kneel above me I see
Tiny red marks on your flanks
Like bites, where the redwood cones
Have pressed into your flesh.
You can find just the same marks
In the lignite in the cliff
Over our heads. *Sequoia
Langsdorfii* before the ice,
And *sempervirens* afterwards,
There is little difference,
Except for all those years.

Here in the sweet, moribund
Fetor of spring flowers, washed,
Flotsam and jetsam together,
Cool and naked together,
Under this tree for a moment,
We have escaped the bitterness
Of love, and love lost, and love
Betrayed. And what might have been,
And what might be, fall equally
Away with what is, and leave
Only these ideograms
Printed on the immortal
Hydrocarbons of flesh and stone.

10-10 Selected Poems – A.R. Ammons

Archie Randolph Ammons (1926–2001) was the Goldwin Smith Professor of Poetry at Cornell. He was the recipient of many honors, including a MacArthur Fellowship, a National Book Award, the Bollinger Prize, and the Ruth Lilly Poetry Prize. Born in North Carolina and educated at Wake Forest and Berkeley, he served with the United States Navy in the South Pacific, and had a varied career as a painter, school principal, glassware manufacturer, business executive, and professor. Intensely involved with nature, many of his poems, including some that are deceptively simple and short, explore both the glory and the brokenness of Earth, its creatures and its objects. Ammons, often described as an Emersonian Transcendentalist, wrote with a conversational style and an engaging skeptical, but generous tone. Our three poems are taken from The Really Short Poems of A.R. Ammons *(1990)*.

Glacials

In the geological rock garden
split boulders
lie about as kinds and ages,

a hundred million years or more in
many frozen solid:
dusty thaws flow

steam-loose from the surfaces and
wind on the way
at last, the wind mixing

old and current time
in mixings beginning and
ended, time unbegun, unended.

Natives

Logos is an engine
myth fuels,
civilization
a pattern,
scalelike crust
on a hill
but the hill's swell
derives from
gravity's
deep fluids
centering elsewhere
otherwise

Undersea

Foraminiferal millennia
bank and spill but
even so
time's under pressure of
diatomaceous event,
divisions a moment
arcs across:
        desperate
for an umbrella, net, longpole,
or fan: so much
to keep for paradigm,
so much to lose.

10-11   Stone – Charles Simic

Charles D. Simic, born in Belgrade, Yugoslavia in 1938, is Emeritus Professor at the University of New Hampshire. A graduate of New York University, he has also taught at Columbia and Boston Universities. He has published more than twenty volumes of poetry and contributed poems to more than one hundred magazines, translated over a dozen books and written another half-dozen books of essays and memoirs. Recipient of the Pulitzer Prize and a MacArthur Award, and a member of the American Academy of Arts and Letters, his work has been described by Victor Contoski as "some of the most strikingly original poetry of our time." The poem, Stone, *which we reproduce, was published in* Dismantling the Silence *(1971).*

Go inside a stone
That would be my way.
Let somebody else become a dove
Or gnash with a tiger's tooth.
I am happy to be a stone.

From the outside the stone is a riddle:
No one knows how to answer it.
Yet within, it must be cool and quiet
Even though a cow steps on it full weight,
Even though a child throws it in a river;
The stone sinks, slow, unperturbed
To the river bottom
Where the fishes come to knock on it
And listen.

I have seen sparks fly out
When two stones are rubbed,
So perhaps it is not dark inside after all;
Perhaps there is a moon shining
From somewhere, as though behind a hill–

Just enough light to make out
The strange writings, the star-charts
On the inner walls.

10-12   Fossils – J.T. Barbarese

Joseph T. Barbarese was born in 1945, completed his PhD at Temple University and teaches English and creative writing at Rutgers. His poetry and fiction appear in numerous magazines and he has four books of original poetry, including Under the Blue Moon *(1984) and* New Science *(1989). We include his short poem* Fossils *which appeared in* The Atlantic Monthly *(2000).*

When he was young he used to spend the whole summer
in the abandoned slag heaps around the old mines
outside the city of Scranton. It would take him hours
to pick through the shale stacks, the sweat writing lines
in the dust on his face, and the old ball peen hammer
slung from his belt pinching his belly button.
Some days there was nothing to read but the signatures
of ice and erosion and tools. Then he'd find one,
a slate unnaturally filigreed with the fright masks
of a trilobite, ferns, the inferior commissures
of ancient clams. He would wrap them in moist newspaper
and carry them carefully home. Once his teacher asked
him to talk to the class about fossils.
    Satan plants them to trick us,
he said. *When I get home I smash them to pieces.*

10-13   Rock – Jane Hirshfield

Jane Hirshfield, a widely published and award-winning poet, was born in 1953 in New York City, and is a devotee of horses, wilderness, gardening, and Zen Buddhism. Graduating from Princeton in 1973 as a member of that university's first class to include women, Hirshfield devoted herself for nearly eight years to study at the San Francisco Zen Center, deferring her writing because "I first had to find out what it meant to live." Lyrical, infused with nature, awareness, the world of everyday, and grace, her poetry, both spiritual and sensuous, is, for her, "an instrument of investigation and a mode of perception." She has taught at Berkeley, the University of San Francisco, and Bennington. Our extract is taken from her poem, Rock, *in* Given Sugar, Given Salt *(2001).*

What appears to be stubbornness,
refusal, or interruption,
is to it a simple privacy. It broods
its one thought like a quail her clutch of eggs.

Mosses and lichens
listen outside the locked door.
Stars turn the length of one winter, then the next.

Rocks fill their own shadows without hesitation,
and do not question silence,
however long.
Nor are they discomforted by cold, by rain, by heat.

The work of a rock is to ponder whatever is:
an act that looks singly like prayer,
but is not prayer.

As for this boulder,
its meditations are slow but complete.

Someday, its thinking worn out, it will be
carried away by an ant.
A *mystrium camillae*,
perhaps, caught in some equally diligent,
equally single pursuit of a thought of her own.

10-14 Poetry Matters: Gary Snyder – W. Scott McLean, Eldridge M. Moores, and David A. Robertson

Poetry can have an intimate connection, not only to landscape, but also to its origin and slow transformation, its conservation and its implication for our self understanding. Nowhere is that connection better illustrated than in the work of Gary Snyder, one of America's fore-most contemporary poets. Born in 1930, raised on small farms in the Pacific north-west, he held a number of jobs – logger, sailor, trail crew worker – which reinforced his concerns for Earth. A graduate of Reed College and University of California at Berkeley, Snyder was involved with the Beatnik movement, and later Zen Buddhism. A Pulitzer Prizewinner, he taught at the University of California, Davis. Three Davis colleagues – W. Scott McLean, a poet, Eldridge M. Moores, a geologist, and David A. Robertson, a Professor of English – discuss Snyder's work and thought in our extract from "Nature and Culture", an essay in Earth Matters *(2000).*

It is a commonplace to speak in the history of science of a sequence of revolutions in thought: from the Aristotelian in the Middle Ages to the Copernican or "scientific" revolution of the sixteenth and seventeenth centuries, the Darwinian in the nineteenth century, and (though one may dispute whether or not Freud's work has the character of science) the Freudian revolution of the twentieth century. But if such cataclysmic upheavals in the philosophy of science mark significant turning points in epistemology and ontology, their real impact is to be found in texts that fall outside the bounds of treatises that delineate philosophic and scientific positions. *Indeed, the real reach of these revolutions is to be found in the poetries written under the aegis of profound cosmological resettings of perspective.* For example, Dante incorporates Aristotle's philosophical rigor into the *Divine Comedy*, and it is in that poem that one can truly feel the impact of Aristotle's ideas on a medieval Catholic. The poets of the late seventeenth and eighteenth centuries incorporate the advances of the new science directly into their works and irrev-ocably alter a generation's perspective (this is most trenchantly detailed in Marjorie Hope Nicolson's *Newton Demands the Muse*). Developments in evolution-ary thought find perhaps their most heartfelt expressions in Tennyson's *In Memor-iam*, a work that gives a truer gauge than any academic analysis of the ways in which the Victorians dealt with the onslaught of Darwin's evolutionary science on their spiritual moorings. And Freud's theories deeply influenced and shaped the work of countless writers in the twentieth century, beginning with Franz Kafka, forever changing our understanding of the mind's dynamics. In each of these cases, the poetic imagination engages experience in ways that mark the depth of the transformation. In each case, as Basil Wiley so incisively notes regarding Tennyson's *In Memoriam*, the poetic text, "like a piece of ritual, enacts what a credal statement merely propounds; 'this', says the poem in effect, 'is a tract of experience lived through in the light of such-and-such a thought or belief; this is what it feels like to accept it.'"

Although geology is not often given pride of place in the list of scientific revolutions referenced above, we believe that at the end of the twentieth century and the second millennium, geology indeed is the most revolutionary of all the sciences. It has, after all, undergone two revolutions in the past 50 years – those of plate tectonics and "Earth in space," – that is, Earth's place in space, its comparison with the other planets, and the role of meteorite and/or comet impacts on processes in Earth history including evolution. These revolutions have extended the work that began with that of James Hutton, Charles Lyell, and Grove Karl Gilbert over a century ago. Today it is the geologic perspectives, concepts, and insights that continue to exercise the most profound hold on our imaginations and will, in the end, have the greatest transformative powers.

Geology does so precisely where it matters most – in the ways in which the findings of geology, as mentioned above, ultimately teach humility. In what is surely one of its most revolutionary implications, the science of unfeeling minerals and rocks leads to reverence, wonder and awe – and, in the end, to compassion. For in disclosing the forces of an inexorable world beyond our daily ken, the geologic sciences disclose the measure of the gods. That is, the geologic sciences provide a nonhuman standpoint for us to view life and experience, helping us look past the interests and perspective of the small self to the larger self that is part of nature. And when they do so, these sciences become an imaginative engagement with the world, an engagement that leads to a profound appreciation for the uniqueness and preciousness of the life of the planet – carrying us, in the end, toward an abiding compassion for all creatures bound up in time's arrow.

At a time when one often hears that the coming century will be the century of biology, one would have to add that this will be the case only if we make our own the lessons of rock's deep time; for in those lessons lies, literally as well as metaphorically, our very bedrock. One could argue just as forcefully, for the reasons outlined below, that the twenty-first century will be the century of geology.

How this is the case one can sketch by referring to the work of an American poet, our colleague Gary Snyder. For many in the sciences, most especially in the biologic sciences, Snyder is the poet of E. O. Wilson's "biophilia" – the poet of a sensibility that takes consistent references to the genetic and species diversity of geology's multi-million-year cycles. Snyder's poems continually address our filiation with that world, seen in the image of the Buddhist "Jeweled Net" of interdependence. Snyder's is a poetry that, in the words of Joseph Baruch's poem "Prayer," is "bright with animals,/images of a gull's wing," asking "the blessing of the crayfish,/the beatitude of the birds." All of Snyder's work circles about a revisioning of that most fundamental of human markers, self-identity; and in his poetry he does nothing less than to define the ways in which our identity is to be found not in terms of our otherness, but in terms of our relations to all of nonhuman nature.

Gary Snyder is a quintessential twentieth-century American poet. Bound to the West and the landscapes of the Great Basin and Pacific Coast, his sense of culture is rooted in those vast spaces Emerson, Thoreau, and Whitman celebrated as the soul of America, an America free of the cultural bones Goethe once decried when he looked to the "New World." But Snyder's poetry is first and foremost of rock, and a full study of Snyder's two most recent books, *A Place in Space* (1995) and *Mountains and Rivers Without End* (1996), sheds crucial light on the ways this poetry is central to an understanding of geology's possible impact on our own times. Indeed, Snyder's verse enacts what the lessons of geology merely propound; this poetry offers "a tract of experience," in effect giving us poems that say what it feels like to accept the geological record.

We are well served by beginning with one of Snyder's early journal entries, this one collected in *Earth House Hold* (1969):

> The rock alive, not barren.
> flowers lichen pinus albicaulis chipmunks
> mice even grass

This is but one of the numerous "geologic meditations" contained in Snyder's early work, the most famous of which are contained in the poems of his first volume, *Riprap*. One of the offerings in this volume, "Piute Creek," is exemplary:

> One granite ridge
> A tree, would be enough
> Or even a rock, a small creek,
> A bark shred in a pool.
> Hill beyond hill, folded and twisted
> Tough trees crammed
> In thin stone fractures
> A huge moon on it all, is too much.
> The mind wanders. A million
> Summers, night air still and the rocks
> Warm. Sky over endless mountains.
> All the junk that goes with being human
> Drops away, hard rock wavers
> Even the heavy present seems to fail
> This bubble of a heart.
> Words and books
> Like a small creek off a high ledge
> Gone in the dry air.
>
> A clear, attentive mind
> Has no meaning but that
> Which sees is truly seen.
> No one loves rock, yet we are here.
> Night chills. A flick
> In the moonlight
> Slips into Juniper shadow:
> Back there unseen
> Cold proud eyes
> Of Cougar or Coyote
> Watch me rise and go.

This poem contains a line that epitomizes Snyder's early work: "A clear, attentive mind/Has no meaning but that/Which is truly seen." As Snyder wrote this while working on trail crews in Yosemite setting stone pathways, this is a rock-setter's jab at Descartes's *cogito ergo sum*. In this and other poems, Snyder answers long and influential traditions in English and American poetics, for example, in Wordsworth's insistence, at the end of "1850 Prelude," that "the mind of man becomes /A thousand times more beautiful than the earth/ On which he dwells."

For Snyder, what Keats called Wordsworth's "egotistical sublime" is a fatal poetic misstep. For this grand poetic "selfhood" denies any genuine engagement beyond a focus on what we might call the "small self." Keats tried, almost singlehandedly, to turn the poetic address back toward the world itself, and in two of his most incisive letters he laid this out plainly:

> "As to the poetical Character itself," He wrote, "(that sort distinguished from the Wordsworthian or egotistical sublime; . . .) It is Not Itself – It Has No Self – It is every thing and Nothing – It has no character – It enjoys light and shade; it lives in gusto, be it foul or fair, high or low, rich or poor, mean or elevated." (27 October 1818)

Keats was talking about what he had earlier called "Negative Capability," when we are

> capable of being in uncertainties, Mysteries, doubts, without any irritable reaching after fact or reason. (December 1817)

A "geologic poetics" demands that we live in uncertainties, mysteries, and doubts, with a wondrous, not irritable, reaching after fact. Snyder grounds his own perspective in an imaginative reaching out to the life of rock in *Left out of the Rain* (1986):

"Geological Meditation"

Rocks suffer,
  slowly,
Twisting, splintering scree
Strata and vein
  writhe
Boiled, chilled, form to form
Loosely hung over with
Slight weight of trees,
  quick creatures
Flickering, soil and water,
Alive on each other.

In this passage we see the poetic enchantment of the observation recorded in an earlier journal entry – for here, in the ways in which inorganic elements take on an emotional life of their own, Snyder imagines himself into the slow suffering, the twisting, chilling life of the planet where the "quick creatures," including us, are "alive on each other."

In these poems, Snyder crafts the foundational blocks of his poetics. One of its most significant accomplishments is a perspective in which the geologic imagination, the life of rocks' million-year cycles, finds a voice. Snyder takes us deep into the landscape of the western United States and shows us how the way to the self is not so much in going deep into the self but in going deep into the world. Snyder watches carefully, attends, knows how far there is to go in learning to know the other and ultimately ourselves. Throughout his great narrative poem of the West, "Mountains and Rivers Without End," Snyder tracks the long process of attention, of listening to the rock, the wind, to write a poetry of the interstices of these contacts in what Jim Dodge has called Snyder's "poised, aggressive receptivity."

"Mountains and Rivers Without End" ends in the desert landscape and vast expanse that was once Lake Lahontan. This is, in a historically complex and abiding way, a nature poetry rooted in place and in the details of the land, with the human community at its center. And it is a geologic poetry in the sense that the landscape of rock, sand, and the million-year processes of geologic formation are active players. It is a poetry that addresses our most fundamental nature, the fact that our relationships and identities, our very own subjectivities, are found in the ephemeral substantiality of all life. Where do we find ourselves? For Gary Snyder, it is looking down the roads and across the passes and along the centuries, here in the world, to be sure – in terms of the ancient mystic traditions, finding in the immanent the transcendent. What sustains us and what really lasts are our love and our compassion, reflected in the beauty and elegance of the songs of the place, of the "foolish loving spaces/full of heart," where "All art and song/is sacred to the real."

10-15 Where Shall Wisdom Be Found? – The Book of Job

The Book of Job, *written probably about twenty-five hundred years ago, contains the thoughts and conversations of a saintly man in the face of adversity. The book is a treasure of mature experience and perspective and is also a gem of literature. Job is an archetype of modern humanity; baffled by death, destruction, disease and loss, a good person in a tragic universe, agonizing over the justice of God and the search for wisdom and understanding. Chapter 28, which we include, contains a moving account of success in mining the Earth and its resources but frustration in the search for wisdom, until it is recognized in a knowledge of the Creator. In Chapter 38, which follows, the Lord answers Job "out of the whirlwind," stressing the limitations of human understanding.*

Book of Job, Chapter 28

"Surely there is a mine for silver, and a place for gold which they refine.
Iron is taken out of the earth, and copper is smelted from the ore.
Men put an end to darkness, and search out to the farthest bound the ore in
    gloom and deep darkness.
They open shafts in a valley away from where men live; they are forgotten by
    travelers, they hang afar from men, they swing to and fro.
As for the earth, out of it comes bread; but underneath it is turned up as by fire.
Its stones are the place of sapphires, and it has dust of gold.

"That path no bird of prey knows, and the falcon's eye has not seen it.
The proud beasts have not trodden it; the lion has not passed over it.

"Man puts his hand to the flinty rock, and overturns mountains by the roots.
He cuts out channels in the rocks, and his eye sees every precious thing.
He binds up the streams so that they do not trickle, and the thing that is hid he
    brings forth to light.

"But where shall wisdom be found? And where is the place of understanding?
Man does not know the way to it, and it is not found in the land of the living.

The deep says, 'It is not in me,' and the sea says, 'It is not with me.'
It cannot be gotten for gold, and silver cannot be weighed as its price.
It cannot be valued in the gold of Ophir, in precious onyx or sapphire.
Gold and glass cannot equal it, nor can it be exchanged for jewels of fine gold.
No mention shall be made of coral or of crystal; the price of wisdom is above
    pearls.
The topaz of Ethiopia cannot compare with it, nor can it be valued in pure gold.

"Whence then comes wisdom? And where is the place of understanding?
It is hid from the eyes of all living, and concealed from the birds of the air.
Abaddon and Death say, 'We have heard a rumor of it with our ears.'

"God understands the way to it, and he knows its place.
For he looks to the ends of the earth, and sees everything under the heavens.
When he gave to the wind its weight, and meted out the waters by measure;
when he made a decree for the rain, and a way for the lightning of the thunder;
then he saw it and declared it; he established it, and searched it out.
And he said to man, 'Behold, the fear of the Lord, that is wisdom; and to depart
    from evil is understanding.' "

Book of Job, Chapter 38

Then the Lord answered Job out of the whirlwind:
"Who is this that darkens counsel by words without knowledge?
Gird up your loins like a man, I will question you, and you shall declare to me.

"Where were you when I laid the foundation of the earth? Tell me, if you have
    understanding.
Who determined its measurements – surely you know! Or who stretched the
    line upon it?
On what were its bases sunk, or who laid its cornerstone,
when the morning stars sang together, and all the sons of God shouted for joy?

"Or who shut in the sea with doors, when it burst forth from the womb;
when I made clouds its garment, and thick darkness its swaddling band,
and prescribed bounds for it, and set bars and doors,
and said, 'Thus far shall you come, and no farther, and here shall your proud
    waves be stayed'?

"Have you commanded the morning since your days began, and caused the
    dawn to know its place,
that it might take hold of the skirts of the earth, and the wicked be shaken out
    of it?
It is changed like clay under the seal, and it is dyed like a garment.
From the wicked their light is withheld, and their uplifted arm is broken.

"Have you entered into the springs of the sea, or walked in the recesses of the
    deep?
Have the gates of death been revealed to you, or have you seen the gates of deep
    darkness?
Have you comprehended the expanse of the earth? Declare, if you know all this."

11. Art

Graham Sutherland (1903–1980) was one of the great British and European painters of the last 200 years. In assessing his career, Robert Waterhouse wrote of the various influences that had shaped his style. "A chance trip to Pembrokeshire in 1934 was a revelation: 'I learnt that landscape was not necessarily scenic,' he wrote later, 'and found that its parts have an individual figurative detachment.' Subjects came together in a clearly resolved sequence of images. In spite of his attachment to the Home Counties and to the Kent coast in particular, Sutherland found his real impetus in the ordered magnificence of Pembroke."

Sutherland was not alone in this, for a host of other artists have been influenced by landscapes from which they drew inspiration, and to which they contributed new vision. Monet and Giverney, other impressionists and Argenteuil, Hornfleur and Auvers-sur-Oise, Gauguin and Tahiti, Constable and southern England, Van Goyen and van Ruysdael and Holland, all spring readily to mind. Sometimes an area can give inspiration and unity to a whole "school" of art. The Hudson Valley School of American landscape painters flourished in the mid-nineteenth century and included such artists as Thomas Doughty, Thomas Cole, and Frederick Church. The artist always draws experience from his subject, and in turn contributes to his subject, often expanding his new perception to other areas and subjects. One artist once exclaimed, "I begin by observing, I end by interpreting."

This creative search has ancient roots. Hand tracings and paintings on the walls of caves and cave paintings and statues represent the Upper Paleolithic period of some 20,000 years ago. Whether magical, symbolic, religious, aesthetic, or all of these things, early humans expressed their innermost feelings with sensitivity and skill in which we can still delight. Nor is landscape the only influence on painting. Materials were of prime importance, for with chalk and charcoal, ocher and oxide, early humans created works of art.

Sculpture and architecture are even more strongly influenced by materials and location, as the selections by Jack Burnham and Jacquetta Hawkes reveal.

11-1   A Land: Sculpture – Jacquetta Hawkes

Jacquetta Hawkes (1910–1996), archeologist, writer, civil servant, lecturer and broadcaster, was educated at Newnham College, Cambridge, where she was the first woman to take honors in archeology and anthropology. Among the better known of her twenty or so books are Early Britain *and* Journey Down a Rainbow, *co-authored with her second husband, J.B. Priestley. In 1951 she received the Kemsley Award for an exceptional book,* A Land *(1951),* in which she shows the close relationship between geology, archeology and landscape, on the one hand, and architecture and art on the other. Chapter 7 of* A Land *is titled, "Digression on Rocks, Soil and Men" from which we include two excerpts, the first here – on rocks, fossils and sculpture and more particularly the work of Auguste Rodin (1840–1917) and Henry Moore (1898–1996) – and the second in the next chapter on page 261.*

Life has grown from the rock and still rests upon it; because men have left it far behind, they are able consciously to turn back to it. We do turn back, for it has kept some hold over us. A liberal rationalist, Professor G.M. Trevelyan, can write of 'the brotherly love that we feel . . . for trees, flowers, even for grass, nay even for rocks and water' and of 'our brother the rock'; the stone of Scone is still used in the coronation of our kings.

The Church, itself founded on the rock of Peter, for centuries fought unsuccessfully against the worship of 'sticks and stones'. Such pagan notions have left memories in the circles and monoliths that still jut through the heather on our moorlands or stand naked above the turf of our downs. I believe that they linger, too, however faintly, in our churchyards – for who, even at the height of its popularity, ever willingly used cast-iron for a tombstone?

It is true that these stones were never simply themselves, but stood for dead men, were symbols of fertility, or, as at Stonehenge, were primarily architectural forms. But for worshippers the idea and its physical symbol are ambivalent; peasants worship the Mother of God and the painted doll in front of them; the peasants and herdsmen of prehistoric times honoured the Great Mother or the Sky God, the local divinities or the spirits of their ancestors and also the stones associated with them. The Blue Stones of Stonehenge, for example, were evidently laden with sanctity. It seems that these slender monoliths were brought from Pembrokeshire to Salisbury Plain because in Wales they had already absorbed holiness from their use in some other sacred structure. There is no question here that the veneration must have been in part for the stones themselves.

Up and down the country, whether they have been set up by men, isolated by weathering, or by melting ice, conspicuous stones are commonly identified with human beings. Most of our Bronze Age circles and menhirs have been thought by the country people living round them to be men or women turned to stone. The names often help to express this identification and its implied sense of kinship; Long Meg and her Daughters, the Nine Maidens, the Bridestone and the Merry Maidens. It is right that they should most often be seen as women, for somewhere in the mind of everyone is an awareness of woman as earth, as rock, as matrix. In all these legends human beings have seen themselves melting back into rock, in their imaginations must have pictured the body, limbs and hair melting into smoke and solidifying into these blocks of sandstone, limestone and granite.

Some feeling that represents the converse of this idea arises from sculpture. I have never forgotten my own excitement on seeing in a Greek exhibition an unfinished statue in which the upper part of the body was perfect (though the head still carried a mantle of chaos) while the lower part disappeared into a rough block of stone. I felt that the limbs were already in existence, that the sculptor had merely been uncovering them, for his soundings were there – little tunnels reaching towards the position of the legs, feeling for them in the depths of the stone. The sculptor is in fact doing this, for the act of creation is in his mind, from his mind the form is projected into the heart of the stone, where then the chisel must reach it.

Rodin was one of the sculptors most conscious of these emotions, and most ready to exploit them. He expressed both aspects of the process – man merging back into the rock, and man detaching himself from it by the power of life and mind. He was perhaps inclined to sentimentalize the relationship by dwelling on the softness of flesh in contrast with the rock's harshness. This was an irrelevance not dreamt of by the greatest exponent of the feeling – Michelangelo. It is fitting that the creator of the mighty figures of Night and Day should himself have spent many days in the marble quarries of Tuscany supervising the removal of his material from the side of the mountain. So conscious was he of the individual quality of the marble and of its influence on sculpture and architecture that he was willing to endure a long struggle with the Pope and at last to suffer heavy financial loss by maintaining the superiority of Carrara over Servezza marble. Michelangelo was an Italian working with Italian marble and Italian light; with us it has been unfortunate that since medieval times so many of our sculptors have sought the prestige of foreign stones rather than following the idiom of their native rock. It is part of the wisdom of our greatest sculptor, Henry Moore, to have returned to English stones and used them with a subtle sensitiveness for their personal qualities. He may have inherited something from his father who, as a miner spending his working life in the Carboniferous horizons of Yorkshire, must have had a direct understanding not only of coal but of the sandstones and shales in which it lies buried and on which the life of the miner depends. Henry Moore has himself made studies of miners at work showing their bodies very intimate with the rock yet charged with a life that separates them from it. (Graham Sutherland in his studies of tin mines became preoccupied with the hollow forms of the tunnels and in them his men appear almost embryonic.)

Henry Moore uses his understanding of the personality of stones in his sculpture, allowing their individual qualities to contribute to his conception. Indeed, he may for a moment be regarded in the passive role of a sympathetic agent giving expression to the stone, to the silting of ocean beds shown in those fine bands that curve with the sculpture's curves, and to the quality of the life that shows itself in the delicate markings made by shells, corals and sea-lilies.

It would certainly be inappropriate to his time if Moore habitually used the Italian marbles so much in favour since the Renaissance. For this fashion shows how a man in his greatest pride of conscious isolation wanted stone which was no more than a beautiful material for his mastery. Now when our minds are recalling the past and our own origins deep within it, Moore reflects a greater humility in avoiding the white silence of marble and allowing his stone to speak. That is why he has often chosen a stone like Hornton, a rock from the Lias that is full of fossils all of which make their statement when exposed by his chisel. Sometimes the stone may be so assertive of its own qualities that he has to battle with it, strive against the hardness of its shells and the softness of adjacent pockets to make them, not efface themselves, but conform to his idea, his sense of a force thrusting from within, which must be expressed by taut lines without weakness of surface.

Moore uses Hornton stone also because it has two colours, a very pale brown and a green with deeper tones in it. The first serves him when he is conscious of his subject as a light one, the green when it must have darkness in it. Differences in climate round the shores of the Liassic lakes probably caused the change in colour of Hornton stone, and so past climates are reflected in the feeling of these sculptures. As for the sculptor's sense of light or darkness inherent in his subjects, it is my belief that it derives in large part from the perpetual experience of day and night to which all consciousness has been subject since its beginnings. The sense of light and darkness seems to go to the depths of man's mind, and whether it is applied to morality, to aesthetics or to that more general conception – the light of intellectual processes in contrast with the darkness of the subconscious – its symbolism surely draws from our constant swing below the cone of night.

It is hardly possible to express in prose the extraordinary awareness of the unity of past and present, of mind and matter, of man and man's origin which these thoughts bring to me. Once when I was in Moore's studio and saw one of his reclining figures with the shaft of a belemnite exposed in the thigh, my vision of this unity was overwhelming. I felt that the squid in which life had created that shape, even while it still swam in distant seas was involved in this encounter with the sculptor; that it lay hardening in the mud until the time when consciousness was ready to find it out and imagination to incorporate it in a new form. So a poet will sometimes take fragments and echoes from other earlier poets to sink them in his own poems where they will enrich the new work as these fossil outlines of former lives enrich the sculptor's work.

Rodin pursued the idea of conscious, spiritual man emerging from the rock; Moore sees him rather as always a part of it. Through his visual similes he identifies women with caverns, caverns with eye-sockets; shells, bones, cell plasm drift into human form. Surely Mary Anning might have found one of his forms in the Blue Lias of Lyme Regis? That indeed would be fitting, for I have said that the Blue Lias is like the smoke of memory, of the subconscious, and Moore's creations float in those depths, where images melt into one another, the direct source of poetry, and the distant source of nourishment for the conscious intellect with its clear and fixed forms. I can see his rounded shapes like whales, his angular shapes like ichthyosaurs, surfacing for a moment into that world of intellectual clarity, but plunging down again to the sea bottom, the sea bottom where the rocks are silently forming.

11-2 Beyond Modern Sculpture – Jack Burnham

In 1934 the English sculptor, Henry Moore (1898–1986), prepared a program for himself, which – almost without exception – he followed throughout his career. Two portions of this program – Truth to Material *and* Observation of Natural Objects *– are reproduced from Jack Burnham's book* Beyond Modern Sculpture *(1968). Both show Moore's concern with pebbles, rocks, shells, bones, and rounding and polishing by erosive processes. Also included is a paragraph relating some of the most powerful works of Henry Moore to the rock formations of Yorkshire and the marine stacks and chimneys off the rugged coast of Brittany. Jack Wesley Burnham was born in 1931, educated at Yale and taught art at Northwestern, Colgate, Williams and Maryland. In his writing he expresses concern for the loss of spiritual qualities in art and the influence of science and technology on the artist.*

No sculptor has brought out with such bluntness, or described with greater precision, the vitalist position than Henry Moore. Where previous sculptors felt content to imbue themselves with a tacit aesthetic based on the organic influence, Moore has always been quite candid about the origins and techniques of his synthesis of natural elements. He has never regarded himself as a magician but simply as a sensitive and observant artist obeying the characteristics of his materials. If, strictly speaking, vitalist doctrine is concerned with the infusion of "life" into the materials of sculpture, then it should be emphasized that Moore has been quite specific in recommending that the word *vital* be used to describe this infusion, rather than *organic*. On the surface the difference may seem small, but *organic* implies a functionalism, an application of materials to various duties, which runs well beyond the range of vitalism.

Vitalism, based as it is on nonphysical substances and states of life, is a metaphysical doctrine concerned with the irreducible effects and manifestations of living things. It was the great discovery of twentieth-century sculpture that these did not have to be appreciated through strict representationalism. Visual biological metaphors exist on many levels besides the obvious total configuration of an animal or human. The aesthetic of true organicism, on the other hand, is not grounded in the appearances of natural forms and their carryover into sculptural materials, but is concerned with the organization of processes and interacting systems.

Moore's vitalism, in no way scientifically analytical, is the vitalism of the naturalist and sensitive craftsman. Besides the sculpture itself, the most obvious example of this lies in Moore's writings. A statement in the anthology *Unit I* (1934) describes the program which Moore set for himself, and which, with a few interruptions, he has followed for the past thirty-five years. Some essential parts read as follows (quoted in Read, 1934, pp. 29–30):

Truth to material. Every material has its own individual qualities. It is only when the sculptor works direct, when there is an active relationship with his material, that the material can take its part in the shaping of an idea. Stone, for example, is hard and concentrated and should not be falsified to look like soft flesh – it should not be forced beyond its constructive build to a point of weakness. It should keep its hard tense stoniness.

Observation of Natural Objects. The observation of nature is part of an artist's life, it enlarges his form-knowledge, keeps him fresh and from working only by formula, and feeds inspiration.

The human figure is what interests me most deeply, but I have found principles of form and rhythm from the study of natural objects such as pebbles, rocks, bones, trees, plants, etc.

Pebbles and rocks show Nature's way of working stone. Smooth, sea-worn pebbles show the wearing away, rubbed treatment of stone and principles of asymmetry.

Rocks show the hacked, hewn treatment of stone, and have a jagged nervous block rhythm.

Bones have marvelous structural strength and hard tenseness of form, subtle transition of one shape into the next and great variety in section.

Trees (tree trunks) show principles of growth and strength of joints, with easy twisting movement.

Shells show Nature's hard but hollow form (metal sculpture) and have a wonderful completeness of single shape.

There is in Nature a limitless variety of shapes and rhythms (and the telescope and microscope have enlarged the field) from which the sculptor can enlarge his form-knowledge experience.

Vitality and Power of Expression. For me a work must first have a vitality of its own. I do not mean a reflection of the vitality of life, of movement, physical action, frisking, dancing figures and so on, but that work can have in it a pent-up energy, an intense life of its own, independent of the object it may represent.

... Not surprisingly, Moore's sculpture has only become devitalized and weakened when he has strayed too far from the tenets of his 1934 statement. This is true of some of his work just prior to the Second World War and particularly during the 1950s. Various half-hearted attempts at formalism, totemism, and realism are in no way consistent with the strengths of Moore's sensibilities.

During the present decade Moore has produced some of the most powerful work of his career – much of it in very large bronze castings. No longer are the plaster models finished in smooth perfection, but instead the plaster rasp and the pick hammer simulate the kinds of graininess and sedimentary stratification which Moore relates to the rock formations in the Yorkshire countryside. At the same time, the tendency of bones to appear smooth-hard or porous, depending upon the function of a bone at a given area, is of more importance to Moore. Some of the figurative conceptions relate to earlier two-piece and three-piece compositions, and, with the aid of polychrome patinas, completely detached sets of shoulders and knees jut up from a plinth's flat surface like the upright rock promontories off the coast of Brittany. ...

Moore's sculpture at the present time involves some of the most literal adaptations to be found in his work. In these new locking pieces the sculptor looks more closely at the various hinge mechanisms and contact points between bones in a skeletal position. Included between the larger segments are thinner members, shock cushions so to speak, resembling the cartilage wrapped around the contacting surfaces of bones. Earlier in the interview with David Sylvester, Moore mentions the truncated surfaces produced by sawed cross-sections of the discovered bones. These truncated surfaces are a new aspect of an idea already adapted by Arp, Brancusi, and others. In sum, the locking forms are Moore's lexicon of touching surfaces, some close together in fine, barely perceivable seams, while others open, tapering toward rounded raised edges. It is interesting that the vitalism in these later pieces has come full circle through abstraction to a kind of transcendent realism: a realism not too far removed from the realm of the naturalist and classical biologist.

11-3 Time's Profile: John Wesley Powell, Art, and Geology at the Grand Canyon – Elizabeth C. Childs

Elizabeth C. Childs, born in 1954, is a graduate of Columbia, from which she holds two master's degrees and a PhD. She teaches art history at Washington University in St. Louis. Childs' major interests are French nineteenth-century visual culture, art, politics, exoticism, history of photography, and caricature. The daughter of a geologist – Orlo Eckersley Childs – her avocational interest is geology. In the following extract from "Time's Profile," a 1996 article published in the Smithsonian Institution's journal American Art, *she explores the relationship between art, geology, and the American west.*

In the years following the Civil War, science and art came together in the exploration of the American West. Ambitious federal surveys, funded between 1867 and 1878, mapped the nation in the common service of scientific knowledge, military control, and industrial expansion. The four largest surveys were led by George Wheeler, Ferdinand Hayden, Clarence King, and John Wesley Powell. It was the golden age of the geographical and topographic survey, of the collection and inventory of natural specimens, and of the development of new theory about geological process and structure. Photographer and painter were often present with the naturalist at the moment of discovery. Expedition photography struggled to match its technical capacities to the challenge of documenting the western spectacle. Exploration called for both the empiricist, who longed to complete the map of the land, and the theorist, who strove to explain land forms the new maps recorded. Questions raised by the West's unique landscape, as it was revealed in the art of the survey artists, forced the scientific mind into new depths of geological time and inspired a new respect for natural processes.

A successful scientist in this era, such as John Wesley Powell (1834–1902), was many things – a topographer and an explorer, a pragmatist, a leader and a politician, as well as a publicist and an entrepreneur. Powell was a largely self-trained geologist with considerable ambition and a flexible mind. In completing his survey of the Colorado River region in 1873, he claimed to have solved one of the last great mysteries of western geography. The task had required years of travel, observation, and mapping throughout the West. It also required astute promotion of the survey's work back East. To this end, Powell collaborated with draftsmen, photographers, engravers, and a professional painter, all of whom made images that recorded the profiles of geological time visible on the walls of the Grand Canyon. With the success they helped him earn, he was catapulted into a distinguished career as a federal administrator of ethnology and geology. . . .

By the end of the 1872 field season, Powell had hundreds of glass negatives, ample material for the documentation and illustration of his work. Not satisfied with this vast resource, he planned to enhance the visual record in a new way: in November 1872, he asked the professional painter Thomas Moran to join the next survey. Powell was well aware that Moran had accompanied Hayden's survey to the Yellowstone and that Hayden had used Moran's sketches to help persuade Congress to pass the National Park Bill in 1872 (Congress subsequently purchased Moran's spectacular canvas of the Yellowstone chasm for ten thousand dollars). Powell, a highly competitive scientist, clearly hoped to gain a similar commercial and political advantage for his survey. A painter of the Grand Canyon could help legitimize Powell's endeavors by linking fine art with that of elite science. Painting offered public celebration, commemoration, poetic commentary, and the aura of uniqueness in a way that survey photography could not. Painting also opened the doors to elite social and political spaces – gallery shows, museums, the halls of Congress – the domains of the educated and the powerful. The right painter could be useful in less elite spheres as well if he could furnish illustrations for journal articles, a concern that had emerged in Powell's publishing plans as early as the spring of 1872. Powell had everything to gain by bringing along a painter at what was becoming a critical time for the survey.

In the late spring of 1873, Powell was losing financial support. For the first time, he failed to get the full appropriation he had requested from Congress. In fact, he received only ten thousand dollars, the lowest appropriation since his survey had been funded. Perhaps as a result of this setback, he pressed his case even more strenuously with Moran, who made the decision only in late June to travel with

Powell after canceling plans to rejoin Hayden. On 8 July Moran arrived in Utah with Jack Colburn, a writer from New York. The two traveled with Powell's survey for the next seven weeks.

Moran's role was not that of a scientific illustrator. He had generated his own professional backing for the trip to the Colorado River region with commissions for illustrations from *Appleton's*, *Aldine's*, and *Scribner's*. With the geologist Thompson and the photographer Hillers, Moran made excursions into the region known today as Zion National Park and along the Vermilion Cliffs. In the main canyon of the Virgin River in Zion he made many pencil drawings, generally working sketches intended only to record plant life and geological detail. Yet even in such a summarizing study as [his] *Canyon of the Rio Virgen* (1873), a view from the Temple of Sinawava, Moran's vista is well composed, extending beyond a dramatic proscenium of rocks, cottonwood trees, and cacti.

Moran must have spent almost as much effort on his annotations as on the drawing itself. In the lower-right corner, pencil notes record plant names, and elsewhere brief color notes describe the red, white, yellow, and gray faces of the cliffs as well as dramatic surfaces stained by iron. The small notation "twice as large," inscribed in the sky in the center background, suggests how Moran had to struggle on his initial visit to capture the scale of this monumental terrain. Similarly, the penciled arrow that shoots above the cliff face in the left middleground records Moran's realization that he had not rendered the altitude correctly. Other notes describe geological features: at the far right the effects of recent rock falls on the face of the cliffs are documented with "radial sweep of large lines"; elsewhere he observed how the rock had been carved into tapering columns. The sheet is a compendium of quick observations to be recalled in the studio.

Moran's party traveled to Mount Trumbull and then to the Toroweap, a dramatic area in the western reaches of the Grand Canyon. Moran was clearly overwhelmed by his first view of the steep inner gorge, declaring it "the most awfully grand and impressive scene" he had ever beheld. A pencil sketch notes that "where water lines enter the Cañon, they are generally white from lime water from the levels above." This observation exemplifies how Powell's team was teaching Moran to see geological action in the landscape. Further evidence of geological theory appears in Colburn's account of the vista from Toroweap. Admitting his desire to see the landscape in terms of catastrophism, Colburn conceded that the shaping of the land was a product of nature's continuous processes and their slow rate of action:

I think the feeling [here] is one of awe and wonder at the evidence of some mighty inconceivable, unknown power, at sometime terribly, majestically and mysteriously energetic, but now ceased. And yet the force that has wrought so wonderfully through periods unknown, unmeasured, and unmeasurable, is a river 3000 feet below.

From Toroweap, Moran's party joined Powell, and all left on 14 August for the Kaibab Plateau, where they camped until at least 25 August. They hiked onto the large Powell Plateau, which Powell claimed offered the greatest vista in the Grand Canyon – a rare, clear view of the Colorado River winding through the canyon's inner gorge (the river is visible from only a few places along the North Rim). From the crest of the Muav Saddle, Moran descended close to a thousand feet to make sketches. Although probably the farthest he ever went into the Grand Canyon, he was still almost four thousand feet above the river level. He hiked down through the layers of Kaibab and Toroweap limestone into the faulted drainage divide, where massive sandstone towers at the top of the Coconino layer are shown looming above him. These towers frame the vista of a creek rapidly descending along the Muav fault, visible at the center; the winding canyon of the tributary draws our eye to an invisible junction with the Colorado. Moran's pencil notation at the top – "The towers should be much higher" – indicates his ongoing effort to capture an appropriate sense of scale. Following Powell's practice of collecting Pai–Ute names for geographical features of the canyon region, Moran noted on the drawing, "Pai Yuni Turn Pui Wi Neuv, name, The Three Standing Rocks in Muav Canyon."

While at Powell Plateau, the party witnessed a dramatic thunderstorm flashing through the canyon. Colburn described the event:

> Here we beheld one of the most awful scenes upon our globe. . . . A terrific thunderstorm burst over the cañon. The lightning flashed from crag to crag. A thousand streams gathered on the surrounding plains, and dashed down into the depths. . . . The vast chasm which we saw before us . . . was nearly seven thousand feet deep.

Moran captured this storm in a number of ambitious sketches. Using modern topographic maps, one can surmise that he worked at or near what is today called Dutton Point on Powell Plateau. Moran was clearly attentive both to the complexity of the rainstorm and to the terrain: the notation "SFM" on the mountains in the distance identifies the peaks of San Francisco Mountain visible on the horizon. Within the canyon itself, he marked the various strata with the letters R, G, and B for the red, gray, and black of the canyon walls. His rendering of the descending profiles of strata is sufficiently accurate to identify distinct layers. Particular buttes and plateaus within the vista are identifiable as well. The Colorado River winds through the left middleground, while the path of a large tributary in Hakatai Canyon is visible on the right. At the right edge of the page, Moran described a tiny butte as the "Sphynx," an apt name for a rock guardian gazing over a New World valley of geological kings. In so doing, Moran reinforced the survey's practice of naming distinctive rock features; besides Native American names, many literary and classical references were grafted onto the land as part of the act of claiming it for science.

October 1873 found Moran back in New Jersey working up compositional studies for his monumental canvas *The Chasm of the Colorado*. The fate of the painting is well known: in July 1874 a joint committee of Congress decided to purchase it for ten thousand dollars and to hang it beside the Yellowstone canvas in the Senate lobby. Powell considered the picture an extension of his survey's enterprise. That he traveled from Washington, D.C., to Newark at Thanksgiving 1873 and again in April 1874 to see the work in progress in Moran's studio indicates his engagement in its outcome. He may well have seen the picture as a key to reviving the government's lagging support for his survey. If so, he was not alone: in December 1873 Hillers had asked Powell about the status of funds and the progress of "our Grand Canyon picture" by Moran, suggesting the proprietary attitude members of the survey took in the artist's work and the hopes they all shared for its success. Moran was so aware of Powell's eagerness to use the painting as a promotional tool that he cautioned him not to prematurely circulate the photograph of the preliminary underdrawing he promised to send Powell.

In that same letter, Moran wrote to Powell that he had "got our storm in good." He was referring to the storm they had witnessed together from Powell Plateau. Rain was clearly as central to Moran's conception of the Grand Canyon as it was to Powell's understanding of the area's geology. As Moran was painting the scene, Powell was stressing in his official report the importance of rain as the great erosive force in this area: "Though little rain falls [here], that which does is employed in erosion to an extent difficult to appreciate. ... A little shower falls, and the water gathers rapidly into streams, and plunges headlong down the steep slopes, bearing with it loads of sand."

Moran's canvas presents an equally comprehensive view of the cycle of water, linking the sunlit pools in the near foreground to the distant rain clouds. The brightest spots at our feet are the little pools that reflect their source, the sky. The standing water and the deeply saturated color of the rocks both suggest the storm's recent passage. In Moran's painting we step into this natural cycle of rain and erosion. Water falling from an amphitheater, or erosion formation, collects in a tributary that descends into the steep, black gorge in the center foreground. A stream cuts around the cliff to an unseen junction with the Colorado River. The sky pours into the heart of the chasm, the tip of the rainbow pointing to the brightly reflective surfaces of the tiny stream. Everywhere the water dashes into the "thousand streams" described by Colburn. Other details point to the erosional cycle. A small cluster of pines in the left foreground is the victim of a recent forest fire. Their ashen white trunks and the charred wood suggest the stand was recently hit by a bolt of lightning – a common partner to the flash storms in the canyon and, of course, a leading cause of the forest fires that denude the plateaus and hasten the removal of softer rocks by rain.

Powell praised Moran for selecting a point of view from a plateau that had been deeply eroded by a stream. A review of the painting in 1874 confirms that it successfully conveyed Powell's uniformitarian theories of erosion and dismissed old notions of catastrophism: "This strange and wild scene ... is not the work of volcanic action or upheaval force, but simply of the quiet and progressive work of erosion by water – thousands and thousands of years have witnessed the progress of the work."

In this painting Powell's impact on Moran is visible both in the grand scheme of erosion and weathering and in the small geological details. A prominent butte in the left foreground, whose wet cliffs glimmer in the sunlight, is a case in point. The formation appears to be based partly on the Holy Grail Temple, a butte named by Powell that appears in the far left of Moran's pencil sketch of the view from Powell Plateau. The surface of the cliff wall is decidedly anthropomorphic, reminiscent of a classical frieze. One imagines the figures are seated and hold lances or staffs; many seem to be wearing helmets like Olympian gods. Although it has been proposed that the details of this butte pay homage to the presence of Native American ruins in the canyon, the architectural frieze suggested here is decidedly Western, even Hellenic, and not Anasazi. Moran's invention probably reflected Powell's tendency to use classical architectural metaphors (primarily in his journalistic writing, not his scientific prose) to describe cliff features in familiar cultural terms – as the dominion won by exploration.

11-4 Thomas Moran: American Landscape Painter – R.A. Bartlett

Thomas Moran (1837–1926), one of the leading landscape artists of the nineteenth century, was a member of two of the great geological and geographical surveys of the West. He accompanied F.V. Hayden to the Yellowstone and, following an extended visit to Yosemite, was a member of Major John Wesley Powell's expedition to southern Utah and

Arizona. Among his best known landscapes are the Lower Falls of the Yellowstone, Glory of the Canyon, Pictured Rocks of Lake Superior, *and* Lower Geyser Basin. *Partly because of the public exposure his paintings provided and because of the authenticity and excitement of his scenes of the American West, a number of his sketching sites have since been made* National Parks. *In our second extract by Richard Adams Bartlett (see page 69 for the first), from* Great Surveys of the American West *(1962), Bartlett gives us a rare glimpse of Thomas Moran. Bartlett's writings on the American West convey the profound influence of a sense of place upon human experience.*

With Hayden that year as a guest was one of the outstanding landscape painters of the nineteenth century, Thomas Moran. Today, the modernists have tried to forget him. His paintings are in storage in art museum basements. It is said that his attempts to paint the realities of the Grand Canyon of the Yellowstone, the Grand Canyon of the Colorado, and the Mount of the Holy Cross were all failures, and anyway, painting has now passed far beyond attempts merely to convey an accurate scene. Yet in the 1870s, Moran's work was judged among the finest, and Congress twice appropriated $10,000 for the purchase of his western scenes. One of them, of Yellowstone Falls and the Grand Canyon, hung for years in the Capitol. "It captured more than any other painting I know, the color and atmosphere of spectacular nature," said Jackson, himself an artist of some ability. But the paintings deteriorated and, though restored, no longer hang in the Capitol.

If painters ever return to reality, to a respect for the man who can make authentic paintings of authentic subjects, then a reappraisal of Moran may bring him back from the limbo of discarded artists. He boldly tried to convey on canvas the most majestic scenes in the entire American West, and some of his work – his painting of Yellowstone Falls and the Grand Canyon of the Yellowstone, for example – is breath-taking in concept, majesty, and color. Writing to Hayden about the Yellowstone painting early in 1872, he expressed his objectives. Most artists, he said, had considered it next to impossible to make good pictures of strange and wonderful scenes in nature, and the most that could be done with such material was to give it topographical or geological characteristics. "But I have always held that the grandest, most beautiful, or wonderful in nature, would, in capable hands, make the grandest, most beautiful, or wonderful pictures," he wrote, "and that the business of a great painter should be the representation of great scenes in nature. All the above characteristics attach to the Yellowstone region, and if I fail to prove this, I fail to prove myself worthy the name of painter. I cast all my claims to being an artist, into this one picture of the Great Canyon and am willing to abide by the judgement upon it."

Moran was thirty-four years old in 1872, tall, gaunt, and cadaverous, but compassionate, helpful, and level headed. He had never mounted a horse before, but he rode the creatures all day long from the very beginning and accepted his sore posterior as part of his sacrifices for the sake of art. He was one of that rare breed of men that have some knowledge about almost everything. He taught the old camp men how to wrap trout in wet paper, bury them beneath the fire, and uncover them baked and delicious half an hour later. He was interested in photography, and he gave unstintingly of his artistic knowledge, especially with regard to the problems of composition.

11-5   Earth Calling – Diane Ackerman

Diane Ackerman, born in 1948 in Illinois, is a graduate of Pennsylvania State and has two masters degrees and a PhD from Cornell. She has had a rich and varied career, serving as a social worker in New York City, a government researcher, a visiting faculty member at Cornell, Pittsburgh, Columbia, New York University and William and Mary University, *a staff writer for the* New Yorker, *a nature writer and a host to a PBS television series. In both her poetry and her prose, Diane Ackerman explores the profound implications of Nature, arguing that to exclude it is to "bankrupt the experience of living." Our extract is taken from* A Natural History of the Senses *(1990).*

We think of music as an invention, something that fulfills an inner longing; perhaps, to be an integral part of the sounds of nature. But not everyone perceives music in that way. About eighty miles north of Bangkok, in the foothills of Wat Tham Krabok, is a Buddhist temple where a group of concerned monks help drug addicts to recover. They use a combination of herbal therapy, counseling, and vocational training. One of the monks, Phra Charoen, a sixty-one-year-old naturalist by disposition, also busies himself in the music room, where, with electronic equipment, he records the electrical phenomena of the earth, which he then translates into musical notation. Charoen and his team of monks and nuns trace the fluctuating sound patterns onto transparent paper, then transfer the graphs to thin strips of cloth that can be catalogued and rolled up for storage. The graphs match up with the traditional eighteen-bar phrases of Thai music. These "pure melodies" are then played on a Thai instrument with an electronic organ as backup, and the result is recorded. Charoen's group are not musicians themselves, but they believe that music is not an imaginary thing, nor even something produced only by people; music falls out of the earth's rocks and roots, its trees and rain. One western woman wrote that "under the temple trees, with birdsong filling the musical pauses, the visitor sits … and hears the earth of ancient Ayuthaya sing, or the stones of the Grand Palace, the sidewalks of Bangkok – or the cracks in the Hua Lampong Railway Station forecourt."

This would no doubt strike a familiar chord with the American composer Charles Dodge, who, in June and September 1970, recorded "the sun playing on the magnetic field of the earth" by feeding magnetic data for 1961 into a specially programmed computer. The performance has a subtitle – "realizations in computed electronic sound" – and three "scientific associates" are prominently mentioned on the album's cover. The result is at times booming, at times squeaking, but consists mainly of shimmering, cascadingly melodic violin and woodwind sounds. Harmonious and breathy, they often create small flourishes and partial fanfares; they don't seem random at all, but rather energized by what, for lack of a better word, I'll call *entelechy*, that dynamic restlessness working purposefully toward a goal we associate with composed music. I also have a recording of Jupiter's magnetic field, a gift from the TRW corporation to visitors to the Jet Propulsion Laboratory during the encounters of Voyager I and II with Jupiter in 1980. An electric-field detector aboard the spacecraft recorded a stream of ions, the chirping of heated electrons, the vibrating of charged particles, lightning whistling across the planet's atmosphere, all accompanied by an aurora we hear as a hiss. Gas from a volcano on the moon Io adds a tinkling and a banshee-like scream of radio waves. Fascinating as this concert is, and useful to scientists, it doesn't sound like music, nor is it supposed to, but music could easily be woven from or around it. Artists have always looked to nature for their organic forms, and so it's not surprising to find a rather pop-sounding composition called "Pulsar." Over four hundred pulsars are known, at various distances from Earth. Using the recorded rhythmic pulses of once-massive stars about 15,000 light-years away, the composer offers Caribbean-like melodies, in which his "drummer from outer space," as he puts it, supplies percussion. The pulsars are identified on the record sleeve by number – 083–45 on side one and 0329+54 on side two – as if they were indeed side men who sat in on the session. On another occasion, Susumu Ohno, a California geneticist, assigned a different note to each of the four chemical bases in DNA (*do* for cytosine, *re* and *mi* for adenine, *fa* and *sol* for guanine, and *la* and *ti* for thymine) and then played the somewhat limited-sounding result. Our cells vibrate; there is music in them, even if we don't hear it. Different animals hear some frequencies better than we do. Perhaps a mite, lost in the canyon on a crease of skin, hears our cells ringing like a mountain of wind chimes every time we move.

When the earth calls, it rumbles and thunders; it creaks. In towns like Moodus, Connecticut, swarms of small earthquakes rattle the residents for months on end. The seismic center of the quake storm is a very small area only a few hundred yards wide near the north end of town. I'm amazed there haven't been horror films about a devil's sinkhole, or some equal abomination. Ground grumblings of this sort are now called "Moodus noises," but long ago, when the Wangunk Indians chose the area for their powwows because it was there the earth spoke to them, they called the spot Machemoodus, which mean "place of noises," and their myths told how a god made the noises by blowing angrily into a cave. Cluster earthquakes can sound as light as corks popping or as relentless as cavalry charging. "Thunder underfoot" is how some have described it. "It's like you got hit on the bottom of your feet with a sledgehammer," one resident complains. The Moodus quakes are noisier than most because they're shallower (only about a mile deep; quakes along the San Andreas Fault are usually six to nine miles deep). Normal deep quakes lose much of their voice to the ground, which dampens and stills it. It may also be that the earth around Moodus simply conducts sound well. Since the town is located between two nuclear power plants, its residents grow anxious when the quakes rage for months, shifting and cracking the earth and sounding like a chronically rattling pantry.

At the Exploratorium in San Francisco, a pipe organ plays the sounds of San Francisco harbor as tide sloshes through its hollows, ringing with a thick brassy murmur. Now that the Russians and the Americans are planning a joint trip to Mars, I very much hope they'll take a set of panpipes along with them, so perfect for the windswept surface of Mars. Pipes would be an especially good choice because, although every culture on our planet makes music, each culture seems to invent drums and flutes before anything else. Something about the idea of breath or wind entering a piece of wood and filling it roundly with a vital cry – a sound – has captivated us for millennia. It's like the spirit of life playing through the whole length of a person's body. It's as if we could breathe into the trees and make them speak. We hold a branch in our hands, blow into it, and it groans, it sings.

Part 4

The Crowded Planet

Planets, so it turns out, are rather less of a rarity, rather less unusual than we had once supposed. They have been discovered orbiting other stars elsewhere in the cosmos. From 1995 to 2007, the number of known exoplanets – planets outside our solar system – has gone from zero to more than two hundred. But Earth is an inhabited planet. How unusual is that? We do not know, though the search for extraterrestrial intelligence (ETI) continues apace. It may well be that in the vast reaches of space other planets support living things. Whether such "life" would resemble living things familiar to us is an open question. If such extraterrestrial organisms do exist, they may have a basic chemistry and means of life quite unlike anything we know on Earth. The flurry of excitement that greeted NASA's announcement in 1996 of life on Mars has now subsided. A number of observers now discredit the evidence on which it was based.

So Earth, as of now, is the only living planet we have. In fact, the terrestrial environment is such a perfect fit for life, that some use it as an argument for design, in an argument that has become known as the Anthropic Principle, though others regard such "fine tuning" of the laws of nature and terrestrial conditions with animal life and human consciousness as a coincidence, an accident of cosmic history.

Whatever conclusion one reaches on this large question – this question of questions – Earth is not only inhabited: it is crowded. When the first edition of this book appeared in 1981, the Earth's population was 4.5 billion. Twenty-seven years later, it is 6.5 billion, and still rising, though the rate of increase has declined.

Two things follow. First, our dependence on Earth resources remains unchanged, even though the particular resources we utilize may change with time (from tallow to whale oil to coal gas to nuclear-fueled electric energy for lighting, for example). Second, growing population produces not only growing demand for Earth resources, but also growing stress upon the delicate systems and balance of Earth itself. For though Earth is a resilient planet, the intricate equilibrium of its various interacting systems is subject to changes, including long-term cycles, that are presently neither predictable nor controllable.

The chapters that follow reflect these various concerns, not only with respect to human history, but also to continuing resources and conservation.

12.   Human History

After a 400-page review of human history, Pulitzer-prizewinning author Jared Diamond concludes in his 1997 book *Guns, Germs, and Steel* that "the striking differences between the long-term histories of peoples of the different continents have been due not to innate differences in the peoples themselves, but to differences in their environment. I expect that if the populations of Aboriginal Australia and Eurasia could have been interchanged during the late Pleistocene, the original Aboriginal Australians would now be the ones occupying most of the Americas and Australia, as well as Eurasia, while the original Aboriginal Eurasians would be the ones now reduced to downtrodden population fragments in Australia." Jared concludes that four features – land area, population size, differences in native plants and animals, and the geographic barriers to distribution and migration, both within and between continents – determine the broad patterns of human history.

That view of history, though perhaps more determinist than some would accept, recognizes the profound impact of topography, climate, natural resources and local plant and animal life upon the development of local societies and ultimately on human history. Geography has determined why military campaigns and naval battles were conducted. Topographic features and the underlying geology that produces them have influenced the sites of particular battles, the tactics involved, and sometimes the outcome of the conflict. Hadrian's Wall, for example, the great Roman defensive rampart that stretches across the width of northern England, was sited in places on the crest of the steep scarp face of the Great Whin Sill, one of a suite of conspicuous East–West intrusions of quartz dolerite, intruded in late Paleozoic times.

Mineral deposits – silver, gold, iron, coal and petroleum – have precipitated wars, from the time of ancient Greece to the Persian Gulf War of 1991. They have disrupted monetary systems, influenced the rise and fall of empires, fueled new economies (iron ore, coal and water in the Industrial Revolution of Victorian England), stimulated exploration (as silver attracted Spanish explorers to the Americas), and decided the lives, well-being and fate of millions of men and women.

We include extracts which discuss the influence of Earth features on the pattern of architecture, military tactics, the location of cities, and the history of ancient Greece. We also include readings in current issues of significance to longer term trends, including water supply, population growth, crime detection, and atmospheric contamination.

12-1   Minerals and World History – John D. Ridge

John Drew Ridge, born in 1909, was for twenty years the Head of the Department of Mineral Economics at Pennsylvania State University, served on the National Academy of Science's panel on Mineral Economics and Mineral Resources, and authored numerous papers in the fields of economic geology and mineral economics. In his 1967 article "Minerals and World History" from the journal Mineral Industries, *he traces the relationship of mineral deposits to historical affairs from the earliest human, who used a pebble for a weapon, to the impact of the mineral wealth of the United States on the outcome of World War II. The following extract from his article covers a brief part of this span.*

Gold and silver became important for ornamentation and decoration early in the history of mankind, but it was not until much later that a monetary value was attached to them. Perhaps the first real effect one of these precious metals had on the history of the world was made possible by the discovery in 483 B.C. of the rich silver-bearing lead deposits at Lavrion, near the tip of the peninsula that runs southeast from Athens. Although silver had been produced at Lavrion since Solon's time (ca. 550 B.C.), the amounts mined had been small until the discoveries of 483. The subsurface rights to minerals in ancient Greece belonged to the state, and the Athenian government leased mining claims to private operations for a rental of about 5,000 ounces of silver per year, plus 1/25th of the silver produced. In those days, of course, 5,000 ounces of silver would buy many times what it would today; thus the return of the Athenian government was enormous. Aeschylus spoke of Lavrion as a "fountain of running silver," and the decision of the Athenians as to what to do with the income from the mines was fundamental to the future not only of Europe but of the world. ...

After silver began to flow from Lavrion in 483, the Greeks gave thought to the use they should make of the purchasing power it gave to the Grecian state. There were those among the Athenians who wished to use the silver further to beautify the city or to buy more grain from Egypt, but Themistocles convinced his fellow citizens that they should spend the money to increase their fleet so that it might compete successfully with the Persians for the mastery of the Aegean. The outcome of his three years of effort was in the 180 triremes and 9 pentaconters, 60 per cent of the total Greek naval force, that the Athenians contributed to the Greek fleet in 480.

The first clash between the Greek and Persian fleets at Atremisium resulted in a drawn battle. The two armies, the Greeks under the Spartan king, Leonidas, met at the pass of Thermopylae. There, on the second day of Atremisium, largely through treachery and panic on the part of certain of the Greek soldiers, their main forces were overwhelmed from flank and rear, and 4,000 men, including Leonidas himself, were killed by the Persians. The land route through Attica lay open, and the Persians poured down the road toward Athens, taking that city quickly and the Acropolis only after fierce fighting, thus clearing the way to the Isthmus of Corinth. When word of this disaster reached the Greek fleet, the decision was made to retreat south to the narrow waters between the island of Salamis and the Greek mainland northwest of the port of Piraeus. In these narrow waters, under the eyes of Xerxes, who watched from Mount Aeguleos, the Greeks defeated the Persians. In a tactical sense the battle was not decisive, but it cut so deeply into Xerxes' confidence in himself and his troops that he returned to Asia Minor leaving his troops, under Mardonius, to suffer a crushing defeat at Platea (near Thebes) on what probably was August 27, 479 B.C.

In that same summer, the Greek troops landed from their fleet on the beach south of Mycale in Asia Minor, marched overland to that port, took the town, and burned the beached Persian fleet. That winter the European terminus of the bridges of boats over the Hellespont was taken by the Greeks, and the bridges were destroyed. The Golden Age of Greece had begun.

Credit for the defeat of the Persians has been given variously to the Greek commanders, such as Themistocles, Aristeides, and Pausanias, as well as to the Greek armor, the Greek organization and discipline, and, above all, to the Greek fleet. But the greatest commanders and the discipline of the Greeks would have been useless had the Greeks not possessed ships, arms, and armor. And none of these would have been available had not the silver of Lavrion paid for them. In short, it was the flowing fountain of silver, even more than the Hellespont or the passes of Thermopylae or the strait of Salamis or the Isthmus of Corinth, that barred the Persians from Greece. Had not the silver of Lavrion belonged to the Athenians, the world of that day after the year 480 would have been under the power of the Persian autocrats and not of the Greek democrats. Whatever the world is today, had the Persians conquered the Greeks, it would be far different from what it is now. Consequently, our present civilization is based on the ships of Salamis and the arms of Platea, and these, in turn, on the silver of Lavrion, and the sound sense of those – foremost among whom was Themistocles – who chose the wooden walls of their ships against the Pentelicon marble walls of buildings the Acropolis was never to know.

12-2 A Land: Architecture – Jacquetta Hawkes

We first encountered Jacquetta Hawkes (1910–1996) in her discussion of Rodin and Henry Moore (see page 245). In this second extract – also from A Land *(1951) – she discusses the influence of geology upon the architecture of Britain, relating both rock type and age to architectural style and mood in the British Isles. One quotation will illustrate her insight and style: "The dour grey and brown rocks of the Carboniferous Age which are so apt an expression of the stubborn civic pride, the puritanical distrust of elegance and light of our northern industrialists are followed by a return of the warm colors that commemorate the Permian and Triassic Ages, the renewed denudation of mountains in desert and heat."*

Building is one of the activities relating men most directly to their land. Everyone who travels inside Britain knows those sudden changes between region and region, from areas where houses are built of brick or of timber and daub and fields are hedged, to those where houses are of stone and fields enclosed by drystone walling. Everywhere in the ancient mountainous country of the west and north stone is taken for granted; where the sudden appearance of walls instead of hedges catches the eye is along the belt of Jurassic limestones, often sharply delimited. The change is most dramatic in Lincolnshire where the limestone of Lincoln Edge is not more than a few miles wide and the transformation from hedges to the geometrical austerity of drywalling, from the black and white, red and buff of timber and brick to the melting greys of limestone buildings, is extraordinarily abrupt.

The distinctive active qualities of the stones of each geological age and of each region powerfully affect the architecture raised up from them; if those qualities precisely meet particular needs then, of course, the stones are carried out of their own region. Since the eighteenth century the value of special qualities in building material has greatly outweighed the labour of transport, and stones of many kinds have not only been carried about Britain to places far from those where they were originally formed, but have been sent overseas to all parts of the world. Men, in fact, have proved immensely more energetic than rivers or glaciers in transporting and mixing the surface deposits of the planet.

Now the process has gone too far; what was admirable when it concerned only the transport of the finest materials to build the greatest buildings has become damnable when dictated by commercial expediency. The cheapness of modern haulage has blurred the clear outlines of locality in this as in all other ways; slate roofs appear among Norfolk reed beds, red brick and tile in the heart of stone country, while cities weigh down the land with huge masses of stone, brick, iron, steel, and artificial marble dragged indiscriminately from far and near.

Nevertheless, there are still regional differences that will hardly disappear. Britain would sink below the sea before a Yorkshireman would buy Scottish granite to build his town hall, or an Aberdonian outrage his granite city with a bank of Millstone Grit. The danger is that Britain will not sink below the sea, but simply into a new form of undifferentiated chaos, when both Yorkshireman and Scot adopt artificial stone and chromium hung on boxes of steel and concrete.

While, on the one hand, it is admitted that even in the twentieth century regional differences still persist, it would, on the other, be false to suggest that even when all transport was by wind or muscle stone was not sometimes moved about the country. If the Blue Stones of Stonehenge are the most startling prehistoric instance, for the early Middle Ages it is the importation of Caen stone. Very many cargoes of this oolitic limestone were shipped from Normandy to build our abbeys and cathedrals. Often it was ordered by the great Norman clerics who, in a hostile land, found reassurance in building with their native rock. The genes of the Norman conquerors are now mingled with those of most of our royal and noble families, and through them also Caen stone has been incorporated in our most sacred national buildings – old St. Paul's, Canterbury Cathedral, Westminster Abbey.

It was of course most usually for ecclesiastical buildings and for castles that stone was shipped and carted about Britain, particularly to those youthful parts of lowland England south-east of the Jurassic belt. The material for Ely Cathedral and other great East Anglian churches came from Barnack in Northamptonshire, as did that for Barnwell, Romsey and Thorney abbeys and many of the early college buildings at Cambridge. The lower courses of King's Chapel at Cambridge, the foundation stone laid by Henry VI, came from the Permian Limestone of Yorkshire, while, after the long interruption in building, the upper courses were constructed of Jurassic stone from Northamptonshire, the personal gift to the college of Henry VII. The fan vaulting of the roof, however, those exquisite artificial stalactites, is again carved from a Permian deposit – the noted Roche Abbey quarries in Yorkshire. So in one building the Permian and Jurassic ages, the north and the Midlands have been made tributary to royal and scholastic pride, the service of God and the imagination of man. I have brought in these facts far from their proper place, to suggest a truth which is perhaps too obvious to need such attention. That the center of gravity of a people in any age may be expected to be found in the objects for which they will transport great quantities of building material. Neolithic communities hauled megalithic blocks to their communal tombs, Bronze Age men did the same for their temples, the Iron Age Celts amassed materials for their tribal strongholds, the Romans for their military works and public buildings; medieval society sweated for its churches, colleges and castles.

12-3   The Geologic and Topographic Setting of Cities – Donald F. Eschman and Melvin G. Marcus

Historians are captivated by those areas where people gather in large numbers, our cities. For it is in population centers where many events which changed the course of humankind occurred and where there is a definitive record of the ebb and flow of civilizations. The site of cities is most often determined by the nature of the topography and also to some extent by the rock type present. These and allied geologic factors also often control the development and ultimate size of population centers. In this extract from the 1972 book Urbanization and Environment edited by Detwyler and Marcus, the authors Donald F. Eschman and Melvin G. Marcus describe the direct relationship of geology to cities. Although the examples cited are for the most part in North America, London, Tokyo, Cairo, Tehran and Rio de Janeiro would have served equally well. Eschman is professor Emeritus of Geology at the University of Michigan and Marcus (1929–1997) was for many years Professor of Geography at Arizona State University.

The geologic and topographic setting of cities plays a major role in their location and growth. First, the physical landscape is a major factor in the initial selection of sites for settlement. Second, topography and landforms strongly influence the early growth and development of settlements, particularly the evolution of their spatial pattern. Last, even though in these days man's technology allows him to move mountains, the economic costs of overcoming geologic and geomorphologic factors continue to impose directional and aerial constraints on urbanization. Thus, although the greatest impact of physical environment on human activities such as urbanization may be found in the historical past or in less developed cultures today, the basic landscape on which cities are situated continues to play an important role in modern urbanization.

This chapter considers the geologic landscape on which cities are built. Because the criteria by which sites for settlement are selected are extremely complex, attention is first directed to the multitude of locational factors in order to place the physical factors in their proper perspective. Basic topographic and geologic questions are then considered. They involve a variety of interactions between cities and their topography, geomorphic processes, and geologic composition. The feedback between human activities and the parent landscape is critical, and therefore examples are given throughout the chapter.

Cities are not simply where you find them. They are located in response to a complex set of interacting processes and forces that encompass a range of factors extending well beyond those presented by the physical landscape. The larger the city, the more complicated are the economic, political, and social factors that influence its location and growth. It is important to recognize, however, that man has only recently acquired the ability to bring a sophisticated technology to bear on the selection and development of his settlements. Although in the long run the survival and growth of urban places must depend upon economic and social factors, success is seldom achieved if rational decisions regarding physical locations are ignored. ...

The parent material and topography on which cities are built are major site factors, although other elements such as water resources, land-water boundaries, and climate commonly are important. In the case of New York City, for example, the original European settlement was on Manhattan Island, then an area of wooded, bedrock hills, interspersed with low-lying marshes and tidal flats. The island's drainage pattern and shoreline were qualities of site. As the population grew and more space was required for human activities, the site was altered by draining and filling the marshes. The dramatic expansion that New York City experienced, however, was not a simple response to its initial site characteristics. Rather, its broader physical situation – expressed in such features as the Hudson River, good harbors, and access to the interior of North America (via the Hudson River and the Mohawk Lowlands) – provided a physical setting favorable to the growth of industry and commerce. The human situation was even more significant in that it provided the social, economic, and political needs and potentials that allowed a great city to evolve upon and spread from the initial site. In short, the city required a physical base, but it was the broader relationships of that site to human activities and distant landscapes that proved most significant in the city's evolution.

The growth of every city can, in large part, be explained in terms of human determinants. It has been suggested by some urban geographers and sociologists that there is little point in attempts to classify the geological and topographical attributes of urban places, because each city responds to a unique set of environmental and human conditions. According to them, generalized explanations of urbanization that are based on a physical typology must inevitably be fruitless exercises that do not address the major social and economic processes at work. Though these arguments are in some part true, the fact remains that cities are built on earth materials that have topographic expression – and the rational use of this physical base is profitable to man. Conversely, topography and geology may place constraints on human activities that require an expenditure of time, money, and effort to overcome.

Historically, man has sought city locations that provide: (1) access to good water transportation or overland routes; (2) protection from natural hazards such as floods, storms, and landslides; (3) security from enemies; (4) water supply; (5) building materials, fuel, and other usable resources; and (6) a stable base for construction. Other factors, some of which may provide commercial and industrial advantages, such as water power, natural breaks in transportation routes, local food sources, and of course historical accident, are also important. Nearly all of these settlement criteria are dependent on geologic and topographic conditions.

The settlement history of North America clearly reflects the need to satisfy some or all of these environmental conditions. New York, Providence, and Boston are good examples of settlements built at the site of a well-protected harbor, while Baltimore and Charleston, South Carolina, are examples of settlements bordering a large estuary (resulting from the drowning of a river valley). In each of these locations, protection from storm waves is provided by deep coastal indentations and offshore islands – landforms created by the recent post-glacial rise of sea level relative to the land. It is interesting that 22 of the world's 32 largest cities are located on estuaries. . . .

Once a settlement has been established, its continued growth commonly is influenced by the geologic and topographic attributes of its site. In an urbanizing area environmental factors such as steep slopes, poorly drained ground, flowing or standing bodies of water, and aesthetic characteristics constrain or bar development of certain areas. Patterns of urban expansion may develop a bias in a particular direction or favor certain landforms; transportation can be similarly affected. Barriers are most apparent during a city's expansion phases, and considerable technological and economic effort is required to overcome them. As long as the least cost and effort can be achieved by building the city on favorable physical locations, problem terrain is avoided. Eventually, if population pressure and economic and political circumstances demand it, urbanization will spread to less desirable landscapes.

It has been economically sensible for cities to expand, especially in their early stages of growth, along the lines of least topographic and geologic resistance. In areas of relatively flat or rolling, well-drained land, cities tend to spread rather evenly in all directions. Examples abound in the Great Plains and Central Lowlands of North America, where only rivers and lakes commonly present major barriers to growth. Houston, Texas, and Saskatoon, Saskatchewan, are good examples of cities where topographic resistance is low and the cities have expanded without interruption and about equally in all directions from the CBD's (central business districts). Chicago and Detroit have sites where only water has been a major obstacle to urban growth; flat, open land has otherwise allowed quite uniform expansion into the rural countryside. In fact, once the lakefront effect has been accounted for in Chicago, the flat terrain so nearly approaches an isotropic surface that that city has become a major focus of studies which attempt to explain patterns of urbanization by various theoretical models. For example, the classical urban studies of Ernest Burgess (1925) and Homer Hoyt (1939) focused on Chicago.

In some regions geologic structure presents formidable obstacles to urban growth. The dramatic relief between ridges and valleys in the Appalachian Mountains has clearly influenced patterns of human settlement. Cities located along the valley floors, on water courses and overland routes, tend to spread longitudinally. Rosary-like strings of towns accompany many rivers, leaving sparsely populated spaces on intervening slopes and ridges. The urban area of Bluefield-South Bluefield, West Virginia, in the Appalachians illustrates the elongation that accompanies urban growth in this region.

The urban pattern of Los Angeles, California, is another example of the constraint that topography can place on urban growth. This city, which has experienced phenomenal population growth in the last half century, exemplifies urban man's determination to subjugate the environment. The population boom seemingly forced the city to expand from the lowlands of the Los Angeles basin onto nearby slopes and hills. The city has become an impressive testomial to man's technological ability to overwhelm his environment – but only if one blindly ignores the risks from landslides, earthquakes, and storms. Even in Los Angeles, though, some steep mountainous areas (those that continue to present physical and economic barriers to urbanization) remain unsettled.

Hills and slopes are not the only kinds of geologic resistance to urban growth. Low, poorly drained areas, such as flood plains, marshes, and tidal flats, also present problems. In the New York metropolitan region, for example, some 15,000 acres (of an original 30,000 acres) of the water-saturated open lands called Hackensack Meadows remain relatively unsettled and unused in the midst of densely populated and industrialized land. Despite dramatic economic pressures, this land is only slowly being drained, filled, and turned to human use. Areas made suitable for some construction by landfilling are not suitable for all types of constructions. Some low-lying fill zones of Manhattan Island, for example, cannot support the skyscrapers that rise on firm bedrock in midtown and lower Manhattan. For many years Chicago was forced to expand horizontally because a means could not be found to soundly support high-rise buildings on the unconsolidated sediments along the lake shore. In recent years this problem has been solved, and vertical growth has become a major feature of Chicago's urbanization.

12-4   Topography and Strategy in the War – Douglas W. Johnson

Douglas Wilson Johnson (1878–1944) was born in West Virginia, studied at Denison, New Mexico, and Columbia, and later taught geomorphology at MIT, Harvard, and Columbia. He made major contributions to the study of shoreline processes and the surface history of the Appalachians. In 1917 he published a study of Topography and Strategy in the War – *from which our extract is taken – and next year joined the United States Army as a major in the Intelligence Division. Written at the height of World War I, the*

passions of the time are evident in the introductory portion of this excerpt. In this book, Johnson discussed the profound influence of topography on World War I. In the present extract, he is concerned with the Paris Basin and the options enforced on defenders and attackers alike. Terrain in western Europe is such that the flat lands of the Netherlands and Belgium invite an attacker from the east to approach Paris from the north. This was true in both of the European conflicts of the twentieth century.

The violation of Belgian neutrality was predetermined by events which took place in western Europe several million years ago. Long ages before man appeared on the world stage Nature was fashioning the scenery which was not merely to serve as a setting for the European drama, but was, in fact, to guide the current of the play into blackest tragedy. Had the land of Belgium been raised a few hundred feet higher above the sea, or had the rock layers of northeastern France not been given their uniform downward slope toward the west, Germany would not have been tempted to commit one of the most revolting crimes of history and Belgium would not have been crucified by her barbarous enemy.

For it was, in the last analysis, the geological features of western Europe which determined the general plan of campaign against France and the detailed movements of the invading armies. Military operations are controlled by a variety of factors, some of them economic, some strategic, others political in character. But many of these in turn have their ultimate basis in the physical features of the region involved, while the direct control of topography upon troop movements in profoundly important. Geological history had favored Belgium and northern France with valuable deposits of coal and iron which the ambitious Teuton coveted. At the same time it had so fashioned the topography of these two areas as to insure the invasion of France through Belgium by a power which placed "military necessity" above every consideration of morality and humanity. The surface configuration of western Europe is the key to events in this theater of war; and he who would understand the epoch-making happenings of the last few years cannot ignore the geography of the region in which those events transpired.

What is now the country of northern France was in time long past a part of the sea. When the sea bottom deposits were upraised to form land, the horizontal layers were unequally elevated. Around the margins the uplift was greatest, thus giving to the region the form of a gigantic saucer or basin. Because Paris today occupies the center of this basin-like structure, it is known to geologists and geographers as "the Paris basin."

Since the basin was formed it has suffered extensive erosion from rain and rivers. In the central area where the rocks are flat, winding river trenches, like those of the

Aisne, Marne, and Seine, are cut from three to five hundred feet below the flat upland surface. To the east and northeast, the gently upturned margin of the basin exposes alternate layers of hard and soft rocks. As one would naturally expect, soft layers like shales have readily been eroded to form broad flat-floored lowlands, like the Woevre district east of Verdun. The harder limestone and chalk beds are not worn so low, and form parallel belts of plateaus, the "côtes" of the French. . . .

The fact that the rock layers dip toward the center of the basin has one striking result of profound military importance. Every plateau belt is bordered on one side by a steep, irregular escarpment, representing the eroded edge of a hard rock layer; while the other side is a gentle slope having about the same inclination as the dip of the beds. The steep face is uniformly toward Germany, the gentle back-slope toward Paris; and the crest of the steep scrap always overlooks one of the broad, flat lowlands to the eastward. The military consequences arising from this peculiar topography will readily appear. It is not difficult to understand why the plateau belts have long been called "the natural defenses of Paris." . . .

Descending the face of the Argonne scarp and crossing the valley of the Aire River, you continue eastward across a minor plateau strip and reach the winding trench of the Meuse. Past immortal Verdun and its outlying forts, you press on to the crest of the next great scarp. What a view here meets the eye! To the north and south stretches the long belt of plateau, cut into parallel ridges by east- and west-flowing streams, – ridges like the Côte du Poivre, whose history is written in the blood of brave men. Below, to the east, lies the flat plain of the Woevre, whose impervious clay soil holds the water on the surface to form marshes and bogs without number. Here the hosts of Prussian militarism fairly tested the strength of the natural defenses of Paris, and suffered disastrous defeat. Moving westward under the hurricane of steel hurled upon them from above, their manoeuvering in the marshes of the plain easily visible to the observant enemy on the crest, the invading armies assaulted the escarpment again and again in fruitless endeavors to capture the plateau. Only at the south where the plateau belt is narrower and the scarp broken down by erosion did the Germans secure a prevarious foothold, thereby forming the St. Mihiel salient; while at the north entering by the oblique gateway cut by the Meuse River, they pushed south on either side of the valley only to meet an equally disastrous check at the hands of the French intrenched on the east-and-west cross ridges. Viewing the battlefields from your vantage point on the plateau crest, you read a new meaning in the Battle of Verdun. You comprehend the full significance of the well-known fact that it was not the artificial fortifications which saved the city. It was the defenses erected by Nature against an enemy from the east, skilfully utilized by the heroic armies of France in making good their battle cry, "They shall not pass." The fortified cities of Verdun and Toul merely defend the two main gateways through this most important escarpment, the river gateway at Verdun being carved by the oblique course of the Meuse, while the famous "Gap of Toul" was cut by a former tributary of the upper Meuse, long ago deflected to join the Moselle. Other fortifications along the crest of the scarp add their measure of strength to the natural barrier.

Once more you resume your eastward progress, traverse the marshy and blood-soaked plain of the Woevre, ascend the gentle back slope of still another plateau belt, and stand at last on the crest of the fifth escarpment. Topographically this is the outermost line of the natural defenses of Paris, and as such might be claimed on geological grounds as the property of France. But since the war of 1870 the northern part of this barrier has been in the hands of Germany, who purposed in 1914 to widen the breach already made in her neighbor's lines of defense. Metz guards a gateway cut obliquely into the scarp, and connects with the Woevre through the Rupt de Mad and other valleys.

Farther south is Nancy, marking the entrance to a double gateway through the same scarp. Here in the first week of September, 1914, under the eyes of the Kaiser, the German armies, moving southward from Metz, where they were already in possession of the natural barrier, attempted to capture the Nancy gateways and the plateau crest to the north and south. Once again the natural strength of the position was better than the Kaiser's best. From the Grand Couronné, as the wooded crest of the escarpment is called, the missiles of death rained down upon the exposed positions of the assaulting legions. The Nancy gateway was saved, and more than three years from that date is still secure in the hands of the French. The test of bitter experience has fully demonstrated to the invading Germans that it was no idle fancy which named the east-facing scarps of northern France "the natural defenses of Paris."

12-5 Geology and Crime – John McPhee

John McPhee's article "The Gravel Page," taken from The New Yorker *of January 29, 1996, describes the use of geologic studies* *in crime detection. For an introduction to McPhee and his work see page 4.*

F.B.I. geologists look first at color, and then at texture. Next they wash the soil and do the mineralogy. They collect "alibi samples," "alibi soils." If you have said that the mud on your skirt came from your back yard, they will collect soil from your yard to prove that it did or did not. About half the work of the Materials Analysis Unit has to do with geology. The rest has to do with things like glass and paint. Bruce Hall, the special agent who is the unit's chief in Washington, points out that forensic geology is broader than the name implies, because it includes chemistry and physics as well. It also includes, in growing numbers, people who testify about environmental impacts and the causal aspects of landslides. For my purposes here, the topic remains concentrated in military puzzles and egregious crimes.

Hall once spent a couple of days on Staten Island collecting alibi samples after a "soldier" in the Bonanno crime family put five dismembered bodies in several graves with a shovel. The shovel was found with bits of soil on its kick plate. Hall collected alibi samples from every place on Staten Island to which an alibi had been – or might be – ascribed. He matched the mineralogy from the bits of soil on the shovel to the mineralogy of the gravesites. With equal care, he unmatched everything else. "You've got to be right every time," he says. "There's no being wrong once in a while." . . .

Some years ago, I asked Chris Fiedler, an F.B.I. geologist, if he could think of a case in which the relevant rock had come in a size class larger than mud, silt, or sand. He remembered a time when the F.B.I. was investigating a group of potential terrorists thought to be moving explosives from safe house to safe house in Eastern states. After a suspect vehicle passed through southern New Jersey, a large rock was observed near an intersection of roads. The F.B.I. wondered if the rock was a marker. They reasoned that the potential terrorists would assume that they were being followed and would use a lead car that was free of incriminating cargo. The lead car would set large rocks in predetermined places to inform a following car that things were so far so safe.

Large rocks are about as common in South Jersey as bent grass on the swells of the ocean. South Jersey, above bedrock, is fifteen thousand vertical feet of unconsolidated marine sand. This rock was out of place, erratic, alien. The F.B.I. took it to Washington. It was a garnet schist, and in the schist was the mineral staurolite in a form sufficiently unusual to eliminate a lot of territory. A metamorphic petrologist at the Smithsonian Institution thin-sectioned the rock and determined that it came from a definable area in the highlands of western Connecticut. F.B.I. agents went there and began asking questions under the outcrops of garnet schist. They found the hideaway they were looking for. In it was evidence that ultimately led them to a safe house in Pennsylvania, full of explosives. The road had been circuitous, and that was the end of the line.

12-6   Tambora and Krakatau – Kenneth E.F. Watt

The eruption in April, 1815 of the volcano Tambora in what was then the Dutch East Indies, and the asking price in shillings for a sack of flour in the commodities market of London in December, 1817 seem to be entirely unrelated events. Yet, in this article written for The Saturday Review *in 1972, Kenneth E.F. Watt documents climatic cooling caused by the addition of solid particles of volcanic origin to the atmosphere and demonstrates the limiting effects on food production. He concludes his short thesis with sobering data on the rate at which the human population is increasing, particulate loading of the atmosphere, and what the complications might be if an eruption comparable to Tambora or Krakatau were to occur in the next 50 years. Watt was born in 1929, received his PhD in Zoology from the University of Chicago, and is presently Professor Emeritus of Zoology at the University of California, Davis. In addition to numerous articles, he has published more than a half a dozen technical and popular books, with his research concentrating on mathematical models of ecosystems and societies.*

You have every reason to be confused as to the possible effects of pollution on the weather. Some scientists have predicted that increasing global air pollution will reduce the amount of solar energy penetrating the atmosphere. This, they say, will cause a decline in the global average air temperature. Other scientists say the earth's weather will warm up. This, it is explained, will come about because of the increase in carbon dioxide concentration in the atmosphere. Since carbon dioxide is a strong absorber of infrared radiation, it will act like the glass roof that prevents loss of heat energy from inside a greenhouse – the infamous "greenhouse effect" – and trap radiation in the atmosphere.

Three types of arguments can be put forth to settle the matter as to which mechanism will be most important. One approach is to argue from mathematical theories of climate determination, using computer-simulation studies. This line of argument is not completely trustworthy, because of the embryonic state of such theories. A second line of argument, pursued by Reid Bryson and Wayne Wendland at the University of Wisconsin, is to discover by statistical analysis the relative strengths of the two processes during recent decades. They found that the impact of the temperature-lowering mechanism will override the impact of the temperature-raising mechanism. Since the temperature has in fact been dropping for more than two decades, their argument is compelling.

However, a third line of argument opens up a most fascinating area of research, both in interpretation of history and in climatological prediction. This argument asserts that the most realistic means of assessing the effect of air pollution is to find some gigantic natural event in the past that operated in a way analogous to modern pollution. Fortunately, we are provided with such historical "experiments" by records of a few particularly gigantic and explosive volcanic eruptions.

A close reading of old newspapers gives evidence that people in the early nineteenth century were not aware of any relationship between unusual weather and volcanoes. To my knowledge, most ancient civilizations thought that the earth's weather was almost totally determined by astronomical phenomena, and this belief explains the spectacular development of observational astronomy in those civilizations. The first man known to have recognized that volcanoes could affect the weather was Benjamin Franklin in 1784, following a series of enormous eruptions in Japan and Iceland the previous year. During the twentieth century many climatologists and geophysicists have come to suspect that volcanoes influenced weather and climate and, more recently, that they had effects similar to the global increase in atmospheric pollution.

There are two obvious sources of information on the impact volcanic eruptions could have on weather and hence on crop production. One is the tables of past weather data recently constructed by historical climatologists such as Gordon Manley and H.H. Lamb. The other is old newspapers, which yield detailed information on local weather, crop growing, planting and harvesting conditions, and agricultural commodity prices.

Consider the effect of the eruption of Tambora in Sumbawa, the Dutch East Indies, in April of 1815. This was the largest known volcanic eruption in recorded history: over the period 1811 to 1818 an estimated 220 million metric tons of fine ash were ejected into the stratosphere. Of this, 150 million metric tons were added to the stratospheric load in 1815 alone, mostly in April. Krakatau, which is much better known because it erupted more recently (1883), ejected only 50 million metric tons of ash into the global stratosphere. It is now known that the ash from certain of these very large and explosive volcanic eruptions spreads worldwide in a few weeks and does not sink out of the atmosphere totally until a few years have elapsed. During all of this time the ash is back-scattering incoming solar radiation outward into space and consequently chilling the earth.

The magnitude of the chilling can be ascertained from Manley's tables for central England. The Tambora volcano erupted most explosively in April of 1815; by November the average temperature of central England had dropped 4.5°F. The following twenty-four months was one of the coldest times in English history. Specifically, 1816 was one of the four coldest years in the period 1698 to 1957; the coldest July in the 259-year period was in 1816; October of 1817 was the second coldest October; May of 1817 was the third coldest May. However, these figures convey little sense of the impact of such chilling on society. For that we must turn to the newspapers of the time. All of the following quotations are from *Evans and Ruffy's Farmers' Journal and Agricultural Advertiser*.

"We had fine mild weather until about the 20th, when it set in cold, with winds at East and North-East, with partial frosts; these together have greatly retarded the operations in Agriculture, and very many cannot purchase seed corn, so that thousands of acres will pass over untilled, and sales of farming stock, and other processes in law, drive many of this useful class in society into a state of despondency. ... The wheats, late sown ... have been partially injured by the frosty mornings. .... Sheep and lambs have suffered from the severity and variableness of the weather. . . . The doing away the Income Tax, and the war duty on Malt, will afford some relief but are wholly insufficient in themselves to restore this country to its former state of happiness and prosperity." (April 8, 1816).

"From about the 9th or 10th of this month, we have never had a day without rain more or less, sometimes two or three days of successive rain with thunder storms. The hay is very much injured; a considerable part of it must have laid on the ground upwards of a fortnight. . . . Wheat is looking as well as can be expected, considering the deficiency of plants in the ground, and those very weakly . . . but still far short of an average crop." (August 12, 1816).

"Throughout the whole month the air has been extremely cold; there has not been more than two or three warm days, being at other times rather cloudy and dark, and the sun seldom seen. The Oats . . . on high situated ground . . . are the most backward and miserable crop ever seen . . . for the greater part of the Wheat, where the mildew did not strike, has been very much affected by the rust or canker in the head. . . . There have been many seizures for rent this month, and many a farmer brought to nothing, and we hear of very few gentlemen who are inclined to lower their farms as yet; it seems they are determined to see the end." (September 9, 1816).

The preceding quotations all refer to agricultural conditions in England. To indicate that this was a worldwide rather than a local phenomenon, the following quotation, from a letter printed in the same newspaper, suggests the state of affairs in other countries.

"Last year was an uncommon one, both in America and Europe: We had frosts in Pennsylvania every month the year through, a circumstance altogether without example. The crops were generally scant, the Indian Corn particularly bad, and frost bitten; the crops, in the fall and in the spring, greatly injured by a grub, called the cutworm. . . ." (November 10, 1817).

Thus, in the case of a volcano, which puts an immense load of pollution into the atmosphere suddenly – unlike modern pollution, which builds up gradually – we have a gigantic experiment, the effects of which can be clearly traced throughout all social and economic systems. For example, the volcano of 1815 had a clear-cut influence on world agricultural-commodity markets, as one would expect from the preceding descriptions of the consequences of weather deterioration on crops. Many measures of market conditions could be used to make this point, but the one I have selected is the highest asking price for best-quality flour within a month of trading on the London commodities market. The following table shows how this price changed, from the period before the volcano, to the peak price in June of 1817, to the normal price, which was finally reached again by the end of December 1818.

Highest Asking Price for a Sack of Flour (in Shillings)

Year	June	December
1814	65	65
1815	65	58
1816	75	105
1817	120	80
1818	70	70*

(*65 by end of month)

The table indicates how volcanic eruptions can serve as the basis for interdisciplinary research, in which a pulse due to a physical event can be tracked through biological, social, economic, and political phenomena.

For example, an interesting feature of this table is the long lag between the time the pollution was introduced into the upper atmosphere and the time that the price elevation was at its peak (April 1815 to June 1817 – twenty-six months). These time lags are characteristic of complex systems and indicate why cause and effect are often not connected in peoples' minds: by the time the effect has occurred, everyone has forgotten the cause.

Could the gradual increase in worldwide air pollution concentrations become serious enough to bring crop production to a halt in high latitudes? To answer this question, we must consider the rate of build-up in pollution now and the likelihood that political power to enforce adequate pollution control will materialize.

The worldwide stratospheric particulate loading due to Krakatau and Tambora was 50 and 220 million metric tons respectively. If worldwide man-caused stratospheric particulate loading continues to build up at recently prevailing rates, . . . by about 2018, the permanent particulate load in the stratosphere would be equal to that produced temporarily by Krakatau, and by about 2039 the permanent load would be equal to that produced by Tambora.

Will they continue? The reader must make his own judgments. There is ample printed evidence testifying to the difficulty of requiring automobile manufacturers to conform to the limits required by the Clean Air Act of 1975. Also, a casual reading of industry journals does not suggest that industry would like to arrest the rate of increase in sales of oil, gas, coal, or other polluting substances. Computer-simulation studies of trends in the use of fossil fuels do not indicate any significant lessening of the rate of increase in pollution prior to about 2004 unless stringent controls are introduced. Thus, without a really massive political and social change of a type that does not seem likely, judging from present attitudes, the quotations from English farm newspapers in 1816 and 1817 may well be read as a scenario for the future.

There are further complications if these are not enough. What if man continues to build up the pollution load in the atmosphere, and then a volcano of the order of Tambora adds still more pollution to the atmosphere? How likely is this to happen? The answer is: very likely. The period since 1835 has been remarkably free of major volcanic eruptions of the explosive, Vesuvian type. But over the historical record an average of five very large volcanoes has occurred each century. Luckily for us, the modern period of great technological activity has been free from major volcanic eruptions, to an almost historically unique extent. But at some time our luck may run out.

Another complication is what is going on in the minds of people. The world population of 1816 had no idea that there was any relationship between the phenomenally bad weather they were experiencing and air pollution. Will we be any wiser? Can we change in time? Will we be able to take the necessary action to ensure a brighter end to the scenario for civilization? Simple extrapolation of present trends does not lead to an encouraging answer, but history teaches us that sudden, surprising changes in political and social attitudes do occur. Perhaps it will occur once more – in time.

12-7 Mortgaging the Old Homestead – Lord Ritchie-Calder

Lord Ritchie-Calder of Bamashannar (1906–1982) was born in Scotland, served during World War II in the British Foreign Office as director of plans for political warfare, and had a distinguished career in journalism, international relations, and public life. He was also a prolific author, writing some thirty books, many of which dealt with social, scientific, ecological or medical topics. Our extract from "Mortgaging the Old Homestead" is taken from the journal Foreign Affairs (1970). Lord Ritchie-Calder argues that the power and carelessness of modern technology pose a threat to the terrestrial homestead we have inherited. Scientific specialization leads to a lack of both consultation and comprehension. This extract concerns an irrigation scheme in Pakistan, and illustrates that major mistakes can result, even from the best of intentions. Those and other hazards are compounded by the continuing population explosion.

Past civilizations are buried in the graveyards of their own mistakes, but as each died of its greed, its carelessness or its effeteness another took its place. That was because such civilizations took their character from a locality or region. Today ours is a global civilization; it is not bounded by the Tigris and the Euphrates nor even the Hellespont and the Indus; it is the whole world. Its planet has shrunk to a neighborhood round which a man-made satellite can patrol sixteen times a day, riding the gravitational fences of Man's family estate. It is a community so interdependent that our mistakes are exaggerated on a world scale.

For the first time in history, Man has the power of veto over the evolution of his own species through a nuclear holocaust. The overkill is enough to wipe out every man, woman and child on earth, together with our fellow lodgers, the animals, the birds and the insects, and to reduce our planet to a radioactive wilderness. Or the Doomsday Machine could be replaced by the Doomsday Bug. By gene-manipulation and man-made mutations, it is possible to produce, or generate, a disease against which there would be no natural immunity; by "generate" is meant that even if the perpetrators inoculated themselves protectively, the disease in spreading round the world could assume a virulence of its own and involve them too. When a British bacteriologist died of the bug he had invented, a distinguished scientist said, "Thank God he didn't sneeze; he could have started a pandemic against which there would have been no immunity."

Modern Man can outboast the Ancients, who in the arrogance of their material achievements built pyramids as the gravestones of their civilizations. We can blast our pyramids into space to orbit through all eternity round a planet which perished by our neglect.

A hundred years ago Claude Bernard, the famous French physiologist, enjoined his colleagues, "True science teaches us to doubt and in ignorance to refrain." What he meant was that the scientist must proceed from one tested foothold to the next (like going into a mine-field with a mine-detector). Today we are using the biosphere, the living space, as an experimental laboratory. When the mad scientist of fiction blows himself and his laboratory skyhigh, that is all right, but when scientists and decision-makers act out of ignorance and pretend that it is knowledge, they are putting the whole world in hazard. Anyway, science at best is not wisdom; it is knowledge, while wisdom is knowledge tempered with judgment. Because of over-specialization, most scientists are disabled from exercising judgment beyond their own sphere. . . .

In the Indus alley in West Pakistan, the population is increasing at the rate of ten more mouths to be fed every five minutes. In that same five minutes in that same place, an acre of land is being lost through water-logging and salinity. This is the largest irrigated region in the world. Twenty-three million acres are artificially watered by canals. The Indus and its tributaries, the Jhelum, the Chenab, the Ravi, the Reas and the Sutlej, created the alluvial plains of the Punjab and the Sind. In the nineteenth century, the British began a big program of farm development in lands which were fertile but had low rainfall. Barrages and distribution canals were constructed. One thing which, for economy's sake, was not done was to line the canals. In the early days, this genuinely did not matter. The water was being spread from the Indus into a thirsty plain and if it soaked in so much the better. The system also depended on what was called "inland delta drainage," that is to say, the water spreads out like a delta and then drains itself back into the river. After independence, Pakistan, with external aid, started vigorously to extend the Indus irrigation. The experts all said the soil was good and would produce abundantly once it got the distributed water. There were plenty of experts, but they all overlooked one thing – the hydrological imperatives. The incline from Lahore to the Rann of Kutch – 700 miles – is a foot a mile, a quite inadequate drainage gradient. So as more and more barrages and more and more lateral canals were built, the water was not draining back into the Indus. Some 40 percent of the water in the unlined canals seeped underground, and in a network of 40,000 miles of canals that is a lot of water. The result was that the watertable rose. Low-lying areas became waterlogged, drowning the roots of the crops. In other areas the water crept upwards, leaching salts which accumulated in the surface layers, poisoning the crops. At the same time the irrigation régime, which used just 1½ inches of water a year in the fields, did not sluice out those salts but added, through evaporation, its own salts. The result was tragically spectacular. In flying over large tracts of this area one would imagine that it was an Arctic landscape because the white crust of salt glistens like snow.

The situation was deteriorating so rapidly that President Ayub appealed in person to President Kennedy, who sent out a high-powered mission which encompassed twenty disciplines. This was backed by the computers at Harvard. The answers were pretty grim. It would take twenty years and $2 billion to repair the damage – more than it cost to create the installations that did the damage. It would mean using vertical drainage to bring up the water and use it for irrigation, and also to sluice out the salt in the surface soil. If those twenty scientific disciplines had been brought together in the first instance it would not have happened. . . .

Always and everywhere we come back to the problem of population – more people to make more mistakes, more people to be the victims of the mistakes of others, more people to suffer Hell upon Earth. It is appalling to hear people complacently talking about the population explosion as though it belonged to the future, or world hunger as though it were threatening, when hundreds of millions can testify that it is already here – swear it with panting breath. . . .

For years the Greek architect Doxiadis has been warning us about such prospects. In his Ecumenopolis – World City – one urban area like confluent ulcers would ooze into the next. The East Side of World City would have as its High Street the Eurasian Highway stretching from Glasgow to Bangkok, with the Channel Tunnel as its subway and a built-up area all the way. On the West Side of World City, divided not by the tracks but by the Atlantic, the pattern is already emerging, or rather, merging. Americans already talk about Boswash, the urban development of a built-up area stretching from Boston to Washington; and on the West Coast, apart from Los Angeles, sprawling into the desert, the realtors are already slurring one city into another all along the Pacific Coast from the Mexican Border to San Francisco. ...

The danger of prediction is that experts and men of affairs are likely to plan for the predicted trends and confirm these trends. "Prognosis" is something different from "prediction." An intelligent doctor having diagnosed your symptoms and examined your condition does not say (except in novelettes), "You have six months to live." An intelligent doctor says, "Frankly, your condition is serious. Unless you do so-and-so, and I do so-and-so, it is bound to deteriorate." The operative phrase is "do so-and-so." We don't have to plan for trends; if they are socially undesirable our duty is to plan away from them; to treat the symptoms before they become malignant.

We have to do this on the local, the national and the international scale, through intergovernmental action, because there are no frontiers in present-day pollution and destruction of the biosphere. Mankind shares a common habitat. We have mortgaged the old homestead and nature is liable to foreclose.

12-8 Breathing the Future and the Past – Harlow Shapley

Harlow Shapley (1885–1972) was born in Missouri, graduated from the University of Missouri, and pursued graduate work in astronomy at Princeton. After several years at the Mount Wilson Observatory, Shapley went to Harvard, where he became director of the Harvard Observatory, a position he held for more than 30 years. An astronomer of substantial achievement, whose work was widely recognized and honored – he pioneered methods for distance measurements that suggested the scale of the Milky Way and discovered the clustering of galaxies – he was also a superb lecturer and a gifted writer, who did much to popularize interest in astronomy. Our extract is taken from his book Beyond The Observatory *(1967).*

Thanks to the mobility of gasses, the winds at the earth's surface keep the atmosphere thoroughly stirred up and in motion. Here today, gone tomorrow, are the atoms that we are now breathing. ...

Since about 1 percent of your breath is argon we can determine approximately the number of atoms in your next argonic intake. The calculations are really rather simple and straightforward, but to some readers this dizzy arithmetic is repulsive and I shall simply state the results. In your next determined effort to get oxygen to your lungs and tissues – that is, in your next breath – you are taking in, besides the nitrogen and oxygen, 30,000,000,000,000,000,000 atoms of argon; in briefer statement 3×10^{19}. (Count the zeros!) A few seconds later you exhale those argon atoms. ...

Now let us follow the career of one argon-rich breath – your next exhalation, let us suppose. We shall call it Breath X. It quickly spreads. Its argon, exhaled this morning, by nightfall is all over the neighborhood. In a week it is distributed all over the country; in a month, it is in all places where winds blow and gases diffuse. By the end of the year, the 3×10^{19} argon atoms of Breath X will be smoothly distributed throughout all the free air of the earth. You will then be breathing some of those same atoms again. A day's breathing a year from now, wherever you are on the earth's surface, will include at least 15 of the argon atoms of today's Breath X.

This rebreathing of the argon atoms of past breaths, your own and others', has some picturesque implications. The argon atoms associate us, by an airy bond, with the past and the future. For instance, if you are more than twenty years old you have inhaled more than 100 million breaths, each with its appalling number of argon atoms. You contribute so many argon atoms to the atmospheric bank on which we all draw, that the first little gasp of every baby born on earth a year ago contained argon atoms that you have since breathed. And it is a grim fact that you have also contributed a bit to the last gasp of the perishing.

Every saint and every sinner of earlier days, and every common man and common beast, have put argon atoms into the general atmospheric treasury. Your next breath will contain more than 400,000 of the argon atoms that Gandhi breathed in his long life. Argon atoms are here from the conversations at the Last Supper, from the arguments of diplomats at Yalta, and from the recitations of the classic poets. We have argon from the sighs and pledges of ancient lovers, from the battle cries at Waterloo, even from last year's argonic output by the writer of these lines, who personally has had already more than 300 million breathing experiences. Our next breaths, yours and mine, will sample the snorts, sights, bellows, shrieks, cheers, and spoken prayers of the prehistoric and historic past.

There was a time when very little argon existed in the earth's atmosphere, and practically no free oxygen at all. That was some billions of years ago. The oxygen has been built up to its present abundance on the earth by the breathing of green plants, and the argon, over the millennia, has grown to its present percentage as the result of the radioactivity of one of the isotopes of the potassium atoms of the rocks. That radioactive decay steadily goes on. In 5 billion years our atmosphere will contain about twice as much argon as it does now.

There ought to be a moral to this story of argon. It tells us of the dramatic smallness of the units of matter. It reminds us of the turbulence in the healthful gaseous envelope which we call our atmosphere. It associates us intimately with the past and the future. It argues that to live long and naturally we want to have in our atmosphere only salutary atoms – oxygen, nitrogen, and argon. We do not want to have this mixture corrupted, polluted, poisoned. Therefore we do not want to have anywhere in our atmosphere the man-made atoms of strontium-90, iodine-131, and similar artifacts produced by ingenious but short-sighted man.

The moral could be: Respect your breath! Keep it decent!

13.   Resources

We are star dust, made of Earth and Earth itself is made of star stuff. We rely on Earth, its airy, watery envelope and its products for our existence, moment by moment and day by day. And throughout human history, we have steadily broadened our dependence on Earth's materials, using a growing range of metals and minerals and exploiting fossil fuels as the basis of industrial development and a continuing technological revolution. Industrialized society runs on Earth's materials. Limit those, and our prospects are bleak.

And that is the challenge we face. The cheap energy – based on coal, natural gas and oil – which has driven our industrial growth is likely to become less cheap as we face the depletion of existing petroleum reserves and the challenge of discovering new sources of supply. Coal, though plentiful, does not yet provide an alternative clean fuel. So we are faced with the need for both conservation and the search for acceptable alternative energy sources. Oil shales, tar sands, coal, solar energy, hydroelectric power and nuclear energy may all be more widely used in the future.

We face a similar situation in the case of some other raw materials. The metals, minerals, chemicals and other materials on which we depend are unevenly distributed in the Earth's crust and exist, for the most part, in low concentrations. Tin, for example, has an abundance of only 0.00015 percent by weight in the Earth's crust, lead 0.00010 percent, and gold 0.0000002 percent. Workable mineral deposits require natural concentration. By contrast, other materials are much more abundant: aluminum has a comparable concentration of 8.00 percent, iron 5.8 percent, and magnesium 2.8 percent. So a workable deposit is a natural concentration of a mineral that may be extracted at a reasonable economic cost. The search for new mineral deposits is an essential continuing activity, as is the search for substitute materials.

It may be argued that recycling will solve all these problems. Certainly, it will help, but recycling, too, involves energy use. Of course, ultimately all the materials we use are recycled into the Earth itself, but the recycling time for many is long, measured not in years, or even centuries, but in aeons.

Although we do not face the short term depletion of most major materials, we do face rising costs of the energy needed to extract them, rising global population, and the rapidly growing industrialization – and thus growing demands – of the developing world. Add to this the environmental impact of increasing industrialization, and it is clear that national and international energy resources and conservation policies are an urgent priority. It is these themes that our extracts explore.

13-1 Wealth from the Salt Seas – Rachel L. Carson

Rachel Louise Carson (1907–1964) was a marine biologist who combined her professional knowledge with a transparent love of nature and a profound sense of stewardship for its protection. Her most successful book, Silent Spring *(1962), not only sold more than half a million copies, but – in warning of the dangers of the indiscriminate use of chemical pesticides – did much to launch the environmental movement of the late 1960s. Our present extract is taken from her book* The Sea Around Us *(1951), which won the John Burroughs Medal and the National Book Award. A brief biography and an assessment of Rachel Carson's influence is given in the next chapter in Gabriele Kass-Simon's essay (see page 293).*

The ocean is the earth's greatest storehouse of minerals. In a single cubic mile of sea water there are, on the average, 166 million tons of dissolved salts, and in all the ocean waters of the earth there are about 50 quadrillion tons. And it is in the nature of things for this quantity to be gradually increasing over the millennia, for although the earth is constantly shifting her component materials from place to place, the heaviest movements are forever seaward.

It has been assumed that the first seas were only faintly saline and that their saltiness has been growing over the eons of time. For the primary source of the ocean's salt is the rocky mantle of the continents. When those first rains came – the centuries-long rains that fell from the heavy clouds enveloping the young earth – they began the processes of wearing away the rocks and carrying their contained minerals to the sea. The annual flow of water seaward is believed to be about 6500 cubic miles, this inflow of river water adding to the ocean several billion tons of salts.

It is a curious fact that there is little similarity between the chemical composition of river water and that of sea water. The various elements are present in entirely different proportions. The rivers bring in four times as much calcium as chloride, for example, yet in the ocean the proportions are strongly reversed – 46 times as much chloride as calcium. An important reason for the difference is that immense amounts of calcium salts are constantly being withdrawn from the sea water by marine animals and are used for building shells and skeletons – for the microscopic shells that house the foraminifera, for the massive structures of the coral reefs, and for the shells of oysters and clams and other mollusks. Another reason is the precipitation of calcium from sea water. There is a striking difference, too, in the silicon content of river and sea water – about 500 per cent greater in rivers than in the sea. The silica is required by diatoms to make their shells, and so the immense quantities brought in by rivers are largely utilized by these ubiquitous plants of the sea. Often there are exceptionally heavy growths of diatoms off the mouths of rivers. Because of the enormous total chemical requirements of all the fauna and flora of the sea, only a small part of the salts annually brought in by rivers goes to increasing the quantity of dissolved minerals in the water. The inequalities of chemical make-up are further reduced by reactions that are set in motion immediately the fresh water is discharged into the sea, and by the enormous disparities of volume between the incoming fresh water and the ocean.

There are other agencies by which minerals are added to the sea – from obscure sources buried deep within the earth. From every volcano chlorine and other gases escape into the atmosphere and are carried down in rain onto the surface of land and sea. Volcanic ash and rock bring up other materials. And all the submarine volcanoes, discharging through unseen craters directly into the sea, pour in boron, chlorine, sulphur, and iodine.

All this is a one-way flow of minerals to the sea. Only to a very limited extent is there any return of salts to the land. We attempt to recover some of them directly by chemical extraction and mining, and indirectly by harvesting the sea's plants and animals. There is another way, in the long, recurring cycles of the earth, by which the sea itself gives back to the land what it has received. This happens when the ocean waters rise over the lands, deposit their sediments, and at last withdraw, leaving over the continent another layer of sedimentary rocks. These contain some of the water and salts of the sea. But it is only a temporary loan of minerals to the land and the return payment begins at once by way of the old, familiar channels – rain, erosion, run-off to the rivers, transport to the sea.

There are other curious little exchanges of materials between sea and land. While the process of evaporation, which raises water vapor into the air, leaves most of the salts behind, a surprising amount of salt does intrude itself into the atmosphere and rides long distances on the wind. The so called 'cyclic salt' is picked up by the winds from the spray of a rough, cresting sea or breaking surf and is blown inland, then brought down in rain and returned by rivers to the ocean. These tiny invisible particles of sea salt drifting in the atmosphere are, in fact, one of the many forms of atmospheric nuclei around which raindrops form. Areas nearest the sea, in general, receive the most salt. Published figures have listed 24 to 36 pounds per acre per year for England and more than 100 pounds for British Guiana. But the most astounding example of long-distance, large-scale transport of cyclic salts is furnished by Sambhar Salt Lake in northern India. It receives 3000 tons of salt a year, carried to it on the hot dry monsoons of summer from the sea, 400 miles away.

The plants and animals of the sea are very much better chemists than men, and so far our own efforts to extract the mineral wealth of the sea have been feeble compared with those of lower forms of life. They have been able to find and to utilize elements present in such minute traces that human chemists could not detect their presence until, very recently, highly refined methods of spectroscopic analysis were developed.

We did not know, for example, that vanadium occurred in the sea until it was discovered in the blood of certain sluggish and sedentary sea creatures, the holothurians (of which sea cucumbers are an example) and the ascidians. Relatively huge quantities of cobalt are extracted by lobsters and mussels, and nickel is utilized by various mollusks, yet it is only within recent years that we have been able to recover even traces of these elements. Copper is recoverable only as about a hundredth part in a million of sea water, yet it helps to constitute the life blood of lobsters, entering into their respiratory pigments as iron does into human blood.

In contrast to the accomplishments of invertebrate chemists, we have so far had only limited success in extracting sea salts in quantities we can use for commercial purposes, despite their prodigious quantity and considerable variety. We have recovered about fifty of the known elements by chemical analysis, and shall perhaps find that all the others are there, when we can develop proper methods to discover them. Five salts predominate and are present in fixed proportions. As we would expect, sodium chloride is by far the most abundant, making up 77.8 per cent of the total salts; magnesium chloride follows, with 10.9 per cent; then magnesium sulphate, 4.7 per cent; calcium sulphate, 3.6 per cent; and potassium sulphate, 2.5 per cent. All others combined make up the remaining .5 per cent.

13-2 Minerals, People, and the Future – Charles F. Park, Jr

Charles Frederick Park, Jr (1903–1990) was born in Wilmington, Delaware, educated at the New Mexico Institute of Mining and Technology, and the Universities of Arizona and Minnesota. During a long career, he worked as a mining geologist, a member of the United States Geological Survey, and as a consultant to the United States Army Corps of Engineers, as well as later serving as Dean of the School of Mineral Sciences at Stanford. His book Affluence in Jeopardy (1968) was one of the most influential books of the late sixties. In it, he argued that exploding world population, growing aspirations of the citizens of the less industrialized nations, and dwindling and non-renewable resources converge to threaten present levels and styles of living. In the present extract, Park reviews the situation, assesses the probabilities of obtaining substitutes and the cost factors, and calls for a threefold program of population control, national mineral policies, and new trading patterns.

"Never look back," advised the philosopher Satchel Paige, also known in the realm of baseball, "something may be gaining on you."

But this philosophical pitcher did not mean that he would ignore the scoring threat of a runner behind him on second base. To the contrary, his record shows that he worked on the future danger in the batter's box with his strategy founded in part on what had already occurred.

As citizens of today face the problems (such as wars, civil unrest, and air pollution) that are lined up in front of us, claiming our immediate attention, we have another problem that is catching up on us from behind, so to speak, because it is largely unobserved and unheeded – the peril of mineral shortages. . . .

Can the earth afford the world's affluence if the philosophy and technology of our society are extended? Can the earth even continue to afford the affluence that exists today? May these questions be dismissed with the assurance that substitutes will be found as they are required, or that the oceans will supply enough minerals for the future, so that our economies may continue to expand? Or must we realize that there are limits to the supplies of minerals that form the basis of a modern industrial civilization? Is affluence in jeopardy?

A few mineral commodities, iron and aluminum for example, exist in sufficient quantities to maintain the industries of the world and to permit considerable expansion. A few other commodities, such as mercury, tin, and silver, are already pressed to maintain their current production and markets; their prices to consumers are rising and, as a consequence, substitutes are being sought (and not always being found).

Between these two extremes lie the great majority of mineral products. They are present in the earth's crust in amounts sufficient to satisfy current demands and to allow for moderate expansion. They cannot, however, over a long period of time keep abreast of the rapid increase in world population. . . . This is especially true of the fossil fuels that provide so much of the energy required by our civilization. . . .

While some substitutions will be possible, general substitutions will not be – substitutes cannot fill all our needs of essential minerals. . . .

And, after all, any substitution must come from the earth's mineral supplies . . .

The mineral industries have made remarkable progress in meeting their needs for the cheaper extraction of lower-grade ores. Still, we cannot look forward to the indefinite utilization of lower and lower grades of any ore. The iron and steel industry, for example, has received the benefit of technological development in the form of pelletizing, which has made economic the mining of lower grades of iron ore. In addition, iron is a common mineral – anyone may take a ton of soil or rock from his back yard and find iron in it. It is doubtful, however, that anyone could afford to extract a pound of iron from a ton of heterogeneous backyard mixture, and to process it, for 7½ cents. If a ton of material contains too little iron, not even the use of pellets can make the extraction of the iron economic at that price.

Although technology is continually improving, the discovery of new deposits and the mining of ever lower grades of ore, both of which must in the future turn to greater depths underground or to the oceans, will reach a point where they are uneconomic. Shortages of essential minerals can come, not only from the ultimate physical exhaustion of the minerals, but also from their unavailability at reasonable cost.

When we speak of costs, we talk in terms of money – of dollars and cents if we are in the United States. But such terms are only symbols of the real cost – that of the energy used. Money is energy, and energy, as has been demonstrated, is one more commodity that is limited in its supply.

Mineral resources and cheap energy are not available in unlimited quantities. Since there are limits to the supplies of minerals that form the basis of a modern industrial civilization, the affluence of our society is in jeopardy. If we are to preserve it, we must answer two questions: How can sufficient quantities of raw materials be obtained over long periods of time? How can they be equitably distributed? In trying to answer these two questions, we must ask a third: How can the emerging nations, which produce large amounts of the world's minerals and energy, be encouraged to develop and to improve their standards of living?

The future of our affluent civilization depends upon the answers to all three of those questions. Since they are so important, their answers cannot come only from a few educated politicians (and one shrinks from thinking of the answers that would come from *uneducated* politicians); it is a knowledgeable public that must everywhere provide the answers. Political leaders must indeed, for the future welfare of the world, be informed about minerals, and so must economists ... The more all citizens know about minerals, however, and the better they recognize their dependence on minerals, the greater the hope that people will in the future have the minerals they need.

[The following are] three recommendations to make in attempting to solve the problem of mineral shortages in the future.

First, the rate of population growth must be controlled. Without such control, all efforts to improve the lot of peoples in the underdeveloped nations, and even to maintain the status quo, can at best achieve only temporary and partial success. Shortages of fertilizers and foods, of energy, and of mineral products of all kinds are inevitable; poverty and starvation are bound to increase. With such conditions at their sources of supply, how will the industrialized nations, with growing needs and growing problems from their own overpopulation, obtain the raw materials that are the basis of their affluence? ...

Second, every nation must formulate a carefully thought out and clearly stated mineral policy that is suited to its particular needs and that recognizes the necessity for international cooperation. Such a policy must not be entrusted to an entrenched, unenlightened, and arbitrary bureau. Such a policy must be under constant review, since the place of minerals in the world economy is seldom static. ...

The emerging nations need help from the industrialized nations to develop the mineral resources which can raise their standards of living. They need foreign capital and foreign personnel to provide the money and the skills required for exploration and for the building and maintaining of mines, mills, and smelters. If they permit foreign capital to earn a fair return, the result will be mutual advantage. If, on the other hand, they overtax foreign investment or resort to shortsighted nationalistic measures of expropriation, they will drive away what they most need – money and trained personnel – to develop their mineral resources and thus to develop their civilization. . . .

Third, the United States should explore the possibilities of establishing – and joining – a common market that might include all peoples in the Western Hemisphere. This would assure the Latin American nations of a market for their raw materials and of a supply of the manufactured articles they need and give them an opportunity to develop the industries for which they are best suited, while this would assure the United States of a source of raw materials for the future and an expanding market for many of its finished products.

13-3   The Bingham Canyon Pit – M. Dane Picard

M. Dane Picard was born in 1927, educated at the University of Wyoming and Princeton, and is now Professor Emeritus in the Department of Geology and Geophysics at the University of Utah. In addition to technical writings in sedimentary petrology and depositional environments, he frequently publishes material for the general public on geologic themes, including Mountains and Minerals, Rivers and Rocks: A Geologist's Notes from the Field *(1993), dozens of essays and articles, and over forty poems in literary journals. In 1997, The National Association of Geoscience Teachers gave him its James Shea Award for outstanding writing about geoscience for the general public and, in 1998, The American Geological Institute honored him with its Award for Outstanding Contribution to the Public Understanding of Geology. In our extract, taken from* Journal of Geoscience Education *(1999), he describes the Bingham Canyon copper mine – the largest human-made excavation on Earth – and located about 25 miles from Salt Lake City, Utah. In 2000, the USA produced 1.45 million tons of copper, about 11% of the world total. The Bingham Canyon mine, owned by Kennecott Utah Copper, has produced more copper than any other mine in history, and presently produces about 20% of the USA's copper.*

At Bingham Canyon most of the rock that is mined is waste. It could be returned to the open pit, but not before the end of mining, because mining would be seriously hampered by the presence of waste. It is probably unreasonable to consider filling the pit after all the mining has been completed, except possibly with municipal waste. Proposing this option would probably give the Utah Department of Health an advanced case of the vapors.

     — Heinrich D. Holland and Ulrich Petersen, 1995, in *Living Dangerously*

Since an autumn afternoon in the early 1950s when I first saw the Bingham Canyon copper pit about twenty miles southwest of Salt Lake City, and walked out to the edge of the rim above the steeply terraced slopes, and stared into Paleozoic and Tertiary worlds that for millions of years had lain undisturbed, but whose rumpled and broken beds were disturbed forever by the mining, the image of the pit has remained in my inner eye.

All of us in the Salt Lake Valley are very much aware of the waste rock and the pit. Bingham is the only place in the world where such an enormous hole in the ground lies so near a large city. A mountain once stood at Bingham; when mining commenced it was called "The Hill." It was at "The Hill" that the brilliant metallurgical engineer Daniel C. Jackling saw the future; open-pit mining and mass production of low-grade porphyry copper ores was possible and could be profitable.

Four centuries ago, copper ores typically held about eight percent metal. Now the average grade of ore mined is less than one percent (at Bingham about 0.6 percent). With the drop in grade, at least thirteen times as much ore (probably a lot more) must be processed to remove the same amount of copper. J.E. Young estimates that 900 million tons of copper ore were mined in the world to produce about 9 million tons of copper in 1990. [Recently], Kennecott applied for a revision of its air-quality permit allowing them to increase yearly production by 31 percent.

The Bingham mine is the word's largest man-made hole, a half-mile deep and nearly two-and-a-half miles across. At the bottom of the pit, the elevation is about the same as the elevation of the Salt Lake valley floor.

The size of the hole and the rock removed from it – more than ten times the amount of material moved for the Panama Canal – are too improbable to be easily contemplated. Bingham is one of only a few man-made features visible from space. It is the great size of the operation that is most worrisome and the past and future effects of mining and smelting on the surroundings and groundwater.

One George B. Ogilvie, while logging in Bingham Canyon during the early part of September in 1863, found the lead-sulfide galena (PbS). He reported the discovery to Colonel Patrick E. Connor, commander of the Third California Infantry, stationed at Camp Douglas on Salt Lake City's eastern foothills. Ogilvie, and the twenty-five others in a picnic and prospecting party organized by Connor, many of them ex-miners, staked on September 17th the first mining claim in the Territory. The mustached, handsome, and dashing Connor organized the area as a mining district.

The mining commenced ninety-three years ago, in 1906, the year of the April 18th San Francisco earthquake. . . . By 1912, the mines at Bingham employed 5,000 workers, 80 percent of them foreign born – Albanians, Armenians, Austrians, Basques, Bulgarians, Chinese, Croats, Czechs, Danes, Dutch, Finns, French, Germans, Greeks. . . . Compared with other work, the rate of accidents and deaths was very high. Frequently, when a miner was hurt, he was not attended to immediately and might lie crushed for hours. As in most mines at the time, working conditions were wretched.

Through the years the miners have removed more than 5 billion tons of Bingham rock. The 1906 miners and all of those connected to the first years of mining are dead. The people now involved with Bingham and working for Kennecott will all become part of exalted or Satanic worlds sometime in the twenty-first century, while mining will still go on at Bingham.

For decades the Bingham hole and the rock removed from it have been a palpable presence. On a clear winter evening, I step onto the deck to feed and water rosy finches that stop by, and the piles of waste rock across the valley shadow the valley floor. In the spring and summer, I walk in the foothills and climb the flanks of the Wasatch Range, and I imagine the bull's-eye of the pit staring uncomprehendingly heavenward.

13-4   The Geological Attitude – John G.C.M. Fuller

John Fuller's article, "The Geological Attitude," from The American Association of Petroleum Geologists Bulletin *(1971), analyzes the relationships between geological activity, industrial and scientific developments, and social pressures. Fuller, a geologist with the Amoco Canada Petroleum Company when* he wrote this article, draws on the history of geology, both early and modern, to clarify the "consumption–pollution and exploitation–injury" dilemma facing our society. The present extract analyzes the problems facing the petroleum industry in the closing years of the twentieth century.

Consider the history of American petroleum geology. Stated simply, drilling for petroleum in Pennsylvania during the 1860s was an outcome of overhunting the whales whose oil served as lamp oil during the first half of the century. Kerosene provided a substitute for whale oil. And although petroleum at that time was already known to yield kerosene, a second competitive answer to whale-oil shortage was found in oil distilled from cannel coal and bituminous shale. The technology of oil-well drilling and casing moreover was also in existence during the first half of the century, having been developed in response to the demand for brine suitable for salt production in the territory west of the Alleghenies.

Petroleum geology thus presents a peculiar feature: the industry possessed a working technology before the geological facts of its commodity were known, and as a consequence its operational and theoretical conceptions diverged wildly.

Forty or fifty years passed in Pennsylvania and the Mid-Continent area before geological principles relating to the origin and accumulation of oil – in particular the "anticlinal" theory – found their way into industrial practice. The reason was quite simply that science was not needed to make a profit from sweet oil lying in huge quantity near the surface of the ground. Drilling mania and greed sustained the momentum until the early 1900s when, by coincidence, wildcat prospecting which had spread southward into Oklahoma encountered proof of mappable geological control over the sites of the oil fields. From that point, between 1913 and 1915, the formal organization of geology in the major American oil companies began. That is one reason why the headquarters of this Association of petroleum geologists, founded in 1917, is in Oklahoma, at Tulsa, and not in Pennsylvania, at Titusville, where the oil industry originated.

Yet it is a fact that an integrated geological conception of the origin of petroleum in black shales and its concentration into oil pools, and of the link between coal rank and indigenous petroleum, went back 50 years to Henry D. Rogers, the first State Geologist of Pennsylvania. Rogers published his findings in 1863, only four years after Drake's discovery well was drilled at Titusville.

> That the volatile hydrocarbons were distilled, as it were, from out the low-lying carbonaceous strata, into the pores and fissures of the over-resting ones, receives strong confirmation from the fact that the elsewhere bituminous shales of the Silurian and Devonian ages, deep under the coal, are altogether as much desiccated and debituminized everywhere in the districts contiguous to the anthracites as the coal beds themselves.

As he wrote this passage did Rogers see in his mind's eye the workings of a refiner's retort, distilling lamp oil from a charge of heated shale? Within a few years such models for metamorphism of organic matter in sedimentary rocks had disappeared from America.

Geology in the petroleum industry serves one function only: to provide eyes in the ground. And all geological activity that is of any use whatever to the industry is directed toward this function. Consequently, petroleum geology clusters round the inventions and innovations which, quite literally, see oil.

The man of action in organized petroleum geology in the United States after the First World War undoubtedly was Everette DeGolyer. What the plane table could do on the surface, a remote sensor could do underground. He succeeded first in Gulf Coast salt-dome terrane, in 1922, with Eötvös torsion balances. Three years later, by converting the principles of German acoustical-seismic apparatus to electrical instrumentation, he assembled what was to be the most effective physical sensing device ever used in geology. Its effect on structural prospecting was immediate and explosive.

The Schlumbergers' invention of "electrical coring" and its introduction to the United States in 1929, later provided the means of detecting by down-hole physical attributes alone the existence of unsuspected closure to potential reservoirs. From this sprang a vogue for "stratigraphic traps."

But the main problem in exploration is now no longer one of simply locating favorable structure or favorable reservoirs, for the fact that there exists only a finite number of pools in a finite amount of continental rock increasingly involves statistical conceptions of hydrocarbon potential and sedimentary volume in the exploratory process. . . .

Concepts generally thought of as belonging to "conservation" are now finding their way into exploration, particularly on the continental shelves, for geologists and the public alike are beginning to visualize the total inventory of prospective rock that extends under the sea to the edge of the continents as final sources of supply, and questions about the meaning of "ownership" are in the newspapers. This is to be expected because social pressure on the industry exceeds anything that it has experienced before. . . .

But the fact remains that the industry faces an immediate necessity of finding and producing unprecedented amounts of petroleum to meet the foreseeable demand: and it is at this point that the petroleum geologist can be confused by the options presented in a pollution-conservation dilemma, of which the essential issues are not at all clear. One can treat the two main options separately.

First, petroleum accounts for something close to three quarters of world energy demand, most of which is concentrated in specific areas of very high consumption. Waste products of this high consumption visibly pollute the environment. Each one of us, for example, produces about 20 times as much pollution by consumption of nonrenewable resources as a so-called underdeveloped person. Problems of this kind involve the petroleum geologist to the extent that he is part of a community which is consuming more energy in its lifetime than the sum total consumed by all preceding generations. Pressure is on the geologist to find oil and gas to satisfy the predicted rates of future consumption, but at the same time he must express his concern for the maintenance of the quality of life in his surroundings, and act on his concerns if he sees any deterioration.

The second part of the dilemma is this: petroleum exploration has now reached areas of the continental shelf and the remote Arctic that were formerly closed through lack of a technology that could operate in those places. The marine and Arctic environments, particularly the latter, are in their own ways peculiarly prone to damage by operations which elsewhere might seem quite normal and inoffensive.

The petroleum geologist, if he has any mind for conservation principles at all, is involved to the degree that the natural environment of the ground in which he hopes to find petroleum is itself a resource, the use of which cannot be simply regarded as a matter of capture, or even of prorated exhaustion.

These alternatives, which we can summarize as consumption-pollution and exploitation-injury are to my mind the horns of the dilemma. The geologist can escape from the dilemma if he wishes by saying that the problems are mainly of recovery, carriage, and use, rather than problems of exploration. But if he does this without first analyzing the whole geological scene he can find himself confused in aim, and very likely suffering from what K. H. Crandall called "motivation mortality." I think many of us suffer bouts of it from time to time, and it is something, as I mentioned before, that the educational and informational resources of the Association can help to correct. Quite a lot depends on how far we align ourselves with the trend that Linn Hoover noted "toward a more humanistic view of the geosciences," because this trend, if it continues for the next few years, he wrote, "may have as much impact as a new scientific concept." That really is the crux of the pollution-conservation problem. It is a lighted fuse.

13-5 Geology – For Human Needs – Michel T. Halbouty

Michel T. Halbouty (1909–2004), "quintessential wildcatter," was born in Beaumont, Texas and attended Texas A & M where he earned three degrees. Widely recognized as an outstandingly successful petroleum geologist, he was involved in the discovery of more than fifty oil and gas fields. He served as an infantry officer and later as Chief of the Petroleum Production Section of the Army–Navy Petroleum Board during World War II and also acted as chairman of President Reagan's Energy Transition Team. He was the author of more than 250 geological articles and several books and was President of the American Association of Petroleum Geologists. In this extract from his article "Geology – For Human Needs" from The Journal of Geological Education (1967), Halbouty discusses the role of geology in ministering to human needs, ranging from the burgeoning demand for raw materials and fuels on the one hand, to water supply and environmental health on the other. Analyzing the growing responsibilities of the geologist in meeting human needs, Halbouty invites enlistment in the "army of scientists" who constitute one of the most vital and essential of the sciences.

Every man who has entered the field of science did so in order to serve his fellow man. The science of geology especially serves mankind, which is the reason that the title "Geology – For Human Needs" was chosen for this article.

Our science, which is based on the ancient studies of astronomy, biology, chemistry, mathematics, and physics, deals with a variety of subjects. Among other things, geology deals with the distribution of natural resources such as ores, rocks and gems, weather and climates of the past, the origin and evolution of animals and plants, history of the rocks, and even the origin of man. Geology is a curious and fascinating mixture of many sciences, yet it is distinct from all others.

It has been the responsibility of geologists to explore for all types of minerals, metals, precious stones, and gems and to examine and evaluate the waters and soils of the earth. In the field of energy supply our profession is responsible for discoveries of raw energy fuels which are so important to the progress and prosperity of this nation and to the remainder of the world. A new age of human progress has been opened in nuclear energy alone, based on results of geological exploration for fissionable materials, such as uranium.

Petroleum geology employs the largest number of geologists, but before a geologist is competent to lead in the hydrocarbon exploration effort he first must be well-grounded in the broad spectrum of science. In other words, a person is a *geologist* first, and a specialist in petroleum geology, second. The petroleum geologist is called upon continually to explore in new provinces, new regions, and new countries. There is no limit to the scientific effort of the petroleum geologist in searching for world-wide deposits of oil and gas to supply energy for the world's needs.

As the space program progresses we are becoming more involved in studies of other planets of the universe. Recently astronaut-geologists have been selected to become members of the space exploration teams of the future. A new branch of geology called astrogeology has developed as a result of our interest in space.

As communications improve, civilization spreads and creates innumerable contacts between once isolated communities. If the American standard of living should spread to only half the world in the next century, it is certain that known supplies of raw materials would be used up far more rapidly than new deposits could be found. All of the scientific potential now available for the exploration for the required new mineral reserves could hardly satisfy the demand.

To meet the growing demand for raw materials we are slowly but surely turning to the seas. The seas make up one of the planet's richest ecological units and they have reservoirs of resources scarcely touched at present. The oceans will absorb an increasing proportion of man's research and development energies for generations to come. Minerals and life of the sea are already challenging to the geologists, biologists, and oceanographers. Marine geology is an important area of specialization for new geologists.

Along with engineers, geologists have been responsible for the development of many major building and construction programs. They have changed the courses of rivers and located dams, harbors, high-rise structures, housing developments, railroads, highways and sites for new cities. However, many large industrial sites, new cities, housing developments, and other types of construction are built without geological advice. But construction undertaken without an understanding of the geology of an area frequently has had tragic consequences.

Tremendous losses have been suffered where homes have fallen into ravines; reservoirs have broken and washed away entire communities, taking the lives of innocent home dwellers. Such tragedies should be a warning to the public to seek geological advice *before* construction, not after a catastrophe.

Geologists and seismologists currently are involved in studying past earthquakes and seeking criteria for the possible predictions of future earthquakes. These studies include investigations of areas vulnerable to earthquakes and could result in recommendations for the removal of major cities, or portions of them, to locations where the likelihood of earthquakes is negligible.

Today a world-wide water shortage demands much of our attention. Lack of sufficient water limits economic growth, undermines and lowers living standards, jeopardizes the security of nations, and above all, and most important, endangers people's health. A supply of water is one of our profession's most serious challenges in meeting human needs for the future. Answers must be found for this baffling and seemingly impossible problem or the world's multiplying population slowly but surely will perish without this vital natural resource.

Modern geology has penetrated even the field of medicine. Recently, a symposium on the subject, "Medical Geology and Geography" was sponsored by the American Association for the Advancement of Science and The Geochemical Society. The symposium revealed that a relation between trace elements and human health may be more than a fascinating hypothesis. Research along this line undoubtedly will enable our science to add biomedical geology to its many fields of endeavor.

Although geology has abandoned none of its old goals, the science is required to meet new challenges and to participate in new projects to help meet the ever expanding needs of mankind. Demands on geologists will continue to increase as human needs increase. An immediate challenge to the profession today is the need to see that enough geologists are trained to meet the growing demands of society. New geologists must be found, trained and put to work.

With world population doubling every 35 years and with per capita consumption of almost every raw material constantly mounting with the progress of civilization, nations of the world are faced with a problem of great magnitude in finding and providing additional mineral reserves. This presents a challenge to all scientists and technologists everywhere, but primarily to geologists and all other earth scientists.

In the more civilized areas of the earth, geologists already have tapped the most obvious sources of raw materials. For example, the oil and gas which were easiest to find have been discovered. The same is true of all other minerals which are now being mined and produced. Now we must find the more difficult and less obvious deposits. We will find these hidden reserves only by employing bold geological deductions, imagination, and relentless enthusiasm: Geologists continuously must plan for the world's increasing multitude of people and therefore the science of geology is important to the future welfare of the world's people.

Challenges to geologists today are greater than ever before but so are the opportunities. Soon people of all nations will demand the type of civilization that we enjoy in America today. Petroleum geologists will be needed to explore for an estimated 700 billion barrels of petroleum which is now waiting to be found along the continental shelves of the world. However, this oil will be only a fraction of that needed to meet demands that seem to have no ending.

Solving the mysteries of space, and finding hidden treasures of the sea will demand the services of hundreds of new geologists. Geologists also are needed in the fields of waste disposal, pollution control, land development, and dam construction aimed at providing new sources of water supply. Growth of the techniques of geologic aerial photographic interpretation has resulted in new importance to this field. Micropaleontology and sedimentology are finding new applications.

Thousands of new geologists will be needed to fill the many assignments in diversified areas of scientific endeavors that fall within the general field of geology. Yet the enrollment of students taking geology in our universities throughout the nation is far less than the demand – not only for the present, but the next ten years.

Teachers who have scientifically inclined or highly imaginative students should give every consideration to leading them into the study of geology. The author belongs to a family of geologists all of whom agree that there is no finer or more rewarding profession. Ours is one of the most vital and essential sciences on earth. It is also one of the most attractive and fascinating.

The profession is not a closed union as many are today. Instead geologists constitute an army of scientists reaching far out for knowledge. We need the help of every brilliant, inquisitive mind that we can attract. The more we learn, the better the men we attract to geology, the more qualified we will be to meet the future challenges of human needs through geology.

14.   Benevolent Planet

We inhabit a benevolent planet. "Surely there is a vein of silver and a place of gold. Iron is taken out of the Earth and copper is smelted from the ore. Men put an end to darkness and search out the opening in a valley away from places where men live," we read in *The Book of Job*. A familiarity with the Earth for early humans bred not only fear, but ultimately an understanding and a growing utilization of the planet on which they lived. Human history is, to some extent, a history of successive exploitation of our tiny planet: the use of accidentally shaped stones for weapons and tools; the discovery of fire; the later crafting of selected stones as weapons; the discovery of the usefulness of certain metals, copper, bronze, and then iron.

We find this progression beginning about 5000 BC, represented by the first copper needles around the Nile Valley. Another 1,500 years later, our ancestors learned to mix a little tin with the copper, to produce bronze. 1,500 years later, the same cultural sequence can be traced in Europe. One way to divide human history is by such inventive stages. The use of fire, the use of different metals, the growing use of language and writing, the domestication of animals and plants, the development of communities, the invention of the wheel – all these were developed as long as 6,000 years ago in parts of Europe and the Middle East. And yet, ancient as are the origins of these foundations of our current prosperity, the remarkable thing is that the provision of this bountiful planet has been such that there is a greater difference between our present life style and that of the early nineteenth century, than there was between that era and the period of the life of Christ. The reason for this is the increasingly successful exploitation of the crust of the planet on which we live. Life expectancy as recorded in one of the older censuses based on Sweden in the middle of the eighteenth century was 19 years in towns and villages. In those same places, it is now 73 years, and the difference that has arisen in that brief time span results from four particular discoveries and applications: medical care, an adequate diet, clean water supplies, and an effective system of sewage disposal. And all those things ultimately depend upon materials provided by the surface of the planet on which we live.

The development during the seventeenth and eighteenth centuries of cheap forms of power and the invention of the blast furnace to produce inexpensive iron and steel have revolutionized the world in which we live. It is somewhat difficult to realize the totality of our dependence on the Earth. The buildings in which we live are built of material manufactured from the Earth's crust. Heating, cooling and lighting systems involve the use of energy derived from materials in the crust, energy stored from the sun by plants and animals millions of years ago. Most of the clothing we wear, the food we eat, gasoline, cars, radios, TV – all these are triumphs of extraction of crude materials from the crust of the Earth: dull, unpromising rocklike materials, discovered, extracted, concentrated, refined, and then manufactured into something of immediate value.

This is not to deny the overriding importance of humanity's intellectual and moral development. Yet even here, there is a measure of interdependence. The supremacy of Greece was established on the discovery around 1,000 BC of the rich silver and lead ores of Laurium, southeast of Athens. Containing up to 120 ounces of silver per ton, these ores were worked by slaves (who had a life expectancy of only four years in the mines) who sunk 2,000 shafts, some 400 feet deep, and produced about two million tons of ore. For three centuries the glory of Athens, the basis of her commerce, and the power of her armies, were financed by this silver.

And so, behind humanity's rise in the intellectual realm, there has also been the increasingly successful exploitation of the crust of the planet on which we live. Each of us today has at our elbow the equivalent of the service of about two hundred workers in the form of the energy supplies that we use. And without energy, all derived from the crust of the Earth, life would be a very different proposition.

There has, however, been a price for this progress. With this transformation has gone a change in attitude. Our ancestors worshipped the Earth. They attempted to placate it. In a literal sense, the Earth was the mother of humanity. But now, we have changed our attitude towards the Earth and we look upon it, not so much as a thing to be worshipped, but as a thing to be exploited. With the increasing success of our exploitation of this bountiful planet, we have changed our attitude towards the planet itself.

For some of our most essential commodities – oil, gas, mercury, tin, silver, amongst them – the end of inexpensive exploitation looms. Reserves dwindle, yet demand and price increase. Our nonreplenishable resources, our burgeoning population, our unequal patterns of prosperity, and our careless technology – all these present both threat and challenge to our survival.

14-1   Gaia – James Lovelock

James Ephraim Lovelock was born in 1919 in Hertfordshire, England and graduated from Manchester and London Universities. He has had a long and distinguished scientific career at the National Institute for Medical Research, London, Baylor University, Harvard, Yale, and the Jet Propulsion Laboratory of NASA, but more recently he renounced "institutional science" to pursue "the solitary practice of science." A Fellow of the Royal Society, his Gaia hypothesis has made him a well-known figure. His work was recognized in 2006 by the award of the Wollaston Medal, the Geological Society's highest honor. Our extract is taken from Healing Gaia: Practical Medicine for the Planet *(1991).*

Gaia is the name the ancient Greeks used for the Earth goddess. This goddess, in common with female deities of other early religions, was at once gentle, feminine, and nurturing, but also ruthlessly cruel to any who failed to live in harmony with the planet.

Such a name seemed particularly appropriate for the new hypothesis, which took shape in the late 1960s. The Gaia hypothesis was first described in terms of life shaping the environment, rather than the other way round: "Life, or the biosphere, regulates or maintains the climate and the atmospheric composition at an optimum for itself." As understanding of Gaia grew, however, we realized that it was not life or the biosphere that did the regulating but the whole system. We now have Gaia theory, which sees the evolution of organisms as so closely coupled with the evolution of their physical and chemical environment that together they constitute a single evolutionary process, which is self-regulating. Thus the climate, the composition of the rocks, the air, and the oceans, are not just given by geology; they are also the consequences of the presence of life. Through the ceaseless activity of living organisms, conditions of the planet have been kept favourable for life's occupancy for the past 3.6 billion years. Any species that adversely affects the environment, making it less favourable for its progeny, will ultimately be cast out, just as surely as will those weaker members of a species who fail to pass the evolutionary fitness test.

14-2   The Web of Life – Fritjof Capra

Fritjof Capra was born in Vienna in 1939, graduating with a PhD in theoretical physics from the University of Vienna. He has held teaching and research positions in France, the UK and the US and is best known for his book The Tao of Physics *(1975), in which he explores the conceptual similarities between the concepts of modern science and Eastern mysticism. Our extract from a more recent book –* The Web of Life: A New Scientific Understanding of Living Systems *(1996) – explores the origins of the controversial Gaia hypothesis in the Romantic movement.*

The Romantic view of nature as 'one great harmonious whole', as Goethe put it, led some scientists of that period to extend their search for wholeness to the entire planet and see the Earth as an integrated whole, a living being. The view of the Earth as being alive, of course, has a long tradition. Mythical images of the Earth Mother are among the oldest in human religious history. Gaia, the Earth Goddess, was revered as the supreme deity in early, pre-Hellenic Greece. Earlier still, from the Neolithic through the Bronze Ages, the societies of 'Old Europe' worshipped numerous female deities as incarnations of Mother Earth.

The idea of the Earth as a living, spiritual being continued to flourish throughout the Middle Ages and the Renaissance, until the whole medieval outlook was replaced by the Cartesian image of the world as a machine. So, when scientists in the eighteenth century began to visualize the Earth as a living being, they revived an ancient tradition which had been dormant for only a relatively brief period.

More recently, the idea of a living planet was formulated in modern scientific language as the so-called 'Gaia hypothesis', and it is interesting that the views of the living Earth developed by eighteenth-century scientists contain some key elements of our contemporary theory. The Scottish geologist James Hutton maintained that geological and biological processes are all interlinked and compared the Earth's waters to the circulatory system of an animal. The German naturalist and explorer Alexander von Humboldt, one of the greatest unifying thinkers of the eighteenth and nineteenth centuries, took this idea even further. His 'habit of viewing the Globe as a great whole' led Humboldt to identifying climate as a unifying global force and to recognizing the coevolution of living organisms, climate, and Earth crust, which almost encapsulates the contemporary Gaia hypothesis.

14-3   Remember the Land – Charles Morgan

Charles Yazzie Morgan (Kin Yaa'áanii) is a Navajo cattle and sheep rancher, living at Dalton Pass near Crowpoint in Navajo Country. Morgan's family have lived in this semi-arid land since the time of his grandfather, but the discovery of uranium in the area scarred the face of the land they loved; mining companies had "dynamited springs, torn up family graves, and planned enough mines and mills to change the place forever." Morgan, refusing to despair, gathered his sixteen children around him in 1978 and addressed them slowly and deliberately. Our account of this speech is taken from Between Sacred Mountains: Navajo Stories and Lessons from the Land *(1982). It reinforces the Navajo tradition that the land and all "things on it and all people that are on it are in our care."*

I grew up knowing the beauty of the land. We used to look at the earth and see it dressed in flowers of all kinds and colors. All the gifts of the earth were thought of as blessings and were kept holy. A long time ago, if rains came at night, mothers and fathers would waken their children and tell them, "You should not be lying down. Sit up!" And fathers would keep the fire going until the rain passed. It was a holy moment. When the rains passed, parents would say, "We are blessed. New plants will grow now. Our water has been replaced."

We think of the Earth as Our Mother. The Sky is Our Father. From the time the sun rises, to the time the sun sets and all through the night, not a day comes out too short. Not a day comes out too long. All is balanced, and this, I think, is what is meant when we say in the Blessingway Ceremony, *"Sa'ah Naagháii Bik'eh Hózhóón."* All is balanced. Without balance there is no life.

We people think of ourselves as something very important. Mother Earth is the same way. She is made of the same things as we are. The way we hurt, that is the way she hurts too. The way we live, that's the way she lives too. Those things that are within us, those things are inside her too.

From her, plants grow. We eat of those plants. Because of her we are able to eat and drink. We are like children who still suck at the mother's breast. Water is like the milk from Mother Earth. We still suckle Our Mother.

Those who live on this earth, the insects, the plants, the animals, and ourselves, should have a place to live. But we bring starvation and thirst to all of these things. To all of them we bring great sorrow and grief. We have not thought about any of them.

What we do to the land can destroy what gives us life. Even now there are springs that no longer run. I am talking not only for myself, but for those who may be born tonight. We have to think of them. How will they drink? How will they live? What kind of life are we giving them?

My father always told us that from the time of the Long Walk our ancestors' strongest advice was never to leave this land again, trade it, or sell it, because they suffered grief, tears, and death for their land.

I say to you, my children: Study! Prepare for a job! Plan for the future! But don't forget the land and the people who went before you. They will be your blessing and will make you strong.

14-4   Rachel Carson: The Idea of Environment – Gabriele Kass-Simon

Rachel Carson (1907–1964) (see page 278) was one of the most influential writers of her time. Gabriele Kass-Simon, a Professor of Zoology at the University of Rhode Island, with degrees from Michigan, Columbia and Zurich, first describes the impact of Carson's work, in an essay she wrote for Women of Science *(1990). This is followed on page 294 by an extract from the closing chapter of Carson's book* Silent Spring *(1962).*

If it is true that the effect and indeed the function of science is to alter our perception of the universe, then it is also true that not all great scientists produce the data from which they create the new perception. Old data is often only differently organized into what then becomes our new vision of the "real" world. Further, one might argue that the importance of a scientific endeavor rests precisely on its ability to create such a new perception.

If, now, we were to compile a list of the great biologists based on this argument, we must immediately include the name of Rachel Carson (1907–1964) – whom, on first consideration, we might have been inclined to overlook, for Carson's achievements as a writer tend to obscure her real contribution as a scientific thinker.

Although a trained biologist – she received her master's degree in zoology from Johns Hopkins in 1932 and worked as a professional biologist both as a teacher at the University of Maryland (1931–1936) and as an aquatic biologist at the United States Department of Fish and Wildlife (1936–1952) – her work was essentially that of instructor and reporter. Indeed, she seems to have seen herself primarily as a writer and an interpreter of nature rather than an original researcher: She once wrote to a friend, "If I could choose what seems to me the ideal existence, it would be just to live by writing," and "biology," she had earlier said, "has given me something to write about."

While with the Fish and Wildlife Department, she produced two large monographs, "Food from the Sea: Fish and Shellfish of New England" (1943), and "Food from the Sea: Fish and Shellfish of the South Atlantic and Gulf Coasts" (1944). Both monographs, written during World War II, were aimed at finding new food sources and preventing the overconsumption of certain species of food fish. They are detailed, accurate descriptions of the anatomy, behavior, and habitat of twenty-six varieties of fish and shellfish as well as an evaluation of their usefulness as food.

In 1941 Carson published her first book, *Under the Sea Wind – A Naturalist's Picture of Ocean Life*. In contrast to the unembellished factual monographs, her first book was a highly romanticized, anthropomorphic portrayal of the lives and habits of a number of aquatic species. Nevertheless, it was well received and, ten years later, it became a best-seller. Two books followed: *The Sea Around Us* (1951) and *The Edge of the Sea* (1955). Like neither the monographs nor *Under the Sea Wind*, these books retain the best characteristics of both; they give a vivid and penetrating account of the physics, chemistry, and biology of the ocean and its shores. Their language is at the same time lyrical and precise and carries none of the melodramatic, sometimes saccharine overtones of her first book. But instructive as they are, these works do not force upon readers a reconstruction of their world. That accomplishment was left to Rachel Carson's last book, *Silent Spring* (1962).

The view that all of nature is interdependent, in the strictest sense, and that human industrialized activity almost invariably causes permanent damage to the earth is a view which today is so commonplace that to reiterate it is almost boring. It is hard to believe, therefore, that when *Silent Spring* first appeared, the idea that modern technology could annihilate us by irretrievably destroying our habitat was an unheard of revolutionary thought – one that provoked debate and discussion throughout the society and was greeted with alarm and controversy in government and industry.

It is further hard to believe that until the publication of *Silent Spring*, not only was it hardly known that small perturbations in the food chain could have severe, far-reaching, and unpredicted ecological consequences, but also that the deleterious genetic and physiological effects of unintended chemical consumption could have been so little suspected. It required some 350 pages (including 55 pages of references) to build an argument that showed how the indiscriminate spraying of pesticides like DDT caused widespread permanent biological destruction. Not only did Carson have to cite reports of such destruction, she also had to present and explain many of the biological and ecological principles which accounted for the destruction. Today we know these principles so well that, as with our own language, we are no longer aware that we had to be taught them.

The immediate effect of Carson's book was the banning of DDT as an insecticide; the long-term effect, as one editor put it, was to change the world: "A few thousand words from her, and the world took a new direction."

In 1964, two years after the publication of *Silent Spring*, Rachel Carson died of cancer at the age of 56.

14-5   Silent Spring – Rachel L. Carson

The influential writings of Rachel Carson (see page 278) made her one of the founders of the modern environmental movement.

We stand now where two roads diverge. But unlike the roads in Robert Frost's familiar poem, they are not equally fair. The road we have long been traveling is deceptively easy, a smooth superhighway on which we progress with great speed, but at its end lies disaster. The other fork of the road – the one "less traveled by" – offers our last, our only chance to reach a destination that assures the preservation of our earth.

The choice, after all, is ours to make. If, having endured much, we have at last asserted our "right to know," and if, knowing, we have concluded that we are being asked to take senseless and frightening risks, then we should no longer accept the counsel of those who tell us that we must fill our world with poisonous chemicals; we should look about and see what other course is open to us.

A truly extraordinary variety of alternatives to the chemical control of insects is available. Some are already in use and have achieved brilliant success. Others are in the stage of laboratory testing. Still others are little more than ideas in the minds of imaginative scientists, waiting for the opportunity to put them to the test. All have this in common: they are *biological* solutions, based on understanding of the living organisms they seek to control, and of the whole fabric of life to which these organisms belong. Specialists representing various areas of the vast field of biology are contributing – entomologists, pathologists, geneticists, physiologists, biochemists, ecologists – all pouring their knowledge and their creative inspirations into the formation of a new science of biotic controls.

"Any science may be likened to a river," says a Johns Hopkins biologist, Professor Carl P. Swanson. "It has its obscure and unpretentious beginning; its quiet stretches as well as its rapids; its periods of draught as well as of fullness. It gathers momentum with the work of many investigators and as it is fed by other streams of thought; it is deepened and broadened by the concepts and generalizations that are gradually evolved."

So it is with the science of biological control in its modern sense. In America it had its obscure beginnings a century ago with the first attempts to introduce natural enemies of insects that were proving troublesome to farmer, an effort that sometimes moved slowly or not at all, but now and again gathered speed and momentum under the impetus of an outstanding success. It had its period of draught when workers in applied entomology, dazzled by the spectacular new insecticides of the 1940s, turned their backs on all biological methods and set foot on "the treadmill of chemical control." But the goal of an insect-free world continued to recede. Now at last, as it has become apparent that the heedless and unrestrained use of chemicals is a greater menace to ourselves than to the targets, the river which is the science of biotic control flows again, fed by new streams of thought. . . .

Through all these new, imaginative, and creative approaches to the problem of sharing our earth with other creatures there runs a constant theme, the awareness that we are dealing with life – with living populations and all their pressures and counterpressures, their surges and recessions. Only by taking account of such life forces and by cautiously seeking to guide them into channels favorable to ourselves can we hope to achieve a reasonable accommodation between the insect hordes and ourselves.

The current vogue for poisons has failed utterly to take into account these most fundamental distinctions. As crude a weapon as the cave man's club, the chemical barrage has been hurled against the fabric of life – a fabric on the one hand delicate and destructible, on the other miraculously tough and resilient, and capable of striking back in unexpected ways. These extraordinary capacities of life have been ignored by the practitioners of chemical control who have brought to their task no "high-minded orientation," no humility before the vast forces with which they tamper.

The "control of nature" is a phrase conceived in arrogance, born of the Neanderthal age of biology and philosophy, when it was supposed that nature exists for the convenience of man. The concepts and practices of applied entomology for the most part date from that Stone Age of science. It is our alarming misfortune that so primitive a science has armed itself with the most modern and terrible weapons, and that in turning them against the insects it has also turned them against the earth.

14-6   Who is El Niño? – S. George Philander

S. George Philander, born in 1942, was educated at the University of Cape Town and Harvard. After research positions in oceanography at MIT and the National Oceanographic and Atmospheric Administration, he joined the faculty at Princeton, where he is Director of the Atmospheric and Oceanic Sciences Program and chair of the Department of Geological and Geophysical Sciences. A recipient of the Sverdrup Gold Medal of the American Meteorological Society, Philander has written on El Niño and on global warming. Our extract is taken from "Who is El Niño?," an essay published in 1998 in the journal Earth in Space.

It is a curious story, about a phenomenon we first welcomed as a blessing but now view with dismay, if not horror. We named it El Niño for the child Jesus, provided it with relatives – La Niña and ENSO – and are devoting innumerable studies to the description and idealization of this family. These scriptures provide such a broad spectrum of historical, cultural, and scientific perspectives that there is now confusion about the identity of El Niño. Scientist K.E. Trenberth summarizes the situation as follows.

The atmospheric component tied to El Niño is termed the "Southern Oscillation." Scientists often call the phenomenon where the atmosphere and ocean collaborate ENSO, short for El Niño–Southern Oscillation. El Niño then corresponds to the warm phase of ENSO. The opposite "La Niña" ("the girl" in Spanish) phase consists of a basinwide cooling of the tropical Pacific and thus the cold phase of ENSO. However, for the public, the term for the whole phenomenon is "El Niño."

The clergy is perplexed, but fortunately, laymen are providing a sensible solution.

El Niño has gained so much publicity over the past year that the term is now part of everyone's vocabulary; it designates a mischievous gremlin. Hence, an erratic stock market or exceptionally bad traffic jams in Los Angeles are attributed to El Niño. This use of the term is entirely consistent with our practice of using atmospheric and oceanic phenomena as metaphors in our daily speech: the President is under a cloud; the test was a breeze. We can now look forward to the term El Niño being used in unexpected and imaginative ways.

Everyone associates El Niño with droughts in Indonesia, torrential rains in Peru and California, and unusually high sea-surface temperatures in the eastern equatorial Pacific. This means that the term as used today refers to a phenomenon with both atmospheric and oceanic aspects. Those who insist that El Niño is strictly an oceanic phenomenon – the term originally referred to a warm, seasonal current that appears along the coast of Peru around Christmas – can take pride in their erudition, but they should realize that their use of the term is becoming archaic. Unless they wish to be part of an elite culture with its own unintelligible argot, they should follow those who accept that El Niño has both atmospheric and oceanic components. By doing so, they will display an awareness of recent scientific results that explain this phenomenon in terms of a tantalizing, circular argument.

The appearance of unusually warm surface waters over the eastern equatorial Pacific does not only cause changes in atmospheric conditions such as the altered rainfall patterns and the relaxation of the trade winds along the equator; the water also warms in response to the relaxation of the trade winds! Intense trades drive the shallow layer of warm surface water westward and expose the cold deep water to the surface in the east. When those winds weaken, the warm water flows back to the east.

This chicken-and-egg argument has an important implication: El Niño is a child of water and air! It emerges from an escalating tit-for-tat (positive feedback) between the ocean and the atmosphere: a random perturbation that amounts to a slight weakening of the trades causes some of the warm water in the western equatorial Pacific to flow eastward, further relaxing the winds and causing more warm water to surge eastward, and so on until El Niño conditions are established. A reversal of the tit-for-tat then invokes La Niña, which restores contemporary conditions of strong trades and cold surface waters in the eastern tropical Pacific.

El Niño is often described as a departure from "normal" conditions, but that is incorrect because normal conditions seldom prevail. The tropical ocean and atmosphere interact to produce a continuous irregular oscillation, known as the Southern Oscillation, between El Niño, and its neglected component La Niña. To ask why El Niño occurs is equivalent to asking why a bell rings or a taut violin string vibrates;

it is part of the spontaneous music of our spheres, the atmosphere and hydrosphere. To cope with El Niño and La Niña we should pay attention to the fascinating, ever-present tune, not only to the high, shrill notes such as the recent flooding in California or the low notes, which we heard when California suffered severe droughts not too long ago.

14-7 Essay on the Earth Sciences – National Research Council

The National Academy of Sciences was established by the Congress in 1863, at the height of the Civil War, to advise the government "upon the subject of science or art." Together with its sister institutions, the National Research Council, the National Academy of Engineering, *and the Institute of Medicine, its elected members provide expert advice on public policy issues involving science, technology, and public health. The following extract is taken from the publication* Solid Earth Sciences and Society *(1993).*

This is a particularly opportune time to assess the state of the earth sciences. New concepts arising from breakthroughs of the past quarter century as well as innovative technologies for gathering and organizing information are expanding the frontiers of knowledge about the earth system at an accelerating pace. Research ranges from the atomic scale of the scanning-tunneling microscope and ion microprobe to the global scale of worldwide seismic networks and images produced from data gathered by orbiting satellites. Earth scientists are constructing a comprehensive picture of the Earth, one that interrelates the physical, chemical, geological, and biological processes that characterize the planet and its history. Scientific advances and opportunities abound at the same time that environmental and resource problems affecting the world's population have become international political issues.

Human societies face momentous decisions in the next few years and decades. Issues such as atmospheric changes, nuclear power, hazardous wastes, environmental degradation, overpopulation, soil erosion, water quality and supplies, and the destruction of species cannot be ignored. The decisions eventually made about these environmental issues, including prediction of changes, must be based on reliable knowledge – knowledge that comes from understanding the earth system and its history.

The more we have learned about the Earth, the more we have come to appreciate the many ways in which it is suited to life. The Earth is just the right distance from the Sun to have surface temperatures optimal to sustain living things. Its vast oceans have remained liquid, neither boiling into the atmosphere nor freezing into a solid block of ice, since shortly after its formation some 4.5 billion years ago. It is the only planet in the solar system that exhibits plate tectonics, which recycles nutrients and other materials essential to life through the interior of the planet and back to the surface. The Earth is unique in sustaining an atmosphere that is one-fifth oxygen – oxygen that was generated over eons by single-celled organisms and that, in turn, spurred the evolution of multicellular organisms.

More particularly, the Earth is a congenial home for humans. All of the materials we use in our daily lives come from the Earth – fuels, minerals, groundwater, even our food (through the intermediaries of soil, water, and fertilizer). We have a strong psychological affinity for certain places on the Earth, for the regions where we grow up and live and for the wild and beautiful tracts, whether preserved in parks or apparent in our everyday surroundings. But we have altered the surface extensively during our occupation – erecting structures, clearing forests, damming rivers. Despite our activities, the Earth continues its normal path of inexorable change – through violent paroxysms such as earthquakes and landslides or through the slow, steady erosion of soils and coastlines.

The Earth is a resilient planet. Long after human beings have vanished from the planet, its basic cycles will persist. The largest man-made structures will erode and disappear, radioactive materials that have been gathered will decay, man-made con-centrations of chemicals will disperse, and new species will evolve and perish. The oceans, atmosphere, solid earth, and living things will continue to interact, just as they did before humans appeared on the scene.

But some earth systems are very fragile. We dispose of our wastes in the same sedimentary basins that supply us with the bulk of our groundwater, energy, and mineral resources. Through our social, industrial, and agricultural activities, we are changing the composition of the atmosphere, with potentially serious effects on climate and on terrestrial and marine ecosystems. The human population is expand-ing into less habitable parts of the world, which increases vulnerability to natural hazards and strains the biological and geological systems that sustain life.

If present trends continue, the integrity of the more fragile systems on which human societies are built cannot be assured. The time scale for the breakdown of these systems may be decades or it may be centuries, but we cannot continue to use the planet as we have been using it. Present trends need not continue, because we are unique among the influences that affect earth systems: we have the ability to decide among various courses, to weigh the pros and cons of alternative actions, and to behave accordingly.

The history of the geological sciences offers many reasons to be optimistic. One of the triumphs of twentieth-century science and technology has been the worldwide identification and extraction of energy and mineral resources, an activity that has brought an increased standard of living to an expanding human population. The geological sciences have demonstrated ways of maintaining water quantity and quality, disposing of wastes safely, and securing human structures and facilities against natural hazards. Essentially, the geological sciences teach us about the nature of the Earth and about our role on it.

The process of understanding the Earth has just begun. If human beings are to survive and prosper for more than a moment in geological time, an understanding of the intricacies of interacting earth systems is a necessity. We must find and develop the resources needed to sustain and improve the human condition. We must also preserve and improve the environment on which essential and aesthetic human needs depend. We need to know enough to predict the beneficial and destructive consequences of our actions.

Among the questions now confronting solid-earth scientists are the following: How did the Earth form, and what accounts for the composition of its various components? How does the core interact with the mantle, and how does the mantle drive plate tectonics? How have the continents evolved? How has the solid earth influenced the course of biological evolution and extinctions, and how have these processes affected the Earth? How have the ocean and atmosphere changed over time, and how can they be expected to change in the future as human influence increases? How can wastes be safely isolated in geological repositories? How are the small-scale movements of the surface related to tectonic processes? None of these questions can be answered without contributions from several subdisciplines of the solid-earth sciences. Thus, interdisciplinary interactions will continue to grow in importance as the earth sciences advance.

Because its concerns are with global features and phenomena, the earth sciences are intrinsically an international undertaking. Basic field data must be gathered from diverse regions of the Earth, and earth scientists must conduct experiments on a global basis to accurately determine its composition and dynamics. International cooperation and exchange among scientists in different countries are essential.

Quantitative models have been developed for a number of earth processes. For instance, numerical computer simulations have been devised for mantle convection, for the evolution of sedimentary basins, for surface processes such as erosion, for fluid flow, and for rupture processes in earthquakes. At the same time, solid-earth scientists are relying on increasingly sophisticated instrumentation to expand and to keep up with the data needed for future discoveries and for models that assess and predict earth processes. Even such traditionally descriptive subdisciplines as mapping and paleontology are becoming increasingly quantitative with the advent of digital analyses and computerized data bases.

Better understanding of the way natural processes operate over time, whether quantitative or descriptive, leads to more accurate prediction of the future effects of those processes. For instance, a greater understanding of surface processes enhances the ability to predict such events as floods, landslides, and subsidence, which assault human structures and facilities. This enhanced ability is the basis for land-use planning, enabling society to minimize the costs and problems associated with geological hazards. The potential losses due to earthquakes and volcanic eruptions have stimulated research on the condition of and changes in the state of the crust before such events, in the hope that similar changes may be used for future predictions.

Prediction in the earth sciences requires an understanding of the role that past events have had in shaping our planet and an accurate projection of data-based interpretations into the future. Field work will remain the principal source of ground truth observations, while laboratory work will calibrate those observations. At the same time, theory will produce generalizations that can be extrapolated to unexplored regions, to different scales, and to events occurring at different times.

Greater predictive abilities should enable earth scientists to tell how human influences will shape the future. Earth systems undergo natural fluctuations on which the changes caused by human influences are superimposed. Distinguishing between natural fluctuations and human-induced changes in earth systems remains an essential task.

An important question is to what extent geological changes, whether driven by natural processes or human intervention, exhibit chaotic behavior. If chaos dominates, the ability to predict many earth phenomena may be limited. But even chaotic systems are subject to statistical predictions related to the size of the events that may be expected and to their average frequency.

14-8   The Round Walls of Home – Diane Ackerman

Diane Ackerman's account from a Natural History of the Senses (1990) is grounded, not ultimately in sensate organisms, in all their startling variety, but in the planet from which they arose and on whose support their

continuing existence depends. Born in 1948, we first saw Diane Ackerman in the chapter on art (page 255). The following extract describes her vision of the home planet – "the big, beautiful, blue, wet ball."

Picture this: Everyone you've ever known, everyone you've ever loved, your whole experience of life floating in one place, on a single planet underneath you. On that dazzling oasis, swirling with blues and whites, the weather systems form and travel. You watch the clouds tingle and swell above the Amazon, and know the weather that develops there will affect the crop yield half a planet away in Russia and China. Volcanic eruptions make tiny spangles below. The rain forests are disappearing in Australia, Hawaii, and South America. You see dust bowls developing in Africa and the Near East. Remote sensing devices, judging the humidity in the desert, have already warned you there will be plagues of locusts this year. To your amazement, you identify the lights of Denver and Cairo. And though you were taught about them one by one, as separate parts of a jigsaw puzzle, now you can see that the oceans, the atmosphere, and the land are not separate at all, but part of an intricate, recombining web of nature. Like Dorothy in *The Wizard of Oz*, you want to click your magic shoes together and say three times: "There's no place like home."

You know what home is. For many years, you've tried to be a modest and eager watcher of the skies, and of the Earth, whose green anthem you love. Home is a pigeon strutting like a petitioner in the courtyard in front of your house. Home is the law-abiding hickories out back. Home is the sign on a gas station just outside Pittsburgh that reads "If we can't fix it, it ain't broke." Home is springtime on campuses all across America, where students sprawl on the grass like the war-wounded at Gettysburg. Home is the Guatemalan jungle, at times deadly as an arsenal. Home is the pheasant barking hoarse threats at the neighbor's dog. Home is the exquisite torment of love and all the lesser mayhems of the heart. But what you long for is to stand back and see it whole. You want to live out that age-old yearning, portrayed in myths and legends of every culture, to step above the Earth and see the whole world fidgeting and blooming below you.

I remember my first flying lesson, in the doldrums of summer in upstate New York. Pushing the throttle forward, I zoomed down the runway until the undercarriage began to dance; then the ground fell away below and I was airborne, climbing up an invisible *flight* of stairs. To my amazement, the horizon came with me (how could it not on a round planet?). For the first time in my life I understood what a valley was, as I floated above one at 7,000 feet. I could see plainly the devastation of the gypsy moth, whose hunger had leeched the forests to a mottled gray. Later on, when I flew over Ohio, I was saddened to discover the stagnant ocher of the air, and to see that the long expanse of the Ohio River, dark and chunky, was the wrong texture for water, even flammable at times, thanks to the fumings of plastics factories, which I could also see, standing like pustules along the river. I began to understand how people settle a landscape, in waves and at crossroads, how they survey a land and irrigate it. Most of all, I discovered that there are things one can learn about the world only from certain perspectives. How can you understand the oceans without becoming part of its intricate fathoms? How can you understand the planet without walking upon it, sampling its marvels one by one, and then floating high above it, to see it all in a single eye-gulp?

Most of all, the twentieth century will be remembered as the time when we first began to understand what our address was. The "big, beautiful, blue, wet ball" of recent years is one way to say it. But a more profound way will speak of the orders of magnitude of that bigness, the shades of that blueness, the arbitrary delicacy of beauty itself, the ways in which water has made life possible, and the fragile euphoria of the complex ecosystem that is Earth, an Earth on which, from space, there are no visible fences, or military zones, or national borders. We need to send into space a flurry of artists and naturalists, photographers and painters, who will turn the mirror upon ourselves and show us Earth as a single planet, a single organism that's buoyant, fragile, blooming, buzzing, full of spectacles, full of fascinating human beings, something to cherish. Learning our full address may not end all wars, but it will enrich our sense of wonder and pride. It will remind us that the human context is not tight as a noose, but large as the universe we have the privilege to inhabit. It will change our sense of what a neighborhood is. It will persuade us that we are citizens of something larger and more profound than mere countries, that we are citizens of Earth, her joyriders and her caretakers, who would do well to work on her problems together. The view from space is offering us the first chance we evolutionary toddlers have had to cross the cosmic street and stand facing our own home, amazed to see it clearly for the first time.

14-9   The Butterfly Effect – Ernest Zebrowski, Jr

Ernest Zebrowski, Jr, holds professorships in science and mathematics education at Southern University in Baron Range and in physics at Pennsylvania College of Technology. Author of several books, he has contributed a comprehensive account of the causes of natural disasters in his book Perils of a Restless Planet: Scientific Perspectives on Natural Disasters *(1997), from which our extract is taken. He concludes that the "secret" of survival is the continuing study of the natural causes of disasters and of the means of remedy and prevention. In our extract, he explores the "nonlinear sciences," butterflies and the subtle linkages of natural events that sometimes make them unpredictable.*

Could a butterfly in a West African rain forest, by flitting to the left of a tree rather than to the right, possibly set into motion a chain of events that escalates into a hurricane striking coastal South Carolina a few weeks later? As bizarre as this premise may sound, research of the last few decades suggests that the answer is yes, and that the effect is hardly limited to butterflies. Many of the large-scale phenomena that threaten human life yet stymie efforts at scientific prediction – tornadoes, earthquakes, volcanic eruptions, and epidemics among them – seem to have one characteristic in common: a sensitive dependence on seemingly innocuous variations in initial conditions. These are phenomena where small disturbances often escalate to larger ones, and if the initial causative agent changes only slightly, the larger effects can differ quite dramatically.

Of course, no one has ever actually observed a butterfly triggering a hurricane. The physical evidence for the butterfly effect is considerably more subtle than this and begins to make sense only through certain theoretical arguments. One fairly compelling line of reasoning runs as follows: If we wish someday to be able to predict future hurricanes, we first need to identify the prerequisite knowledge that would have permitted us to predict today's hurricane. And there ought to be a simple way of doing this: Just run the documentary in reverse. Take lots of data as a hurricane develops, then run it backward in time to see how the event got started in the first place. When we identify how its started, and we know how step C follows from step B and that from step A, we ought to have a program for predicting future events of the same type.

In fact this research strategy has been followed, many times, and it always ends in dismal failure when applied to complex natural phenomena. The problem is that we can only see so much detail, we can measure only to limited precision, and we can compute only to finite accuracy. The swarm of satellites now circling the globe indeed provides us with vast quantities of highly detailed information: in the technical limit, temperature variations of a few thousandths of a degree, and geometrical resolutions approaching 30 centimeters (1 ft) horizontally and a few centimeters vertically. But suppose that a tropical storm develops, and that we play back the data record of the previous few days. What do we find as we go back in time? A smaller storm, and yet a smaller disturbance, then a warm moist windy spot, then a set of atmospheric conditions that looks no different from that at many other locations in the tropics. What is it that whips some of these minor atmospheric fluctuations into full-blown hurricanes, while others disperse after causing no more annoyance than blowing off someone's hat in Africa? It's impossible to say for sure. All we know is that the fundamental agent must be very small, because all of our expensive and sophisticated instrumentation can't detect it. When scientists run up against something very curious like this, they often become whimsical; hence, the "butterfly effect." The premise is that of extremely sensitive dependence on initial conditions. Hummingbirds or flying squirrels would do as well.

The butterfly effect is not limited to a storm's birth but also applies to its future development (which, as we saw in the detailed case of Hurricane Emily, can be quite complicated indeed). Why, for instance, do hurricanes frequently change storm speed and direction? If we could understand this, we might at least improve the accuracy of our predictions of where and when a specific hurricane will make landfall. We now program our computers with dozens of equations to predict the future course of a storm after it is born, and we feed these equations hundreds of thousands of data values that originated in direct readings of sophisticated instruments, yet our predictions just one day into the future are still only marginally accurate at best. What is going wrong?

An insight into the answer has been gained by comparing computer simulations of hypothetical storms whose initial conditions differ in only the tiniest ways. Hypothetical initial data is supplied, and a simulated storm develops on the computer screen, with columns of numbers telling how the wind speed, storm speed, temperature, barometric pressure, and other measurable variables change as time passes and the storm evolves. Here "initial data" does not refer to data from the time of birth of the storm (a time that is, after all, unknown), but to data from some arbitrary instant in time during the storm's development; the data are therefore "initial" only in that they are used to initiate the computation. By itself, of course, such a single computer simulation tells us nothing of any value. But now conduct a second computer run, and a third, and a fourth, with the same computer program but slightly different initial data. How different? Change the temperature profile by just a few millionths of a degree here and there (which is far beyond the limits of physical measurement), or change the wind-speed distribution in the fourth or fifth decimal place (which might mimic a flock of birds alternately flying into and against the wind). Now compare the results of these different computer runs. What do we find? In the first few hours, as one might expect, these simulated storms differ very slightly. But as time passes, it turns out that their behaviors always begin to diverge, and eventually they often develop in quite different ways. One simulated storm may veer northward while another continues westward, one may intensify while another is dying, or one may stand stationary while another gallops toward a shoreline.

Such experiments with computer simulations suggest that the future of a storm is always extremely sensitive to tiny fluctuations in what goes on within it – so sensitive, in fact, that even variations that are too tiny to measure may seriously affect the future course of the event.

Variations of this type need not come from butterflies or birds. Going back three centuries to Isaac Newton's law of action and reaction (One system cannot affect another without the second exerting an effect back on the first), we realize that if a hurricane strikes even a small island, for example, the island itself will affect the future of the hurricane. Newton himself viewed action and reaction as always arising in equal and opposite force pairs, and at any instant in time they do. In light of the butterfly effect, however, a very small reaction by the island at one instant may initiate a larger action-reaction pair at the next instant, and the process may cascade until it ultimately sends a hurricane off on a significantly different future course. Moreover, the butterfly effect suggests that it may take a lot less than a whole island to do this; even a few newly built hotels may have the capacity to affect the future of a particular storm.

Today's scientists have generally accepted this fundamental premise, and within the last decade a number of new scientific journals have appeared that are devoted to the study of "nonlinear dynamics." The term "nonlinear" refers to situations were two agents combine to produce an effect that is quite different from the sum of the two separate effects. As a simple analogy, take two children (boys, in particular), put them in separate rooms full of toys, watch their behavior, then try to use your observations to predict what will happen when you put them together in the *same* room full of toys. Can such a prediction, even in principle, be made with any degree of reliability? Is there any way to anticipate whether the children will cooperate, fight, ignore each other, smash the toys to smithereens, or alternate between these behaviors? At best, it's a tall order. Regardless of your prediction, you wouldn't want to put the little guys together and then walk away in confidence.

Nonlinear natural phenomena – such as storms, earthquake clusters, and epidemics – tend to be wildly unstable. They may seem to be behaving themselves at one instant in time, following at least some statistically predictable pattern; then suddenly they swing into a drastically different mode of behavior, for reasons that are not apparent because the agent responsible for the change was too small to measure or even to notice. Add a little glitch, a metaphorical butterfly, to a complex process, and sometimes you get an outcome no rational person would ever have expected.

In social and political arenas, the counterpart of the butterfly effect has long been recognized and written about. A well-known fable that apparently originated in pre-Elizabethan England runs as follows:

> For the want of a nail, the shoe was lost.
> For the want of a shoe, the horse was lost.
> For the want of a horse, the rider was lost.
> For the want of a rider, the battle was lost.
> For the want of a battle, the kingdom was lost.
> All for the want of a nail.

It's worth noting that the moral to this story is ambiguous. On the one hand, the writer may have been quite pessimistic about our human abilities to plan for anything, given that even the tiniest glitch can lead to the most drastic consequences. On the other hand, the writer may have been advising us to inspect carefully for loose horseshoe nails.

There are many metaphorical butterflies (and loose horseshoe nails) in the world, and at present there is no way of telling which of them may send a large-scale nonlinear system into upheaval. Mother Nature's message, however, remains just as ambiguous as the horseshoe nail fable. Is she telling us that we've bumped up against the limits of what it is possible for humans to know and predict? Or, maybe what she's really telling us is that we need to explore paths to knowledge other than deterministic or statistical numerical prediction. Then again, the real messages may be this: Find the right metaphorical butterfly at the right time, snag it and send it on a different course, and perhaps we can occasionally *prevent* a natural disaster.

14-10   Pale Blue Dot – Carl Sagan

Carl Sagan (1934–1996) was, perhaps, the most widely known scientist of his generation. A quadruple graduate of the University of Chicago, which he entered at age sixteen, he spent six years on the faculty at Harvard and the rest of his career at Cornell, where he was David Duncan Professor of Astronomy and Space Sciences and Director of the Laboratory for Planetary Studies. He was known to millions throughout the world for his PBS television series Cosmos – *said to have been watched by three percent of the world's population – and wrote over twenty books on topics ranging from astronomy to science fiction, UFOs, nuclear war, and missile defense. He was also widely recognized for his research in astronomy, especially in planetary development, the search for extraterrestrial intelligence, and exobiology. Our extract is taken from his book* Pale Blue Dot: A Vision of the Human Future in Space *(1994). Pale blue dot is a description of the view of Earth from the perspective of the* Voyager *spacecraft.*

Look again at that dot. That's here. That's home. That's us. On it everyone you love, everyone you know, everyone you ever heard of, every human being who ever was, lived out their lives. The aggregate of our joy and suffering, thousands of confident religions, ideologies, and economic doctrines, every hunter and forager, every hero and coward, every creator and destroyer of civilization, every king and peasant, every young couple in love, every mother and father, hopeful child, inventor and explorer, every teacher of morals, every corrupt politician, every "superstar," every "supreme leader," every saint and sinner in the history of our species lived there – on a mote of dust suspended in a sunbeam.

The Earth is a very small stage in a vast cosmic arena. Think of the rivers of blood spilled by all those generals and emperors so that, in glory and triumph, they could become the momentary masters of a fraction of a dot. Think of the endless cruelties visited by the inhabitants of one corner of this pixel on the scarcely distinguishable inhabitants of some other corner, how frequent their misunderstandings, how eager they are to kill one another, how fervent their hatreds.

Our posturings, our imagined self-importance, the delusion that we have some privileged position in the Universe, are challenged by this point of pale light. Our planet is a lonely speck in the great enveloping cosmic dark. In our obscurity, in all this vastness, there is no hint that help will come from elsewhere to save us from ourselves.

The Earth is the only world known so far to harbor life. There is nowhere else, at least in the near future, to which our species could migrate. Visit, yes. Settle, not yet. Like it or not, for the moment the Earth is where we make our stand.

It has been said that astronomy is a humbling and character-building experience. There is perhaps no better demonstration of the folly of human conceits than this distant image of our tiny world. To me, it underscores our responsibility to deal more kindly with one another, and to preserve and cherish the pale blue dot, the only home we've ever known.

Sources

The editors and publishers gratefully acknowledge the permission granted to reproduce the copyright material in this book. They apologize for any errors or omissions.

Chapter 1: Eyewitness Accounts of Earth Events

1–1 **LOS ANGELES AGAINST THE MOUNTAINS**: John McPhee, from *The Control of Nature*. (Farrar, Straus & Giroux LLC) 1989, pp. 183–186. Copyright © 1989 John McPhee. Reprinted by permission of Farrar, Straus & Giroux LLC. [The text of this book originally appeared in *The New Yorker*.]

1–2 **THE NIGHT THE MOUNTAIN FELL**: Gordon Gaskill, from *Reader's Digest*. May 1965, **86**, pp. 59–67. Copyright © 1965 by The Reader's Digest Association, Inc.

1–3 **THE TURTLE MOUNTAIN SLIDE**: R.G. McConnell and R.W. Brock, from *Report on the Great Landslide at Frank, Alberta*. Canada Department of the Interior: Annual Report, 1902–1903. 1904, **VIII**, pp. 6–7.

1–4 **CANDIDE**: Voltaire, from "Candide", in *The Portable Voltaire*. (New York: Viking) 1968, pp. 239–243. Edited by Ben Ray Redman, copyright © 1949 renewed © 1977 by Viking Penguin Inc. Used by permission of Viking Penguin, a division of Penguin Group (USA) Inc. [The novel *Candide* was originally published in 1759].

1–5 **THE LISBON EARTHQUAKE**: James R. Newman, from *Science and Sensibility*. (Garden City, New York: Doubleday and Company, Inc.) 1970, pp. 56–66. Reprinted with permission.

1–6 **THE TEMBLOR**: Mary Austin, "The Temblor: A Personal Narration" in David Starr Jordan's *The California Earthquake of 1906*. (San Francisco: A.M. Robertson) 1907, pp. 341–350.

1–7 **THE ALASKAN GOOD FRIDAY EARTHQUAKE**: Jonathan Weiner, from *Planet Earth*. (New York: Bantam Books) 1986, pp. 6–8. © Jonathan Weiner. Reprinted with kind permission of the author.

1–8 **TSUNAMI**: Francis P. Shepard and H.R. Wanless, from *Our Changing Coastlines*. (New York: McGraw Hill Book Company) 1971, pp. 521–522.

1–9 **NOT A VERY SENSIBLE PLACE FOR A STROLL**: Haroun Tazieff, from *Craters of Fire*. (New York: Harper and Brothers) 1952, pp. 11–21. Reprinted by permission of Harper & Row, Publishers, Inc.

1–10 **LAST DAYS OF ST PIERRE:** Fairfax Downey, from *Disaster Fighters*. (New York: G.P. Putnam's Sons) 1938, pp. 74–78, 83–86, 88–91. Copyright 1938 by Fairfax Downey.

1–11 **BEACONS ON THE PASSAGE OUT:** Hans Cloos, from *Conversation with the Earth*. (New York: Alfred A. Knopf) 1953, pp. 14–21. Translated from the German by E. B. Garside. Edited and slightly abridged by Ernst Cloos and Curt Dietz. Reprinted with the permission of Random House Inc. [Originally published as *Gespräch mit der Erde*, 1947.]

1–12 **ERUPTION OF THE ÖRAEFAJÖKULL, 1727:** Jon Thorlakson. Account taken from Ebenezer Henderson, 1819, *Iceland; or the Journal of a Residence in that Island during the Years 1814 and 1815*. 2nd edn. (Edinburgh: Waugh and Innes) pp. 208–212.

Chapter 2: Exploration

2–1 **THE VOYAGE OF THE BEAGLE:** Charles Darwin, from *The Voyage of the Beagle* (New York: P.F. Collier & Son Corporation) 1969, pp. 380–384. 62nd printing. [First published as *Journal of Researches into the Geology and Natural History of the Various Countries Visited by H.M.S. Beagle* (1839)]

2–2 **THE MAP THAT CHANGED THE WORLD:** Simon Winchester, from *The Map That Changed the World: William Smith and the Birth of Modern Geology*. (New York: HarperCollins Publishers) 2001, pp. xv–xix. Copyright © Simon Winchester 2001. Reprinted with permission of Sterling Lord Literistic and HarperCollins Publishers USA.

2–3 **THE EXPLORATION OF THE COLORADO RIVER**: John Wesley Powell, from *The Exploration of the Colorado River*, 1957, abridged edition by the University of Chicago Press, pp. 77–79. [Originally published 1875, *The Exploration of the Colorado River of the West*]

2–4 **MONO LAKE–AURORA–SONORA PASS:** William H. Brewer, from *Up and Down California, 1860–1864*. (New Haven: Yale University Press) 1930, pp. 416–418. Reprinted 1949, University of California Press.

2–5 **THRILLS IN FOSSIL HUNTING:** George F. Sternberg, from *The Aerend*. Summer 1930, **1**(3), pp. 139–153. Extract reprinted in the *Journal of Geological Education*, 1967, **15**, pp. 78–79.

2–6 **THE CREATIVE EXPLOSION:** John E. Pfeiffer, from *The Creative Explosion: An Inquiry into the Origins of Art and Religion*. (New York: HarperCollins Publishers) 1982, pp. 19–20. Reprinted with permission of HarperCollins Publishers USA.

2–7 **ATTENDING MARVELS: A PATAGONIAN JOURNAL:** George Gaylord Simpson, from *Attending Marvels: A Patagonian Journal*. (New York: The Macmillan Company) 1934, pp. 58–59. Reprinted with permission of The George Gaylord Simpson Trust for Great-Grandchildren.

2–8 **EXPLORATIONS**: Robert D. Ballard (with Malcolm McConnell), from *Explorations*. (New York: Hyperion) 1995, pp. 184–189. Copyright © 1995 by Robert D. Ballard, Ph.D. Reprinted by permission of Hyperion. All rights reserved.

2–9 **THE BLUE PLANET:** Louise B. Young, from *The Blue Planet*. (New York: Meridian Book/New American Library) 1983, pp. 255–256. Copyright © 1983 by Louise B. Young. Reprinted by permission of Little, Brown and Co and the author.

Chapter 3: Geologists are also Human

3–1 **STEPPING STONES:** Stephen Drury, from *Stepping Stones: the Making of our Home World.* (Oxford: Oxford University Press) 1999, pp. 1–2. © 1999 by Stephen Drury. Reprinted with permission of Oxford University Press.

3–2 **WILLIAM BUCKLAND:** Elizabeth O.B. Gordon, from *The Life and Correspondence of William Buckland, D.D., F.R.S., sometime Dean of Westminster, twice president of the Geological Society and first President of the British Association* (London: John Murray) 1894, pp. 31–33, 41–42, 142–143.

3–3 **THE OLD RED SANDSTONE:** Hugh Miller, from *The Old Red Sandstone.* (Boston: Gould and Lincoln) 1851, pp. 3–11, 13, 14. [First appeared serially during 1840 in Witness]

3–4 **A LONG LIFE'S WORK:** Sir Archibald Geikie, from *A Long Life's Work.* (London: Macmillan and Company) 1924, pp. 55–58.

3–5 **LIFE, TIME, AND DARWIN:** Frank H.T. Rhodes, Inaugural Lecture of the Professor of Geology delivered at University College of Swansea (Jan 23, 1958) from *Life, Time and Darwin.* (Swansea, England: University College of Swansea) 1958, pp. 3–7.

3–6 **KING'S FORMATIVE YEARS:** R.A. Bartlett, from *Great Surveys of the American West* (Norman, Oklahoma: University of Oklahoma Press) 1962, pp. 127–140. Reprinted with permission of University of Oklahoma Press.

3–7 **WITH SHACKLETON IN THE ANTARTIC:** Mary Edgeworth David, from *Professor David, The Life of Sir Edgeworth David* (London: Edward Arnold & Company) 1937, pp. 129–130, 139–140, 143–148, 151, 175–177. Reprinted with the kind permission of Anne Edgeworth.

3–8 **THE GREAT DIAMOND HOAX:** William H. Goetzmann, from *Exploration and Empire.* (New York: Alfred A. Knopf, Inc.) 1966, pp. 452–457. © 1966 by William H. Goetzmann. Reprinted by permission of Alfred A. Knopf, Inc.

3–9 **SAND, WIND, AND WAR:** Foreword by Luna B. Leopold, Paul D. Komar, and Vance Haynes, to Ralph Bagnold's *Sand, Wind, and War: memoirs of a desert explorer.* (Tucson: University of Arizona Press) 1990, pp. ix–xii. © 1990 Arizona Board of Regents. Reprinted with permission of University of Arizona Press.

3–10 **SHIP'S WAKE:** Hans Cloos, from *Conversation With the Earth.* (New York: Alfred A. Knopf) 1953, pp. 22–32. Translated from the German by E.B. Garside. Edited and slightly abridged by Ernst Cloos and Curt Dietz. Reprinted with the permission of Random House Inc. [Originally published as: *Gespräch mit der Erde,* 1947.]

Chapter 4: Celebrities

4–1 **BENJAMIN FRANKLIN AND THE GULF STREAM:** H. Stommel, from *The Gulf Stream.* (University of California Press) 1965, pp. 4–5. Reprinted with permission.

4–2 **LEONARDO DA VINCI AS A GEOLOGIST:** Thomas Clements from *Language of the Earth,* (New York: Pergamon Press Inc) 1st edn, 1981, pp. 310–314.

4–3 **MINERALOGY, GEOLOGY, METEOROLOGY:** Rudolf Magnus, from *Goethe as a Scientist.* (New York: Henry Schuman) 1949, pp. 200–205, 209–213. [Translated by

Heinz Norden from the German, *Goethe als Naturforscher* (Leipzig, Germany, 1906)]. [Original substantially abridged for this edition.]

4–4 **MEGALONYX, MAMMOTH, AND MOTHER EARTH:** Edwin T. Martin, from *Thomas Jefferson: Scientist*. (New York: Henry Schuman) 1952, pp. 107–114. [Original substantially abridged for this edition.]

4–5 **THREE SHORT, HAPPY MONTHS:** William A. Stanley, from *ESSA World*, US Department of Commerce/Environmental Science Services Administration. Jan 1968, **3**(1), pp. 14–15.

4–6 **MOUNTAIN-WORSHIP:** W.G. Collingwood, from *The Life of John Ruskin*. (Cambridge: Houghton, Mifflin and Co., The Riverside Press) 1902, pp. 41, 60–61, 246. [*The Life and Work of John Ruskin*, by W.G. Collingwood, M.A., in two volumes, of which this volume is a revised and abbreviated edition, was published and copyrighted in 1893, by Houghton, Mifflin & Co.]

4–7 **STANFORD UNIVERSITY, 1891–1895:** Herbert C. Hoover, from *The Memoirs of Herbert Hoover: Years of Adventure, 1874–1920*. (New York: Macmillan) 1951, pp. 16–20, 23–24. Reprinted with permission of The Herbert Hoover Presidential Library Association.

Chapter 5: Philosophy

5–1 **CONCERNING THE SYSTEM OF THE EARTH, ITS DURATION AND STABILITY:** James Hutton, from *Abstract of a Dissertation Read in the Royal Society of Edinburgh, upon the Seventh of March, and Fourth of April, M,DCC,LXXXV, Concerning the System of the Earth, Its Duration, and Stability*, in James Hutton's *System of the Earth*, 1785, pp. 3–30. [(Hafner Publishing Company) 1970, Facsimile of the original 1785 version].

5–2 **THE METHOD OF MULTIPLE WORKING HYPOTHESES:** T.C. Chamberlin, from *Science*. 1965, **148**, pp. 754–759. [Originally published 1890, *Science* (old series), **15**, pp. 92–97]

5–3 **HISTORICAL SCIENCE:** George Gaylord Simpson, from *The Fabric of Geology*. Edited by Claude A. Albritton, Jr (London: Addison-Wesley Publishing Company) 1963, pp. 24–25, 31–33. Reprinted with permission of The George Gaylord Simpson Trust for Great-Grandchildren.

5–4 **WHAT IS A SPECIES?** Stephen Jay Gould, from *Discover* magazine. Dec 1992, **13**, pp. 40–44.

5–5 **MESSAGES IN STONE:** Christine Turner, from *Earth Matters: The Earth Sciences, Philosophy, and The Claims of Community*. Edited by Robert Frodeman. (Englewood Cliffs, New Jersey: Prentice Hall) 2000, pp. 53–54. Reprinted with permission of Pearson Education USA.

5–6 **NATURAL SCIENCE, NATURAL RESOURCES, AND THE NATURE OF NATURE:** Marcia G. Bjørnerud, from *The Earth Around Us: Maintaining a Livable Planet*. Edited by Jill S. Schneiderman (New York: W. H. Freeman & Co.) 2000, pp. 59, 66–69. Copyright © 2000 Jill S. Schneiderman. Reprinted with permission of Henry Holt and Company, and the author.

5–7 **DOES GOD PLAY DICE?** Ian Stewart, from *Does God Play Dice? The New Mathematics of Chaos*. Extract from 2nd edn. (Penguin Books) 1997, pp.xi–xii. Copyright © Ian Stewart 1997. Reprinted with permission of Penguin Books UK and Blackwell Publishing UK.

Chapter 6: The Fossil Record

6–1 **EARTH AND MAN:** Frank H.T. Rhodes, from *Wooster Alumni Magazine*. Oct 1972, pp. 5–7.

6–2 **THE FLOWERING EARTH:** Donald Culross Peattie, from *Flowering Earth*. (New York: G. P. Putnam's Sons) 1939, pp. 22–26, 132–143. Reprinted with permission of Curtis Brown Ltd.

6–3 **HABITS AND HABITATS:** Robert Claiborne, from *Climate, Man and History*. (W.W. Norton & Company Inc) 1970, pp.170–178. Copyright © 1970 by Robert Claiborne. Used by permission of W.W. Norton & Company Inc and IMG Literary Agency.

6–4 **DIPLODOCUS, THE DINOSAUR:** James A. Michener, from *Creatures of the Kingdom: Stories of Animals and Nature*. (New York: Random House, and UK: Martin Secker & Warburg) 1993, pp. 43–47, copyright © 1993 by James Michener. Reprinted with permission of The Random House Group Limited and Random House Inc. [Originally written for *Centennial* (1974).]

6–5 **A WINDOW ON THE OLIGOCENE:** Berton Roueché, from *The New Yorker*. Nov 13, 1971, pp. 141–148. Reprinted with permission of David Higham Associates Limited.

6–6 **A FISH CAUGHT IN TIME:** Samantha Weinberg, from *A Fish Caught in Time: the search for the Coelacanth*. (New York: HarperCollins Publishers) 2000, pp. vii–xx. Ogden Nash, "The Coelacanth" from *Verses from 1929 On* (Boston: Little, Brown and Company, Inc., 1959). Used with permission.

6–7 **APE-LIKE ANCESTORS:** Richard E. Leakey, from *The Making of Mankind*. (New York: E. P. Dutton) 1981, p. 40. Copyright Sherma B.V.

6–8 **THE RELIC MEN:** Loren Eiseley, from *The Night Country*. Copyright © 1971 by Loren Eiseley. Copyright © 1947, 1948, 1951, 1956, 1958, 1962, 1963, 1964, 1966 by Loren Eiseley. Reprinted with permission of Scribner, an imprint of Simon & Schuster Adult Publishing Group. All rights reserved.

Chapter 7: Geotectonics

7–1 **FROM THE BOUNDLESS DEEP and THE BIRTH OF THE ROCKIES:** James A. Michener, from *Creatures of the Kingdom: Stories of Animals and Nature*. (New York: Random House, and UK: Martin Secker & Warburg) 1993, pp. 5–6, 25–28. Copyright © 1993 by James Michener. Reprinted with permission of The Random House Group Limited and Random House Inc.

7–2 **WHEN PIGS RULED THE EARTH**: Anna Grayson, from *Equinox: The Earth*. (London: Channel 4 Books/Pan Macmillan) 2000, pp. 19–26. Copyright © Anna Grayson, 2000.

7–3 **THE LIVING PLANET**: David Attenborough, from *The Living Planet: A Portrait of the Earth*. (Boston: Little, Brown, and Co.) 1984, pp. 13, 15, 308. Reprinted with permission of David Attenborough Productions Limited.

7–4 **THE ROAD TO JARAMILLO:** William Glen, from *The Road to Jaramillo: Critical Years of the Revolution in Earth Science*. (Stanford: Stanford University Press) 1982, pp. 1–4. Copyright © 1982 by the Board of Trustees of the Leland Stanford Jr University. Reprinted with permission of Stanford University Press.

7–5 **MAO'S ALMANAC: 3,000 YEARS OF KILLER EARTHQUAKES:** J. Tuzo Wilson, "Mao's Almanac: 3,000 Years of Killer Earthquakes," *Saturday Review.* Feb 19, 1972, pp. 60–64. Copyright © 1972 *Saturday Review.* All rights reserved.

7–6 **GEOLOGIC JEOPARDY:** Richard H. Jahns, from *The Texas Quarterly.* 1968, **11**(2), pp.69–83. Copyright © 1968 by The University of Texas Press. All rights reserved. Reprinted with permission of University of Texas Press.

Chapter 8: Controversies

8–1 **APES, ANGELS, AND VICTORIANS:** William Irvine, from *Apes, Angels, and Victorians: The Story of Darwin, Huxley, and Evolution.* (New York: McGrawHill Book Company, Inc) 1955, pp. 3–7. Reprinted with permission of McGrawHill Education.

8–2 **THE GREAT PILTDOWN HOAX:** William L. Straus, Jr, from *Science.* 1954, **119**, pp. 265–269. Copyright © 1954 AAAS. Reprinted with permission.

8–3 **FOSSILS AND FREE ENTERPRISERS:** Howard S. Miller, from *Dollars for Research: Science and Its Patrons in Nineteenth-Century America.* (Seattle: University of Washington Press) 1970, pp. 133–143. Reprinted with permission of University of Washington Press.

8–4 **THE K-T EXTINCTION:** Charles Officer and Jake Page, from *Tales of the Earth: Paroxysms and Perturbations of the Blue Planet.* (Oxford: Oxford University Press) 1993, pp. 140–144. Reprinted with permission of Oxford University Press.

8–5 **THE FOUNDERS OF GEOLOGY:** Sir Archibald Geikie, from *The Founders of Geology.* (New York: Dover Publications, Inc.) 2nd edn., 1962, pp. 144–153. [This new Dover edition is an unabridged and unaltered republication of the second (1905) edition of the work first published by Macmillan and Company in 1897.]

8–6 **TO A ROCKY MOON:** Don E. Wilhelms, from *To a Rocky Moon: A Geologist's History of Lunar Exploration.* (Tucson: The University of Arizona Press) 1993, pp. 7–10. © 1993 Arizona Board of Regents. Reprinted by permission of University of Arizona Press.

8–7 **PROPERTIES AND COMPOSITION OF LUNAR MATERIALS: EARTH ANALOGIES:** Edward Schreiber and Orson L. Anderson, from *Science.* 1970, **168**, pp. 1579–1580. Copyright 1970 by the American Association for the Advancement of Science.

8–8 **CFCs:** Joel L. Swerdlow, from "Making Sense of the Millennium", *National Geographic.* 1998, **193**(1), pp. 5–6. Reprinted with permission of National Geographic.

Chapter 9: Prose

9–1 **OUT OF AFRICA:** Isak Dinesen, from *Out of Africa.* (New York: Random House) 1938, pp. 3–7. Copyright © 1937 by Random House Inc and renewed 1965 by Rungstedlundfonden. Used with permission of Random House Inc.

9–2 **SEVEN PILLARS OF WISDOM:** T.E. Lawrence, from *Seven Pillars of Wisdom.* (New York: Garden City Publishing Co.) 1938, pp. 183–185. Copyright © 1926, 1935 by Doubleday, a division of Random House Inc. Used by permission of Doubleday, a division of Random House Inc.

9–3 **GREEN HILLS OF AFRICA:** Ernest Hemingway, from *Green Hills of Africa.* (New York: Charles Scribner's Sons, and Jonathan Cape) 1935, pp. 170, 250–251. © 1935 by Charles Scribner's Sons. Copyright renewed © 1963 by Mary

Hemingway. Reprinted with permission of The Random House Group Limited and Simon and Schuster Inc.

9–4 **WIND, SAND AND STARS:** Antoine de Saint-Exupéry, from *Wind, Sand, and Stars*. (Harcourt Brace Jovanovich, Inc.) 1968, pp. 99–104. Translated from the French by Lewis Galantière. Reprinted with permission of Harcourt Inc. [Originally published in 1939, Paris: Gallimard.].

9–5 **THE FRENCH LIEUTENANT'S WOMAN:** John Fowles, from *The French Lieutenant's Woman*. (New York: Signet Books/New American Library) 1970, pp. 111–114. Copyright © 1969 by John Fowles. Reprinted by permission of Little, Brown and Co and Gillon Aitken Associates.

9–6 **TRIP TO THE MIDDLE AND NORTH FORKS OF SAN JOAQUIN RIVER:** John Muir, from *The Yosemite*. (New York: The Century Company) 1912, pp. 153–158.

9–7 **ROUGHING IT:** Mark Twain, from *Writings of Mark Twain*, VII. (Hartford, Connecticut: The American Printing Company) 1903, pp. 294, 297–303. [First published 1872].

9–8 **A PLACE ON THE GLACIAL TILL:** Thomas Fairchild Sherman, from *A Place on the Glacial Till: Time, Land, and Nature Within an American Town*. (Oxford: Oxford University Press) 1997, pp. 35–39. Reprinted with permission of Oxford University Press.

9–9 **BASIN AND RANGE:** John McPhee, from *Basin and Range*. (New York: Farrar Straus & Giroux LLC) 1981, pp. 78–83. Copyright © 1980, 1981 by John McPhee. Reprinted by permission of Farrar Straus & Giroux LLC.

9–10 **NEANDERTHAL:** John Darnton, from *Neanderthal*. (New York: Random House, and Hutchinson), 1996, pp. 35–38. Copyright © 1996 by John Darnton, Endpaper illustrations copyright © 1996 by Mikhail Iventisky. Reprinted by permission of The Random House Group Limited and Random House, Inc.

9–11 **ANTARCTICA:** Kim Stanley Robinson, from *Antarctica* (New York: Bantam Books) 1998, pp. 238–242. Copyright © Kim Stanley Robinson, 1998. Reprinted with permission of HarperCollins Publishers and Random House Inc.

9–12 **THE LOST WORLD:** Sir Arthur Conan Doyle, from *The Lost World*. (London: John Murray) 1912, pp. 173–176.

Chapter 10: Poetry

10–1 **LANDSCAPE AND LITERATURE:** Sir Archibald Geikie, from *Landscape in History and Other Essays*. (London: Macmillan & Company, Ltd.) 1905, pp. 124–127.

10–2 **THE EXCURSION:** William Wordsworth, "The Excursion", being a portion of *The Recluse*, a poem. (London: Longman, Hurst, Rees, Orme, and Brown), 1814. This extract is lines 173–189 from Book Three (out of nine) of the poem.

10–3 **THE LISBON EARTHQUAKE:** Voltaire, from "The Lisbon Earthquake", in *The Portable Voltaire*. (New York: Viking) 1968, pp. 561–564. Edited by Ben Ray Redman, copyright © 1949 renewed © 1977 by Viking Penguin Inc. Used by permission of Viking Penguin, a division of Penguin Group (USA) Inc. [Poem originally published in 1755].

10–4 **THE FOUNTAINS OF THE EARTH:** C.S. Rafinesque, from *The World, or Instability. A Poem. In Twenty Parts, With Notes and Illustrations*. (J. Dobson) 1836, pp. 89–90. [This extract is lines 791-814 in Part IV, "The Earth and Moon, Water, Fire and Land".]

10–5 **TO A TRILOBITE:** Timothy A. Conrad, quoted by P.E. Raymond, *Prehistoric Life*. (Harvard University Press) 1939, p. 47. [Originally written in 1840.]

10–6 **A SHROPSHIRE LAD:** A.E. Housman, in *The Collected Poems of A.E. Housman*. (Holt, Rhinehart and Winston) 1968, ch. XXXI, p. 60. Copyright © 1939, 1940, 1965 by Holt, Rhinehart and Winston. Copyright © 1967, 1968 by Robert E. Symons. Reprinted by permission of Holt, Rhinehart and Winston, Publishers. [This poem was originally published in 1896, *A Shropshire Lad*, Kegan Paul.]

10–7 **MENTE ET MALLEO:** Andrew C. Lawson, in *Andrew C. Lawson, Scientist, Teacher, Philosopher*. Edited by Francis E. Vaughan. (Glendale, California: Arthur H. Clark Company) 1970, p. 356. [This poem was originally published in 1888.]

10–8 **SELECTED POEMS: 'STRATIGRAPHICAL PALAEONTOLOGY', 'COAL MEASURES', 'ICHTHYOSAURUS', AND 'FROZEN MAMMOTHS':** John Stuart Blackie, from *Selected Poems*. Edited by A.S. Walker. (London: John MacQueen) 1896, pp. 133–138. [These four poems were quoted in Archie Lamont, "Palaeontology in Literature", *The Quarry Managers' Journal*, **xxx** (Jan- Mar) 1947, pp. 432–441, 542–551.]

10–9 **LYELL'S HYPOTHESIS AGAIN:** Kenneth Rexroth, from *The Signature of All Things* (1950). Later collected in *The Collected Shorter Poems* (New York: New Directions) 1967, pp. 180–181. Copyright © 1949 by Kenneth Rexroth. Reprinted with permission of New Directions Publishing Corporation.

10–10 **SELECTED POEMS: 'GLACIALS', 'NATIVES' AND 'UNDERSEA':** A.R. Ammons, from *The Really Short Poems of A.R. Ammons*. (W.W. Norton & Company Inc) 1990, pp. 22, 34, 150. Copyright © 1990 by A.R. Ammons. Used by permission of W.W. Norton & Company Inc.

10–11 **STONE:** Charles Simic, from *Dismantling the Silence*. (New York: George Brazillier, Inc) 1971. Reprinted by permission of George Brazillier Inc.

10–12 **FOSSILS:** J.T. Barbarese, from *The Atlantic Monthly*. Sept 2000. Copyright © J.T. Barbarese 2000. Reprinted with kind permission of the author.

10–13 **ROCK**: Jane Hirshfield, from *Given Sugar, Given Salt*. (New York: HarperCollins) 2001, and in *Each Happiness Ringed by Lions* (Bloodaxe Books) 2005. Reprinted with permission of Bloodaxe Books and HarperCollins New York.

10–14 **POETRY MATTERS: GARY SNYDER:** W. Scott McLean, Eldridge M. Moores, David A. Robertson, "Chapter 11: Nature and Culture" from *Earth Matters: the Earth Sciences, Philosophy, and the Claims of Community*. Edited by Robert Frodeman. (Englewood Cliffs, New Jersey: Prentice Hall) 2000, pp. 53–54, 143–147. © 2000. Reprinted by permission of Pearson Education Inc, Upper Saddle River, NJ. Gary Snyder, "Journal Entry" from *Earth House Hold* (New York: New Directions) 1969. Copyright © 1969 by Gary Snyder. Reprinted by permission of New Directions Publishing Corp.
Gary Snyder, "Piute Creek" from *Riprap* (Shoemaker & Hoard Publishers) 2003. Copyright © 2003 by Gary Snyder. Reprinted with permission of Shoemaker & Hoard Publishers USA.
Gary Snyder, "Geological Meditation" from *Left Out in the Rain* (Shoemaker & Hoard Publishers) 2005. Copyright © 2005 by Gary Snyder. Reprinted with permission of Shoemaker & Hoard Publishers USA.

10–15 **THE BOOK OF JOB: WHERE SHALL WISDOM BE FOUND?** The Book of Job, Chapter 28, Verses 1–28; Chapter 38, verses 1–18. The Holy Bible. Revised Standard Version (New York: Thomas Nelson & Sons) 1952.

Chapter 11: Art

11–1 **A LAND: SCULPTURE:** Jacquette Hawkes, from *A Land* (London: The Cresset Press) 1951, pp. 100–105.

11–2 **BEYOND MODERN SCULPTURE:** Jack Burnham, from *Beyond Modern Sculpture: The Effects of Science and Technology on the Sculpture of this Century.* (New York: George Brazillier) 1968, pp. 94–95, 98. Reprinted with permission of George Brazillier Inc.

11–3 **TIME'S PROFILE: JOHN WESLEY POWELL, ART, AND GEOLOGY AT THE GRAND CANYON:** Elizabeth C. Childs, from *American Art.* 1996, **10**(1), pp. 7–8, 20–27. Reprinted with permission of University of Chicago Press and the author.

11–4 **THOMAS MORAN: AMERICAN LANDSCAPE PAINTER:** R.A. Bartlett, from *Great Surveys of the American West.* (Norman, Oklahoma: University of Oklahoma Press) 1962, 41–42. Reprinted with permission of University of Oklahoma Press.

11–5 **EARTH CALLING:** Diane Ackerman, from *A Natural History of the Senses.* (New York: Random House) 1990, pp. 222–225. Copyright © 1990 by Diane Ackerman. Used by permission of Random House Inc.

Chapter 12: Human History

12–1 **MINERALS AND WORLD HISTORY:** John D. Ridge, from *Mineral Industries.* 1967, **36**(9), pp. 3–4.

12–2 **A LAND: ARCHITECTURE:** Jacquette Hawkes, from *A Land.* (London: The Cresset Press) 1951, pp. 105–108.

12–3 **THE GEOLOGIC AND TOPOGRAPHIC SETTING OF CITIES:** Donald F. Eschman and Melvin G. Marcus, from *Urbanization and Environment.* Edited by Thomas R. Detwyler and Melvin G. Marcus. (Belmont, California: Duxbury Press), 1972, pp. 27–34.

12–4 **TOPOGRAPHY AND STRATEGY IN THE WAR:** Douglas W. Johnson, from *Topography and Strategy in the War.* (New York: Henry Holt & Co.) 1917, pp. 1–8.

12–5 **GEOLOGY AND CRIME:** John McPhee, from the article "The Gravel Page" in *The New Yorker.* 29 Jan 1996, pp. 46, 48. Reprinted by permission. © 1996 John McPhee. Originally published in *The New Yorker.* All rights reserved.

12–6 **TAMBORA AND KRAKATAU:** Kenneth E.F. Watt, from "Tambora and Krakatau: Volcanoes and the Cooling of the World" from *Saturday Review.* 23 Dec 1972, pp. 43–44. Reprinted with the kind permission of the author.

12–7 **MORTGAGING THE OLD HOMESTEAD:** Lord Ritchie-Calder, from *Foreign Affairs.* 1970, **48**(2), pp. 207–208, 216–218, 220. Copyright © 1970 by the Council on Foreign Relations Inc. Reprinted by permission of Foreign Affairs. [Original was substantially abridged for this edition.]

12–8 **BREATHING THE FUTURE AND THE PAST:** Harlow Shapley, from *Beyond the Observatory.* (New York: Charles Scribner's Sons) 1967, pp. 44–50. Copyright © 1966, 1967 by Harlow Shapley. Reprinted with permission of Scribner, an imprint of Simon & Schuster Adult Publishing Group. All rights reserved.

Chapter 13: Resources

13–1 **WEALTH FROM THE SALT SEAS:** Rachel L. Carson, from *The Sea Around Us.* (New York: Oxford University Press) 1951, pp. 188–191. Copyright © 1950 by Rachel L. Carson. Reprinted with permission of Pollinger Limited and Frances Collin, Trustee.

13–2 **MINERALS, PEOPLE, AND THE FUTURE:** Charles F. Park, Jr., from *Affluence in Jeopardy: Minerals and the Political Economy.* (San Francisco: Freeman, Cooper and Company) 1968, pp. 321, 323–328, 330–331, 334.

13–3 **THE BINGHAM CANYON PIT:** M. Dane Picard, from *Journal of Geoscience Education.* 1999, **47, pp.** 369–373. Reprinted with permission of NAGT.

13–4 **THE GEOLOGICAL ATTITUDE:** John G.C.M. Fuller, from *AAPG Bulletin.* 1971, **55**(11), pp. 1927–1938. Reprinted with permission of AAPG whose permission is required for further use.

13–5 **GEOLOGY - FOR HUMAN NEEDS:** Michel T. Halbouty, from *Journal of Geological Education.* 1967, **XV**(2), pp. 80–82. Reprinted with permission of NAGT.

Chapter 14: Benevolent Planet

14–1 **GAIA:** James Lovelock, from *Healing Gaia: Practical Medicine for the Planet.* (New York: Harmony Books) 1991, p. 25. Now publishing as *Medicine for an Ailing Planet* (London: Hamlyn). Reprinted with kind permission of the author.

14–2 **THE WEB OF LIFE:** Fritjof Capra, from *The Web of Life: A New Synthesis of Mind and Matter.* (London: Flamingo/HarperCollins Publishers) 1997, pp. 22–23. Originally published (London: HarperCollins Publishers) 1996. Copyright © 1996 by Fritjof Capra. Reprinted with permission of HarperCollins Publishers and Doubleday, an imprint of Random House Inc.

14–3 **REMEMBER THE LAND:** Charles Morgan, from *Between Sacred Mountains: Navajo Stories and Lessons from the Land.* (Tucson: Sun Tracks/University of Arizona Press) 1982, p. 245. Reprinted with permission of Rock Point Community School.

14–4 **RACHEL CARSON: THE IDEA OF ENVIRONMENT:** Gabriele Kass-Simon, ''Biology is Destiny'' from *Women of Science.* Edited by G. Kass-Simon and Patricia Farnes. (Bloomington: Indiana University Press) 1990, pp. 257–258. Reprinted with permission of Indiana University Press.

14–5 **SILENT SPRING:** Rachel L. Carson, from *Silent Spring.* (Boston: Houghton Mifflin Company) 1962, pp. 277–279, 296–297. Copyright © 1962 by Rachel L. Carson. Copyright renewed 1990 by Roger Christie. Reprinted with permission of Pollinger Limited and Houghton Mifflin Company. All rights reserved.

14–6 **WHO IS EL NIÑO?** S. George Philander, from *Earth in Space.* 1998, **10**(8), pp. 6–7. Reprinted with permission of American Geophysical Union and the author.

14–7 **ESSAY ON THE EARTH SCIENCES:** National Research Council, from *Solid-Earth Sciences and Society.* Commission on Geosciences, Environment, and Resources. (Washington, D.C.: National Academy Press) 1993, pp. 13–16. Reprinted with permission.

14–8 **THE ROUND WALLS OF HOME:** Diane Ackerman, from *A Natural History of the Senses.* (New York: Random House) 1990, pp. 283–285. Copyright © 1990 by Diane Ackerman. Used by permission of Random House Inc.

14–9 **THE BUTTERFLY EFFECT:** Ernest Zebrowski, Jr., from *Perils of a Restless Planet: Scientific Perspectives on Natural Hazards.* (Cambridge: Cambridge University Press) 1997, pp. 263–266. Reprinted with permission of Cambridge University Press.

14–10 **PALE BLUE DOT:** Carl Sagan, from *Pale Blue Dot: a Vision of the Human Future in Space.* (New York: Random House) 1994, pp. 8–9. Copyright © 1994 Carl Sagan. Reprinted with permission of Democritus Properties LLC.

Names index

For subject index, see p. 320.

Subject index

For names index, see p. 315.